Systems and Synthetic Immunology

Shailza Singh

Editor

Systems and Synthetic Immunology

 Springer

Editor
Shailza Singh
National Centre for Cell Science
Pune, Maharashtra, India

ISBN 978-981-15-3352-5   ISBN 978-981-15-3350-1   (eBook)
https://doi.org/10.1007/978-981-15-3350-1

This Springer imprint is published by the registered company Springer Nature Singapore Pte Ltd.
The registered company address is: 152 Beach Road, #21-01/04 Gateway East, Singapore 189721, Singapore

Contents

About the Editor

Shailza Singh is a scientist at the Bioinformatics and High Performance Computing Facility, NCCS, Pune, India. Her research chiefly focuses on the systems and synthetic biology of infectious diseases such as leishmaniasis. In this regard, her group is working to integrate the action of regulatory circuits, cross-talk between pathways, and non-linear kinetics of biochemical processes through mathematical modeling. Dr. Singh has been honored with the DBT RGYI, DST Young Scientist, and INSA Bilateral Exchange Programme awards and was selected by the DBT for a SAKURA EXCHANGE Programme in Science in the field of artificial intelligence and machine learning to Tokyo in 2018. She serves as a reviewer for prestigious international grants such as the RCUK; for national grants from the DBT, DST, and CSIR; and for several prominent international journals, e.g., *Parasite and Vectors, PLOS One, BMC Infectious Disease, BMC Research Notes, Oncotarget, and the International Journal of Cancer.*

Vaccine Design, Nanoparticle Vaccines and Biomaterial Applications

Pragya Misra and Shailza Singh

Abstract

Leishmaniasis is a neglected tropical disease subverting the immune system of the infected individual. Most of available treatment regimens are associated with various drawbacks such as drug resistance, toxicity, and cost. Development and implementation of vaccines seem to be the only rationale to eradicate the disease. However, various traditional approaches for vaccine development have been implicated against leishmaniasis, but till date, no vaccine is available for humans in the market. It has been observed that vaccination strategy including live or attenuated vaccines is mainly due to their ability to deliver the antigens to the appropriate immune cells for generating an immune response. This indicates that pan-Leishmania vaccine packaged into a suitable delivery system could not only increase the stability of the vaccine candidate but also lead to its targeted delivery which will mimic the natural infection and recognition of the antigen by the desired antigen-presenting cells. Various natural and synthetic polymers have been used as delivery vehicles encapsulating the vaccine components against leishmaniasis. Herein, we have tried to summarize such attempts, along with our insight on using synthetic circuits as delivery system, not only for targeted but also controlling the expression dynamics of antigen as needed.

Keywords

Leishmania · Vaccine · Synthetic circuit · Biomaterials

P. Misra · S. Singh (✉)
National Centre for Cell Science, Pune, Maharashtra, India

1.1   Introduction

In the modern era, infectious diseases have become a major cause of health threat across the globe [1]. Many new infectious diseases have been identified, and old ones have reemerged, becoming a major concern for human health. Leishmaniasis and tuberculosis are two of the most important infectious diseases. Among 16 categories of neglected tropical diseases, during the period of 2005–2013, leishmaniasis ranks second in age-standardized DALYs (disability-adjusted life years), next to malaria [2], and in 2017, 20,792 out of 22,145 (94%) new cases reported to WHO occurred in seven countries: Brazil, Ethiopia, India, Kenya, Somalia, South Sudan, and Sudan. Tuberculosis accounts for death of nearly five thousand people every day. For both these diseases, there is need of global, multi-sectorial approach. Since drug resistance is a common problem associated with both the abovesaid diseases, vaccines appear as a safe and better treatment strategy.

Area of vaccine development holds much importance in today's arena of drug resistance and toxicity associated with drugs. Significant work has been done toward development of new vaccines and improves the efficacy of existing ones, and the global efforts toward vaccine development have improved the health status universally. It has been inferred that by improving the vaccination program, nearly 1.5 million lives could be saved annually [1]. On the contrary, various deadly infectious diseases do not have an approved vaccine, although theoretical strategies confer that vaccine could be an effective therapeutic strategy [2, 3]. Hindrances on various levels are responsible for this, which include legal and ethical reasons.

The major reasons are associated with the link between nature of pathogen and vaccination technologies evolved for it [4]. When it comes to vaccination strategies for diseases like tuberculosis and leishmaniasis, the most important point to be considered is that both these pathogens are intracellular and vaccines based on humoral immune response will be of no use [5, 6] along with the fact that these pathogens have high antigenic diversity and various immune-evasion strategies to combat the host immune response [2, 7].

In spite of enormous efforts and strategies followed to develop the vaccine, effective vaccines against both the abovesaid infectious diseases are still a distant variable. Herein this review, we would focus on various vaccination strategies of major infectious disease, namely, leishmaniasis, along with the loopholes in vaccine development program. The important discussion in the present study would be on use of biomaterials in improving the vaccination and other immunotherapies. Biomaterials hold importance in vaccine development because they allow controlled responses to antigens, adjuvants, or immunomodulators and have also been explored for targeted delivery of vaccine candidates to specific cell/tissue.

1.2   Leishmania and Vaccines Overview

Leishmaniasis is a neglected tropical vector-borne disease which is transmitted by bite of infected sand fly, afflicting nearly 900,000 ± 1.3 million people annually with 30,000 deaths per year [8]. The disease is spreading by natural phenomenon and the man-made condition for which efforts are being taken using technology, knowledge, and communication to effectively control it.

1.3 Challenges Associated with Control Program for Leishmaniasis

Leishmaniasis being a zoonotic and vector-borne disease encounters various challenges for its control. Since it is a neglected tropical disease afflicting mainly poor population [9, 10], the magnitude and stability of research funds for it are limited as compared to diseases such as cancer, HIV, or diabetes. Animal reservoir, of which few are wild and inaccessible, is another big challenge to be addressed for leishmaniasis elimination program. Another major challenge is that since it is a vector-borne disease, aspects of sand fly as well as human behavior need to be thoroughly understood to dissect the transmission dynamics of disease and vector control as well [11].

Many other challenges involve availability of drugs, cost of treatment (drugs and hospitalization), efficacy, adverse effects, and growing parasite resistance. Therefore, there is need for new therapeutic interventions which can become truly accessible to the population in endemic regions.

1.4 A Brief Overview Why and How Vaccines Could Work Against Leishmaniasis

A complex relationship exists between the host, vector, and reservoir for *Leishmania* parasite, and this makes treatment strategies for leishmaniasis a bit complicated. Of all the available treatment regimens, most of the drugs have shown cases of drug resistance, and they all require long-term hospitalization which is quite challenging. Assessing the present status of antileishmanial treatments along with the fact that once infected, the individual develops long-lasting immunity against the infection, vaccine fits for the best way forward to cure leishmaniasis [12–15].

The causative parasite for different form of leishmaniasis is different, but the sequence homology is more than 90%. The next question that rises is that vaccine against which form is the need of the day. It would be an ideal condition that vaccine against leishmaniasis should have broad spectrum of protection, showing protective response against all the leishmanial species. Selection of antigens for such vaccine should be based on the fact that highly conserved antigens should be selected for designing cross-protective vaccine. Another point to be taken in consideration is that parasite resides in two hosts, and the relationship between host-vector reservoirs is not well studied, which might be a problem in assessing the efficacy of a vaccine candidate. For example, the antigenic candidates which could enhance the susceptibility of host to the infection need to be avoided [16].

Not only for leishmaniasis but for many infectious diseases, there is a big debate on how the vaccine should work, either prophylactically or therapeutically alone or in combination with some type of adjuvant. Therapeutic vaccines development has gained less importance; however, it would be of more use to patients with active infection by modulating their immune response [17].

Another important aspect which needs to be considered is the potential strategy for designing a vaccine, which has been initiated from leishmanization, which is still practiced in few regions of the Middle East to first-generation vaccines to second and third generation vaccines [14]. In the next section, we would discuss about various vaccination strategies against leishmaniasis, with detailed discussion on biomaterial-based vaccines.

1.5 Vaccination Strategies Adapted Against Leishmaniasis

Till date, no vaccine is available against human leishmaniasis, but it has been widely accepted by researchers across the globe that vaccine against this disease is feasible, and the most important point in consideration is that vaccination is the only viable option to achieve disease elimination [18].

1.6 Leishmanization: A Traditional Practice

In leishmanization, live *Leishmania* parasite was introduced in infected individuals in covered part of the body to protect against lesion development. In simple terms, leishmanization was controlled induction of disease to prevent the consequences of natural infection [19, 20].

Later on, the virulent parasites which were harvested from cell-free cultures were used for this. This practice using *Leishmania major* parasites was used in former USSR, Israel, and Iran [21] but discontinued due to loss in infectivity by repeated subculturing or freezing, complications at the inoculation site, or major complications due to immunosuppression [22]. Many cases of nonhealing lesions in Iran further complicated the possibility of widespread use of leishmanization for treating the disease [23–26]. However, leishmanization is still performed in Uzbekistan [14].

Although the practice of leishmanization trials is there, the information it gave at the end of century is very important, proving the feasibility of vaccines against leishmaniasis and for defining strategies to develop vaccine against leishmaniasis. Both C57BL/6 and BALB/c experimental mouse models have shown that a key factor in the efficacy of leishmanization is the persistence of the parasite following inoculation indicating that persistent antigen presentation drives T-cell immunity [27, 28].

Various studies are ongoing to develop killed, naturally attenuated, or genetically modified live parasites or subunit vaccines which are based on replicating the immune protection as in case of leishmanization. Vaccine based on various strategies can be categorized as live vaccines ("leishmanization like"); first-, second-, and third-generation vaccines; and vector-derived vaccines.

1.7   Live Vaccine Candidates

Vaccination principles postulate that the vaccine which is more similar to the natural infection will evoke a better immune response. Thus, live attenuated vaccine can be good strategy for treating leishmaniasis. Moreover, against various intracellular organisms, live attenuated vaccines have served as a gold standard for treatment of diseases such as smallpox, measles, mumps, and rubella. Success of leishmanization also supports the research going on to develop vaccines based on live parasites, and it has also been termed as leishmanization revisitation [29], because these have advantage of partially mimicking the natural course of infection [30].

Various efforts have been taken since years to develop attenuated strains in in vitro cultures [31] by selecting for temperature sensitivity, chemical mutagenesis [32], and γ-attenuation [33] or by keeping parasite culture under drug pressure [34]. Specially in the era when genetic engineering was not so flourished, these chemically and physically attenuated parasites showed effectiveness in preclinical trials against cutaneous leishmaniasis (CL), mucocutaneous leishmaniasis (MCL), and visceral leishmaniasis (VL) [15, 35–37]. All these methods have further shown that these attenuated strains have shown remarkable protection in murine models against challenge by virulent *Leishmania* parasite; however, as a drawback, a clear genetic profile and probability of reversal to virulent parasite could not be predicted restricting their human use. There is one more possibility that the continuous presence of such asymptomatic *Leishmania* infection can increase the risk of subsequent reactivation, particularly in case of HIV/*Leishmania* coinfection. There is another drawback that such undefined attenuation can also cause loss of effectiveness for protective immunity which can be due to failure of such strains to establish a subclinical infection or due to the loss in expression of antigenic epitopes. One such study carried out using *L. chagasi* has shown that inoculating high dose of *L. chagasi* subcutaneously caused subclinical infection, and this induced protective immune response in mice. On the contrary, when attenuated *L. chagasi* parasite obtained either by long-term passage or knockout was inoculated in mice, no protective response was elicited because these parasites failed to establish a subclinical infection or no expression of immunogenic antigen epitopes [38].

In the post-genomic era, these strategies for attenuation were replaced by the genetically modified parasites. For preclinical studies, genetically modified *Leishmania* parasites were generated using two approaches: loss-of-function mutants (knockout) and gain-of-function mutants (knock-in). For knockout studies, defined genetic alterations of the *Leishmania* genome are usually generated using a gene-targeted disruption strategy through homologous recombination which allows selection and long-term survival and virulence of parasite lacking gene of interest in selection medium.

Targeted deletion of an essential metabolic gene, DHFR-TS (dihydrofolate reductase thymidylate synthase), was the first attempt for developing knockout parasites for vaccination studies. Homozygous null mutant auxotrophic for thymidine was created by a two-step process as *Leishmania* is a diploid parasite. These *DHFR-TS* were able to survive in vivo but did not induce infection or cause disease even in most susceptible mice. When used for immunization against *L. major* challenge, these induced a potential protection against the virulent parasite [39]. Another group studied the efficacy of these genetically mutated parasites in comparison to inactivated autoclaved promastigotes (ALM) with bacillus Calmette-Guérin (BCG) for protection in *Rhesus macaques* (Macaca mulatta) against *L. major* infection. Protective immunity was not observed in monkeys post vaccination as all the monkeys exhibited skin lesions in all the study groups. Moreover, another striking observation was these attenuated parasites were not pathogenic in monkey model. Therefore, further studies on these knockout parasites were stopped with a conclusion that although the vaccine protocol is safe in primates, for clinical use, it needs more modifications [40]. Many other studies have been conducted using various target genes for developing knockout vaccine candidates. These included *L. mexicana* cysteine proteases+(CPA/CPB−/−), *L. major* lipophosphoglycan 2 (LPG2−/−), *L. major* phosphomannomutase (PPM−/−), *L. donovani* Centrin (Cen−/−), *L. infantum* heat shock protein 70 type II (HSP70-II−/−), *L. donovani* arabino-1,4-lactone oxidase (ALO−/−), and *L. donovani* biopterin transporter 1 (BT1−/−)] and a single knockout [*L. infantum* silent information regulatory protein 2 (SIR2+/−)] which showed protective responses in primate models of CL, MCL, and VL [37, 41–46]. One amastigote-specific protein p27 (Ldp27) was knocked out using *L. donovani*, and it was found that the parasite had reduced virulence in vivo. Further studies carried out by the same group had shown that these knockout parasites did not survive for long in BALB/c mice and hence could serve as an immunogen. When mice with Ldp27$^{-/-}$ were challenged with virulent parasites, immunized mice showed significantly lower parasite burden in liver and spleen along with anti-inflammatory cytokine and NO production. Long-term memory response was proven by adoptive transfer of T cells from immunized mice to naive mice against *L. donovani* challenge. These knockout parasites also demonstrated cross-protection against the *Leishmania major* and *Leishmania braziliensis* infection [47, 48]. Recently, a *Leishmania major* p27 gene knockout (*Lmp27$^{-/-}$*) strain was developed that was safe and immunogenic in BALB/c mice [49]. In this study, protective immunity and efficacy of *Lmp27$^{-/-}$* were evaluated against homologous (*L. major*) and heterologous (*L. infantum*) *Leishmania* species. Results showed a significant Th1 response along with smaller skin lesions and lower parasite burdens following a *L. major* challenge. These mutant also showed cross-protection against *L. infantum* infection [50].

Another strategy of gain-of-function mutants also showed efficacy as potential vaccines against cutaneous and visceral leishmaniasis. These were termed as "suicidal mutants" as they would be completely eliminated from the immunized host either by the action of chemotherapeutics [*L. major* thymidine kinase (herpes simplex virus), cytosine deaminase (*Saccharomyces cerevisiae*) knock-in: tk-cd+/+] or by photodynamic therapy (*L. amazonensis* δ-aminolevulinate dehydratase, porphobilinogen deaminase knock-in: alad-pbgd+/+) [36, 51, 52].

Most of the studies show exciting data, but live attenuated vaccine is a long road to be covered. These attenuated parasites are associated with many safety constraints such as possibility for reversal of virulence, reactivation in immunosuppressed individuals, and manufacturing considerations which majorly include stability/viability. Most of these attenuated parasites are made by inserting an antibiotic resistance gene to be used as a selection marker during the steps of gene deletion which again strongly restrain their clinical studies.

To overcome such hindrances, a new approach has come into picture using *Leishmania tarentolae* parasite, which infects reptiles but did not cause sustained infection in mice, and most importantly, it shares more than 90% gene homology to other *Leishmania* species [53]. Based on this strategy, a live recombinant *Leishmania tarentolae* which expressed lipophosphoglycan 3 (LPG3) antigen was tested against *L. infantum* infection in Balb/c mice. It caused enhanced expression of IFN-γ along with decreased expression of IL-10 when compared to control group with virulent parasites, indicating its Th1 stimulatory role [54]. Further studies are needed to explore this approach of live vaccination.

1.8 First-Generation Vaccines

First-generation vaccines comprised of whole-killed pathogen or their fractions along with live attenuated vaccines, and many of these are approved for human use. In Latin America, since early part of twentieth century, these first-generation vaccines have been under experimentation. After the era of leishmanization, killed/fractionated vaccines were developed to assure the safety issues associated with leishmanization as well as attenuated counterparts. These vaccines elicit a specific memory response without any expected pathology even in immunocompromised individuals [55]. However, this has a disadvantage as well, since the antigen would be needed to be administered more than once to boost the primary response or coadministered along with some agent which can act as an immune enhancer, not required for generation of immune response by live vaccines [56, 57]. Many first-generation vaccines have gone to the clinical trial, and this outnumbers the other vaccines. The concept of designing these vaccines is quite simple, and production cost is also low which makes these vaccines as an attractive candidate to be developed for human use. *Killed Leishmania* vaccine because of abovesaid merits along with stable biochemical composition and antigenicity gained a lot of attention, but these could not confer significant protection against human leishmaniasis [14]. Leishvaccine, which was prepared from whole-killed promastigotes of *Leishmania amazonensis* and bacillus Calmette-Guérin (BCG), showed protective efficacy against canine leishmaniasis by inducing a mixed cytokine response. This was taken successfully to phase I and II clinical trials, wherein it showed good safety and immunogenicity, but it failed to give similar immunoprotective results in randomized phase III clinical trial [58].

Most of the first-generation vaccines are focused on CL, and no clinical trials have been done for visceral leishmaniasis. Alum-precipitated autoclaved *L. major* was given along with BCG which showed promising efficacy as vaccines for VL and PKDL [22]. Psoralen compound amotosalen-treated *L. infantum* and *L. chagasi* along with treatment with UV radiation were used as whole cell vaccine. Such treatment caused the generation of permanent covalent DNA cross-links within parasites which resulted in parasite termed as killed but it was metabolically active (KBMA). Initial data with this approach was quite promising [59].

Efficacy of *L. mexicana* along with BCG was also tested both prophylactically and as an immunotherapy. Application of this vaccine showed low levels of leishmanin skin test (LST), and the participants which showed LST conversion had low incidences of leishmaniasis [60]. The strategy of immunotherapy has shown success in the patients afflicted by mucocutaneous and diffuse forms of CL treated with pasteurized *Leishmania braziliensis (L. braziliensis)* promastigotes along with BCG. This treatment cured patients of nonhealing CL which did not respond to three courses (2 months) of antimonial treatment [60].

L. major has also been well explored for its immunogenicity in various clinical trials for Leishmania treatment [60]. Autoclaved *L. major* (ALM) was used along with BCG in phase I and II clinical trials within healthy individuals in non-endemic areas of CL, and it was observed that LST conversion was observed only in 36% of healthy individuals with low levels of IFN-γ production on stimulation with soluble Leishmania antigen. This vaccine was also assessed in healthy individuals of endemic area, and similar results were obtained [15, 61].

Alum was optimized by adsorption of antigenic fraction to alum and, along with BCG, was used in combination with sodium stibogluconate for treating post-kala-azar dermal leishmaniasis (PKDL), and data suggested that this combination was more effective than sodium stibogluconate (Stb) alone [36, 60].

In case of *L. donovani*, this type of parasite-killing approach was tested as a vaccine in preclinical trial against visceral leishmaniasis (VL), and significant potential was observed in protecting against the disease [62].

Total or soluble antigens of *L. donovani* obtained after sonication of the parasite have also been used along with MPL-A, BCG, or liposomes as vaccine candidates for VL in preclinical trials with promising results in all models tested (mice, hamsters, and monkeys) [43, 63–65].

As a vaccination strategy, fractionation of soluble proteins of *L. donovani* was carried out based on molecular weight, and different fractions were tested for their prophylactic efficacy in hamster model. It was observed that fraction within the range of 97–68 KDa showed nearly 90% protection, which was further characterized by proteomics studies [66, 67].

Two of the approved Leishmania vaccines Leishmune® in Brazil and CaniLeish® in Europe licensed for veterinary use to protect dogs belong to fractionated vaccines only. Leishmune is a vaccine of a purified fraction named as fucose mannose ligand (FML), which is a glycoproteic complex isolated from *Leishmania donovani* plus a saponin adjuvant which include QS21 and two deacylated saponins. It showed more than 90% efficacy in Brazil] [12, 22 clinical trial paper].

CaniLeish® comprises of purified excreted-secreted proteins (ESP) of *Leishmania infantum* (LiESP) produced by means of a patented cell-free, serum-free culture system [68] and adjuvanted with QA-21, a highly purified fraction of the *Quillaja saponaria* saponin. Dogs vaccinated with this vaccine showed Th1-type immune response within three weeks. However, both of these vaccines were never tested for human use which might be due to more stringent and lengthy process for human approval as well since this vaccine consists of heterologous antigens which again is harder to standardize.

1.9 Second-Generation Vaccines

This category of vaccines comprises of further refined products, such as recombinant proteins, which are produced by genetically engineered cells along with adjuvant or expression in heterologous microbial vector. Since these proteins can be produced in a large scale, are reproducible, and have low cost, these represent a more feasible vaccination strategy. The response elicited by them can be further enhanced by formulation with adjuvant [69, 70].

In animal models, defined antigens which are delivered as plasmid DNA/vector DNA or as recombinant protein with adjuvant have shown promising efficacy, but for human use, only recombinant proteins are licensed.

Various attempts have been made to develop second-generation vaccine against leishmaniasis. In a study of recombinant stage-specific hydrophilic surface protein of *Leishmania donovani*, recombinant hydrophilic acylated surface protein B1 (HASPB1) was evaluated for its prophylactic efficacy and it was able to control parasitic burden in spleen along with production of IL-12 and IFN-γ [71].

To investigate the immune response generated against amastigote antigens, three stage-specific antigens, namely, A2, P4, and P8, purified from in vitro-cultured amastigotes of *L. pifanoi* were evaluated. It was found that along with *Corynebacterium parvum* as an adjuvant, P4 and P8 showed partial to complete protection of BALB/c mice challenged with *L. pifanoi* promastigotes. P8 showed complete protection against *L. amazonensis* infection of CBA/J mice and partial protection of BALB/c mice [72]. A hypothetical *Leishmania* amastigote-specific protein (LiHyp1) also showed protective response in mice [73].

KMP-11 is a highly conserved surface membrane protein present in all members of the family Kinetoplastidae. This protein is differentially expressed in amastigote and promastigote stage of *Leishmania* parasite. A construct containing KMP-11 was tested in susceptible golden hamsters against challenge by both pentavalent antimony-responsive (AG83) and antimony-resistant (GE1F8R) virulent *L. donovani*. It showed substantial magnitude of protection as evident by decreased parasitic load and increased IFN-γ, TNF-α, and IL-12 levels [74].

LCR1 antigen of *L. chagasi* was found to stimulate the production of IFN-γ from T cells isolated from infected BALB/c mice and, when used for immunization, showed partial protection. To enhance its immunogenicity, BCG expressing LCR1 (BCG-LCR1) was engineered which showed better protection than LCR1 alone promoting Th1 immune response which strengthened its potential as a component for Leishmania vaccine [75].

Similar strategy was adapted for another Leishmania surface protein gp63. Gp63 of *L. major* was cloned and expressed in BCG using two different expression systems. It was found that BALB/c mice immunized with recombinant BCG producing

Gp63 as a hybrid protein with the N-terminal region of the beta-lactamase stimulated significant protection against *L. major* challenge [76].

Along with membrane proteins, various antigens identified in soluble fraction of Leishmania proteins were also evaluated for their antigenicity and feasibility as vaccine candidates [77–83].

Most of the proteins in such studies have been identified by immune-proteomics approach. Amastigote stage considering that this is the form which resides as intracellular parasite in humans has been explored for identifying vaccine candidates [73, 84–86]. Data shows that many categories of proteins, including ribosomal proteins, metabolic enzymes stress-related proteins, antioxidant-machinery components, and even hypothetical proteins, have been evaluated for their efficacy as vaccine candidates. One such vaccine, *L. donovani* A2, has been licensed as veterinary vaccine against leishmaniasis in Brazil—LeishTec® [87]. Peptide vaccines and many combinations of immunogenic peptides/multi-epitopes and/or multi-specific vaccines have been tested against leishmaniasis [88–94].

In spite of so many efforts, only very few antigens have gone to clinical or veterinary trials. One of the important drawbacks associated with recombinant vaccines is that these generally induce weak T-cell response, which could be overcome by addition of adjuvant or a delivery vehicle, which we will study in detail in next section.

Few of the recombinant vaccines along with suitable adjuvants have reached to clinical trials. One such vaccine is LEISH-F1 produced by the Infectious Disease Research Institute (IDRI, Seattle, WA, USA), which was previously known as Leish-111f. It is comprised of a recombinant artificial protein encoded by three genes: *L. major* homologue of eukaryotic thiol-specific antioxidant (TSA), *L. major* stress-inducible protein-1 (LmSTI1), and *L. braziliensis* elongation initiation factor (LeIF). This vaccine was emulsified with an adjuvant called "monophosphoryl lipid A which stimulates Toll-like receptor (TLR)" (MPL-SE) and reached up to phase II of clinical trials. This vaccine not only protected individuals affected with cutaneous leishmaniasis (CL) or mucocutaneous leishmaniasis (ML) but also induced the production of protective immunity in healthy volunteers [95–97].

IDRI launched another vaccine called LEISH-F2. This has all the constituents similar to LEISH-F1; the only modification was removal of N-terminal histidine tag, which made protein resemble more to the natural protein. This vaccine also reached up to phase II clinical trial associated with MPL-SE adjuvant (25 µg) [95].

Another multicomponent vaccine is LEISH-F3, which includes two proteins, namely, nucleoside hydrolase (NH) and sterol 24-c-methyltransferase (SMT), derived from *L. donovani* and *L. infantum,* respectively [98]. This was formulated with an adjuvant, which is a ligand for TLR-4, glucopyranosyl lipid A-stable oil-in-water nanoemulsion (GLA-SE) [98]. This vaccine was tested in healthy and adult individuals, living in Washington (US), and showed promising results, which was evident by robust immune response against VL [98, 99].

One such multivalent vaccine is Protein Q which has been tested for canine VL and has showed more than 90% in combination with various adjuvants [100].

1.10 Third-Generation Vaccines

Concept of generating an antigen-specific immune response by intramuscular injection of plasmid in animal model brought up the idea of new arm of vaccine research, DNA vaccines. These vaccines were not well accepted due to the ethical implication that this foreign DNA might integrate in the human genome along with the possibility of generation of autoimmune pathology which could be generated by anti-DNA immune response [101]. These shortcomings were later on ruled out by various preclinical and clinical trials for DNA vaccines suggesting that these vaccines are safe and immunogenic. However, till date, no third-generation vaccine has been approved for human use. Membranous and soluble, both antigens were studied for DNA vaccination strategy in animal models. Immunological response induced by DNA vaccination with LACK (Leishmania analogue of the receptor kinase C), TSA (thiol-specific antioxidant) genes alone, or LACK-TSA fusion was studied against cutaneous leishmaniasis by assessing cellular and humoral immune responses after challenge with *L. major*. Partial immunity was shown by all the groups with IFN-γ/interleukin (IL)-4 and IgG2a/IgG1 ratios showing that fusion of LACK-TSA produced highest IFN-γ and IgG2a. Overall data suggested that a bivalent vaccine can induce stronger immune responses [102].

Efficacy of a synthetic DNA vaccine encoding *Leishmania* glycosomal phosphoenolpyruvate carboxykinase (PEPCK) delivered by electroporation by intradermal route was found to be superior to the intramuscular route for generating skin-resident PEPCK-specific T cells. It was observed that mice immunized intradermally, when challenged with *Leishmania major* parasites, exhibited significant protection, while mice immunized intramuscularly did not [103]. Hemoglobin receptor (HbR) of *Leishmania* was found to be conserved across many strains of *Leishmania*, and anti-HbR antibody was detected in kala-azar patients' sera. Based on this, immunization with HbR-DNA was carried out, and data suggested that it induced complete protection against virulent *Leishmania donovani* infection in both BALB/c mice and hamsters with production of Th1 type of immune response [104].

Recently, a first-in-human dose-escalation phase I trial was conducted in 20 healthy volunteers to assess the safety, tolerability, and immunogenicity of a prime-only adenoviral vaccine for human VL and PKDL. ChAd63-KH is a replication-defective simian adenovirus which expresses a novel synthetic gene (KH) encoding two *Leishmania* proteins, KMP-11 and HASPB. Synthetic *haspb* gene was designed to reflect repeat diversity and repeat domain structure of the gene product as known from clinical isolates of *L. donovani* from India and East Africa which represented a novel approach. Innate immune response was seen by whole blood RNA-Seq and antigen-specific CD8+ T-cell responses by IFN-γ ELISpot and intracellular flow cytometry. It was found that ChAd63-KH was safe at intramuscular doses of 1×10^{10} and 7.5×10^{10} vp. Transcriptomic profiling of whole blood showed that ChAd63-KH induced innate immune responses characterized by IFN-γ and the presence of activated dendritic cells. Robust CD8+ T-cell response was induced in all the subjects in the study [105].

1.11   Role and Use of Biomaterials for Leishmania Vaccines

1.11.1 Introduction to Biomaterials and Their Use in Vaccine Delivery

Various advances have been made in the area of vaccine development against infectious diseases. Numerous first-generation, subunit second-generation, and RNA or DNA (third-generation) vaccines have been developed to elicit immune response against the disease.

Some of the important points to be taken care of while developing a new vaccine include (i) safety, (ii) stability, and, (iii) the most important one, generation of disease-specific immune response to combat the disease with a minimum dose [106, 107].

Although RNA/DNA vaccines are associated with various advantages with minimum risk, the delivery of these vaccine molecules to the target site is a big challenge along with requirement of booster dose. There is a strong probability of premature degradation of these molecules, and another challenge is, in some cases, their inability to translate into a functional immunogen [108–110]. Proteins-based vaccines although in use for various infectious diseases are associated with few drawbacks such as need for an adjuvant to potentiate their immunogenicity, and they are more prone to early degradation when exposed to hostile milieu [111, 112]. These shortcomings associated with various vaccines indicate the need for some efficient vaccine delivery system which, along with doing targeted delivery, help in evoking a stronger immune response with minimum dose and side effects. Biomaterials which include natural or synthetic polymers, lipids, nanostructures, and engineered artificial cells cannot only control the required immune response for combating the disease but also help in targeted delivery. Some of the biomaterials used include nanoparticles and microparticles prepared from polymers or lipids. Many scaffolds are also prepared which are either stable or degradable for implantation and devices like microarray needles for targeted delivery to skin [113–117]. Along with these, an array of protein and peptide biomaterials has been used to improve the efficacy and delivery of subunit vaccines for various diseases including infectious diseases, cancer, and autoimmune disorders. Merit associated with these delivery vehicles is that they are biodegradable and have control over both material structure and immune function. These are sometimes made from engineering self-assembling proteins which occur naturally for loading vaccine components [118].

1.11.2 Physical and Chemical Properties of Biomaterials Affecting Their Efficacy

Various properties such as physical and chemical associated with biomaterials affect the final outcome, such as it has been observed that ellipsoidal particles improve the pharmacokinetics better than the spherical molecules which enhances the circulation time and thus promotes immunity [119]. Studies related to immune responses have been facilitated by reductionist system of which acellular artificial antigen-presenting cell (aAPC) is one approach [120]. These aAPCs are being generated by coupling proteins that deliver first signal of TCR signaling, that is, binding of MHC complex or anti-CD3 to TCR, and proteins that deliver second signal which includes binding of co-stimulatory receptors on the APC to the surface of particles prepared from variety of materials such as PLGA microparticles, polystyrene particles, etc. This approach has shown efficacy against tumor. One such study with aAPCs has shown that ellipsoidal PLGA particles functionalized with peptide-MHC complex and anti-CD28 on the surface-mimicked antigen presentation and stimulation of T cells in a better way, thereby increasing the efficacy [121].

In order to elucidate how immune system differentiates various shapes and sizes of antigen, role of morphological features of particles of various sizes and shapes was assessed in antigen presentation and processing by immune cells. It was found that among particles of different types, small spherical particles generated a stronger Th1 and Th2 response when compared with other particle types. Particles of spherical and rod shape were internalized by dendritic cells. This data suggested that modulation of immune response is dependent on size of particle along with shape [122]. In another study in the same row to evaluate the effect of size of particles, lung macrophages and dendritic cells were studied. Inert nontoxic polystyrene nanoparticles 50 nm in diameter (PS50G) and 500 nm in diameter (PS500G) were studied for immunological responses. It was observed that 50 nm *particles* were taken up preferentially by alveolar and nonalveolar macrophages, B cells, and CD11b(+) and CD103(+) DC in the lung. In case of dendritic cells in draining lymph nodes, PS50G were exclusively uptaken. Frequency of antigen-laden DCs was also decreased with PS50G being more efficient. Differential modulation of induction of acute allergic airway inflammation was done by both these particles with PS50G but not PS500G significantly inhibiting adaptive allergen-specific immunity.

Overall data suggested that particles with distinctive sizes differentially modulate the immune response [123].

The immunogenicity of the biomaterial used in the vaccine delivery system is highly impacted and controlled by the chemistry of the material used. The effect of porous silicon nanoparticles with different surface chemistries was evaluated on human monocyte-derived macrophages and lymphocytes. It was observed that thermally oxidized and thermally hydrocarbonized nanoparticles induced very high rate of immunoactivation by increasing the expression of surface co-stimulatory markers. Undecylenic acid-functionalized nanoparticles as well as poly(methyl vinyl ether-alt-maleic acid) conjugated to (3-aminopropyl)triethoxysilane-functionalized thermally carbonized porous silicon nanoparticles and polyethyleneimine-conjugated undecylenic acid-functionalized porous silicon nanoparticles showed moderate immunoactivation. On the contrary, thermally carbonized porous silicon nanoparticles and (3-aminopropyl)triethoxysilane-functionalized porous silicon nanoparticles did not induce any immunological responses [124].

This data concluded that nanoparticles which have more nitrogen or oxygen on the outermost backbone layer are less immunogenic than nanoparticles with higher C-H structures on the surface, suggesting that chemistry plays an important role in immunogenicity of nanoparticles. In another study, hydrophobicity of gold nanoparticles having specific functional groups altered the expression profile of cytokines in splenocytes. In vivo studies also established a direct, quantitative correlation between hydrophobicity and immune system activation, an important determinant for nanomedical and nanoimmunological applications [125].

1.12 Biomaterial-Based Vaccine Delivery Systems Used Against Leishmaniasis

Figure 1.1 illustrates how biomaterial-based vaccine delivery systems can be used against leishmaniasis. In this section, we briefly discuss those.

1.13 Liposomes-Based Vaccine Delivery

1.13.1 Liposomes and How They Work

Liposomes are vesicles of spherical shape and are composed of natural amphiphilic phospholipids which are nontoxic and non-immunogenic. Based on number of lipids in bilayer, liposomes are classified as multilamellar vesicles (MLVs), small

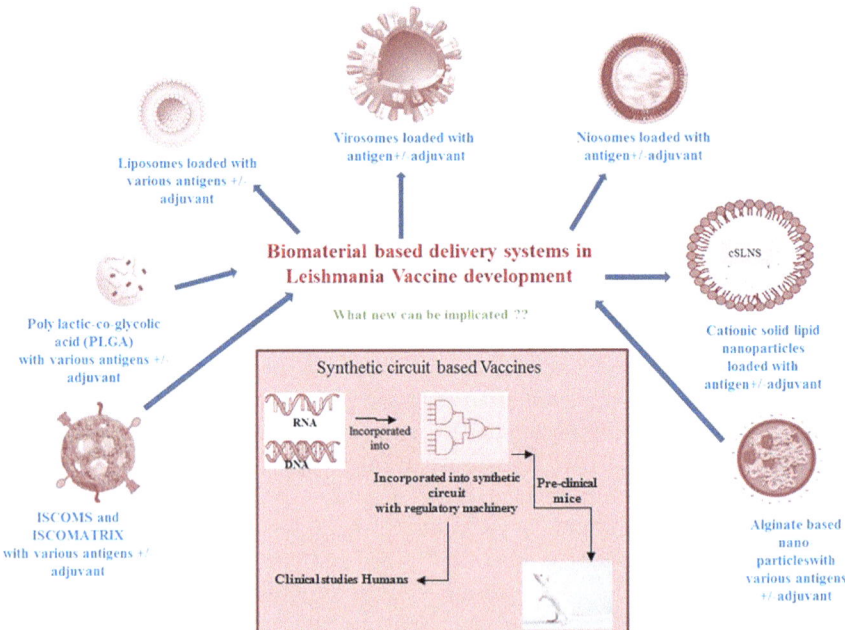

Fig. 1.1 Leishmania vaccine development through biomaterials

unilamellar vesicles (SUVs), or large unilamellar vesicles (LUVs). Liposomal delivery systems are associated with various advantages for vaccine development such as safety, and these are biodegradable because they are often composed of lipids which are found naturally in cell membranes such as phosphatidylcholine (PC) and cholesterol [126]. These liposomes protect the antigens from being cleared from the body and their targeted delivery to the respective antigen-presenting cells. Depending on the nature of antigen, whether it is lipophilic or hydrophilic, it would be incorporated either in lipid layer or inside the aqueous core [127]. This organization enables all types of antigens such as peptides, proteins, carbohydrates, nucleic acids, and small molecules to be encapsulated in liposomes, whenever needed adjuvants such as lipid A, muramyl dipeptide and its derivatives, and various interleukins can also be inserted along with these antigens in liposomes. Liposomes have the ability to channel the proteins and peptide antigens to MHC class II pathway of antigen presentation. This increases the induction of antigen-specific humoral and adaptive T-cell response. These also serve as delivery vehicles for exogenous proteins and peptide antigens to the MHC class II pathway for efficient presentation and induction of cytotoxic T-cell response [128]. The antigenic proteins which are delivered by conventional liposomes are processed via MHC class II pathway and those by pH-sensitive liposome carriers via MHC I presentation. Liposomes have the property of increasing the expression of various chemokine genes such as CCL2 (chemokine (C-C motif) ligand 2), CCL3, and CCL4 by dendritic cells. They increase the longevity of antigens inside the APCs, and since the exposure to antigen is increased, it prolongs the primary activation of T cells.

Not only this, the parameters of liposomal formulations also affect the immune response generated. These parameters include the composition of phospholipids, fluidity of bilayer, charge present on the surface, size of the particle, lamellarity, liposome preparation, antigen attachment, and lamellar-hexagonal bilayer phase transition ability.

Composition of lipids affects in the way that few lipids have main-phase transition temperature (T_m) below 37 °C and will be in the liquid crystalline state in the body, and those with a Tm higher than this will be in gel state. A correlation has been established between the Tm of the phospholipids and the immune response generated with membrane antigens [129]. The state in which bilayer is present physically can also affect many things which influence the immunogenicity, such as endocytosis, intracellular trafficking, and processing of the vaccine components [130]. Phospholipid composition might induce better immune response for a particular antigen and not for the other. Fluidity of bilayer may affect in a way that it affects the rate of release of antigen from vesicles and the interaction of liposome with APCs [126].

The other major thing which comes into play is the size of the vesicle which not only affects the uptake and trafficking of antigens but also influences the draining kinetics of liposomes from the site of the injection [131–133].

Contents of liposomes are delivered by passive or active targeting mechanisms in the cells. The physiochemical properties discussed above such as size and surface charge also affect this targeting [134].

1.14   Various Liposomal Vaccine Formulations Against Leishmaniasis

In this section, we would focus on various liposomal formulations studied against leishmaniasis along with vaccine candidates and the parameters associated with them in consideration.

Soluble antigen of *L. donovani* promastigotes was encapsulated in non-phosphatidylcholine (non-PC) liposomes (escheriosomes) and tested for their protective prophylactic efficacy. Stronger and protective immune response was generated by escheriosomes when compared with soluble antigens alone [135]. A 63-kDa leishmanial glycoprotein gp63 has been used and shown partial protection against visceral leishmaniasis in Balb/c mice without adjuvant. However, when this antigen vaccine was entrapped in cationic DSPC liposomes, it showed significant efficacy as evident by decreased parasitic burden with enhancement of antigen-specific IFN-γ response and downregulation of IL-4, demonstrating a Th1 bias. Results showed that cationic liposomes loaded with gp63 showed long-term protection against *L. donovani* infection [136].

Another study formulated recombinant gp63 either within monophosphoryl lipid A-trehalose dicorynomycolate (MPL-TDM) or entrapped within cationic liposomes or both. It was observed that combined formulation showed better protection both in vitro by restricting the replication of amastigotes and reducing parasitic burden in spleen and liver [137].

In another study, effect of bilayer composition with different phase transition temperature of liposomes on T-cell response was evaluated. Three different liposomes with different bilayer composition were taken, namely, egg phosphatidylcholine (EPC, $T_c < 0$ °C), dipalmitoylphosphatidylcholine (DPPC, T_c 41 °C), and distearoylphosphatidylcholine (DSPC, T_c 54 °C), and were prepared, all loaded with recombinant gp63. Mice immunized with these liposomes in the same dose schedule showed different immunological responses which indicated that these were influenced with the bilayer composition of the liposomes. Liposomes with egg phosphatidylcholine induced Th2-type immune response in mice, and DPPC or DSPC induced Th1 type of immune response signifying that liposomes with higher value of Tm are suitable and induce Th1 type of immune response and protection when used with antigenic Leishmania proteins [138].

Another study using rgp63 emphasized the importance of size of liposomes on their efficacy. Liposomes of different sizes including 100, 400, and 1000 nm were loaded with rgp63 and evaluated for their efficacy against *L. major* challenge in Balb/c mice. It was found that larger size of liposomes induced better production of IFN-γ, highest IgG2a/IgG1 ratio, thereby inducing Th1 type of immune response, whereas small size one of 100 nm induced Th2 response. The data inferred that size of liposomes also plays a significant role in generation of immune response [139].

Role of the charge present on the liposomal surface such as positively charged, negatively charged, or neutral liposome formulations has also been explored in context of efficacy. It has been shown that liposomes with positive charge target antigens for endocytosis more efficiently because of electrostatic interactions between positively charged particles (such as cationic liposomes or cSLN) and negatively charged cell surface of APC and can therefore improve the induction of immune responses even at lower doses. On the contrary, anionic liposomes have low percentage interactions with APCs.

Various negatively and positively charged lipids have been used in liposomal formulations, and net surface charge on the surface of liposomes can be changed by combination of positive- or negative-charged lipids.

Leishmanial antigens isolated from the membrane of *Leishmania donovani* promastigotes were encapsulated in positively charged liposomes (consisting of egg lecithin/stearylamine/cholesterol) when used for immunization, significantly enhancing the protective efficacy of these antigens in comparison to when used alone in BALB/c mice and hamsters model of infection [140]. Similar studies carried out with negatively charged liposomes (consisting of egg lecithin/phosphatidic acid/cholesterol) showed that the level of protection by Leishmania membrane Ag-liposome was not significantly different from that induced by free LAg. It was found that stimulation of insufficient cellular response, as reflected by DTH and potentiation of IgG1 over IgG2a, IgG2b, and IgG3, suggested a dominance of Th2 response with this liposome-antigen formulation, resulting in weak protection against visceral leishmaniasis [141]. Neutral liposomes also showed average protection. Overall data suggests that protection induced by liposomes varied depending on the charge of the vesicles, with maximum induction by positively charged liposomes, followed by neutral liposomes and last negatively charged liposomes. Further studies were done on characterization of Leishmania antigens and antigens entrapped in liposomes of different charges which showed that gp63 was immunodominant in all the vaccine preparations. In addition to gp63, 72-, 52-, 48-, 45-, 39-, and 20-kDa components showed strong reactivity in neutral and positively charged liposomes in contrast to reactivity of a greater number of leishmanial antigenic components in negatively charged liposomes. Data indicated that resistance to VL depended on the immune response induced by gp63 and few other selective antigens with appropriate liposomes [142].

Another study carried out to show role of charge in liposomal formulation prepared liposomes containing rgp63 by dehydration-rehydration vesicle (DRV) method. Composition of liposomes was neutral liposomes with dipalmitoylphosphatidylcholine and cholesterol, and positively charged ones and negatively charged were prepared by adding dimethyldioctadecylammonium bromide (DDAB) or dicetyl phosphate (DCP), respectively, to the neutral liposome formulation. Contrary to above study, it was observed that mice immunized with neutral liposomes showed smaller footpad swelling, significantly lowest splenic parasite burden, the highest IgG2a/IgG1 ratio and IFN-gamma production, and the lowest IL-4 post challenge as compared to other immunized groups. It was evident from the data that Th1 response was induced more efficiently by neutral liposomes than positively charged liposomes, whereas negatively charged liposomes induced a Th2 type of immune response [143].

It was observed that susceptible BALB/c mice when immunized with recombinant stress-inducible protein 1 (rLmSTI1) encapsulated in cationic liposomes induced a significant protection against challenge with *L. major* with significant reduction in parasite burden in spleen and significantly smaller footpad thickness after challenge, again indicating that cationic liposomes increase the efficacy of Leishmania antigens as a vaccine [144].

Liposomal formulation of *Leishmania major* stress-inducible protein 1 (LmSTI1) antigen was evaluated for its efficacy against *L. major* challenge by co-encapsulating CpG ODN in a liposome (Lip-rLmSTI1-CpG ODN) along with other control groups. It was observed that mice immunized with Lip-rLmSTI1-CpG ODN showed a significant decrease in infection as compared to mice immunized with recombinant protein with CpG ODN without liposomal form [145]. Similarly, cationic liposomes containing soluble protein of *L. major* along with CpG ODNs showed a significantly smaller footpad swelling, lower spleen parasite burden, higher IgG2a antibody, and lower IL-4 level compared to the control groups post challenge [146].

Leishmanial elongation factor-1α (EF1-α) has been identified as an immuno-dominant component of soluble leishmanial membrane antigens showing cytokine response in PBMCs of cured VL subjects. 36 kDa truncated and cloned recombinant EF1-α of the *L. donovani* were formulated in cationic liposomes and induce strong resistance to parasitic burden in liver and spleen of BALB/c mice through induction of DTH and a IL-10- and TGF-β-suppressed mixed Th1/Th2 cytokine responses. Multiparametric analysis of splenocytes for generation of antigen-specific IFN-γ, IL2, and TNF-α producing lymphocytes indicates that cationic liposome facilitates expansion of both CD4$^+$ as well as CD8$^+$ memory and effector T cells. Liposomal EF1-α is a novel and potent vaccine formulation against VL that imparts long-term protective responses. Moreover, the flexibility of this formulation opens up the scope to combine additional adjuvants and epitope-selected antigens for use in other disease forms also [147].

Majumdar et al. also showed the effect of composition of phospholipid on the adjuvanticity and efficacy of liposomes carrying *Leishmania donovani* antigens. They used liposomes prepared with distearoyl derivative of L-a-phosphatidylcholine (DSPC) having liquid crystalline transition temperature (T_c) of 54 °C and liposomes prepared from dipalmitoyl (DPPC) (T_c 41 °C) and dimyristoyl (DMPC) (T_c 23 °C) derivative. All these liposomes entrapped *Leishmania donovani* membrane antigen with equal efficiency. However, strong DTH response was shown by Leishmania antigen in DSPC liposomes, whereas other two showed the inconsistent response. Moreover, in terms of protection, also DSPC liposomes showed significantly high protection, with other two formulations showing no protectivity [148]. The authors defended the results by the fact that liposomal structural versatility helps in designing the vesicles for the optimum efficacy, and herein this study, they took the fluidity of bilayer into consideration to improve the stability of the formulation and thereby enhance the efficacy of vaccine. High-melting phospholipid DSPC in the vesicles reduce the bilayer destabilization promoted by plasma, and therefore, the liposomes prepared from it are more rigid and resistant to particle adsorption. Therefore, the significant potential DSPC liposomes observed against experimental VL might be due to their prolonged stay in circulation which enabled more effective delivery of antigens to the antigen-presenting cells [149, 150].

A liposomal formulation coated with neoglycolipids containing oligomannose residues (OMLs) has been explored for better potency as an adjuvant to induce Th1 immune responses and CTLs specific for the encased antigen. These OMLs are uptaken by the phagocytic cells in the periphery, and it has been observed that cells uptaking OMLs secrete IL-12 selectively, enhance the expression of co-stimulatory molecules, and migrate into lymphoid tissues from peripheral tissues [151]. A study conducted using intraperitoneal administration of soluble leishmanial antigen (SLA) entrapped in liposomes coated with neoglycolipids containing oligomannose residues (mannopentaose or mannotriose) showed a strong antigen-specific immune response against *L. major* challenge with high production of IFN-γ and IL-2 and lower IL-4 and IL-5. This immune response generated is thought to be triggered by peritoneal CD11b-positive cells (macrophages) which take up SLA-OML [152].

Above studies indicate that liposomal formulation used for vaccine delivery should be prepared by keeping various parameters in mind for an effective formulation which also depends on the type of antigen being loaded/encapsulated, and such liposomes would not only increase the efficacy but also stability of Leishmania vaccine.

1.15 Virosomes

Spike glycoproteins present on the viral membrane mediate the binding and fusion of membrane-enveloped viruses with the cell surfaces. Lipids vesicles were generated consisting of viral spike proteins derived from influenza virus firstly. Preformed liposomes were used along with hemagglutinin (HA) and neuraminidase (NA), purified from influenza virus to generate membrane vesicles with spike proteins protruding from the vesicle surface and named as virosomes. After that, many other protocols were developed for generating virosomes [153–157].

In short, virosomes are reconstituted viral envelope which resembles intact virus in antigenicity without genetic material and can be used for delivery of vaccines. Viral membrane is reconstituted with help of a detergent used to solubilize the viral envelope, followed by removing the viral nucleocapsid by ultracentrifugation and removal of detergent from the supernatant [158]. These spherical, unilamellar vesicles in contrast to liposomes contain functional viral envelope glycoproteins: influenza virus hemagglutinin (HA) and neuraminidase (NA) intercalated in their phospholipid bilayer membrane. Presence of HA and NA contributes to the unique properties of the virosomes. The viral proteins not only confer structural stability and homogeneity to virosomal formulations but significantly contribute to the immunological properties of virosomes, which are clearly distinct from other liposomal and proteoliposomal carrier systems. The unique properties of virosomes partially relate to the presence of biologically active HA from influenza virus in the virosome membrane.

These virus like particles were loaded with three different recombinant proteins, namely vector-derived (VD) component LJL143 obtained from *Lutzomyia longipalpis*'s saliva, parasite-derived components KMP11 and LeishF3+, and a TLR4 agonist, GLA-SE, taken as an adjuvant, were assessed for in vivo safety and immunogenicity. This vaccine was found to be safe during the treatment time frame. Antigen-specific cellular and humoral responses confirmed the immunogenicity of the vaccine formulation. There was an interesting and noticeable finding that VD proteins induced a more robust immune response, and these were not due to immunodominance of the VD antigen. Moreover, priming with VD antigen alone and then using complete vaccine candidate as booster improved the immune response remarkably [159].

In another study, novel virosomal formulations of a synthetic oligosaccharide were prepared and evaluated as vaccine candidates against leishmaniasis. A synthetic tetrasaccharide antigen related to lipophosphoglycan was conjugated to a phospholipid and to the influenza virus coat protein hemagglutinin. Lipid membrane of reconstituted influenza virus virosomes was used to embed these glycan conjugates. It was observed that this virosomal formulation showed both IgM and IgG anti-glycan antibodies in mice, indicating an antibody isotype class switch to IgG. Along with this, the antisera cross-reacted with the corresponding natural carbohydrate antigens in vitro expressed by leishmanial cells. Overall, experimental observations suggest that virosomes can be used as a universal antigen delivery platform for synthetic carbohydrate vaccines [160].

1.16   Niosomes

These are weakly immunogenic nonionic surfactant vesicles which consist of one or more bilayers of lipid encapsulating an aqueous core which can encapsulate both lipophilic and hydrophilic content and protect them from acidic environment in gastrointestinal tract and their enzymatic degradation [161].

A study was conducted using different positively charged niosomal formulations with the composition of sorbitan esters, cholesterol, and cetyl trimethyl ammonium bromide. These delivery vehicles were prepared by film hydration method for the entrapment of autoclaved Leishmania major (ALM). Stability and size distribution of these niosomes were evaluated by laser light scattering method. Percentage of encapsulated ALM was quantified by bicinchoninic acid method. Based on above observations, the selected niosomes were assessed for their efficacy to induce an immune response against Balb/c mice model of cutaneous leishmaniasis. Data showed that niosomes with ALM delayed the lesion development and reduced their size as compared to ALM alone, but this formulation did not show complete protection. It was inferred that the delay in lesion development might be due to the slow release of antigens from niosomes, thereby evoking a strong immune response. Using a more refined strategy for antigen selection along with improving the niosomal formulation could be a prevention strategy for CL [162].

Another group encapsulated gp63 protein in niosomal formulation. These niosomes were prepared following the procedure of Baillie. Vaccination data in C57BL/10 mice indicated that group vaccinated with purified gp63 entrapped into niosomes induced considerable resistance to disease, whereas other group vaccinated with liposomal formulation did not. In the niosomal group, at the time of termination of experiment, mice presented only ulcerated lesions that started to heal [163].

Both the studies indicated that niosomes have a potential to enhance the immune response when used with appropriate antigens. But the drawback summarized from these studies emphasizes the need for better standardization of niosomal formulation along with the antigenic combination to be selected.

1.17 Cationic Solid Lipid Nanoparticles

These nanoparticles are an efficient alternative to the available traditional colloidal carriers such as emulsions and liposomes. Solid lipid nanoparticles have an advantage over the colloidal carrier with the use of biocompatible lipids such as triglycerides, fatty acids, free fatty acids, steroids, fatty alcohols, or waxes [164]. Another distinct advantage of these nanoparticles over polymeric nanoparticles is their production without any organic solvent, by the use of high-pressure homogenization (HPH) method which is well implemented in pharma industry [165].

Cationic solid lipid nanoparticles (cSLNs) have at least one cationic lipid and have been implicated as nonviral vectors of gene delivery, and these have been found to bind effectively with nucleic acids protecting them from degradation by DNAse I and their delivery to live cells [166–169].

They act as delivery systems via two mechanisms, either by encapsulating the antigen inside the lipid matrix or by absorbing antigen on their surface by electrical interaction.

Based on these properties, cationic liposomes were tested in vitro for delivery of cysteine proteinases *cpa*, *cpb*, and *cpb*CTE. For this, melt emulsification method was used followed by HSH method to prepare cSLNs. Plasmids having type I and II cysteine proteinases were anchored on the cationic surface of these nanoparticles. This strategy was found to be efficient enough to deliver the immunogenic CP genes which were evident by expression of CPs in vitro for 72 h after their COS-7 cells treatment. It also had an advantage to overcome the drawback of degradation of naked DNA delivery in the circulation [170]. In another study, three pDNAs encoding cysteine proteinase type I (cpa), II (cpb), and III (cpc) of *L. major* were formulated using cSLNs and used for immunizing BALB/c. It was observed that group vaccinated with SLN-pcDNA-cpa/b/c showed significantly high production with Th1 type of immune response [171].

This delivery system was also compared with electroporation in administering DNA vaccine containing A2 gene of *L. donovani* along with cysteine proteinases [CPA and CPB without its unusual C-terminal extension (CPB_CTE)] of *L. infantum*. It was observed that cSLNs were equally efficient as electroporation delivery system in protecting Balb/c mice against L. donovani challenge by evoking an immune response with high levels of IFN-γ [172]. The protective efficacy of these two vaccine delivery systems containing abovesaid DNA vaccine was further evaluated against *L. infantum* challenge in outbred dogs. The results indicated the efficacy of cSLNs as carrier systems to increase the efficacy of DNA vaccines against canine visceral leishmaniasis [173].

1.18 Poly Lactic-Co-Glycolic Acid (PLGA) Delivery Systems

These biodegradable poly(D,L-lactide-co-glycolide) (PLGA) nanoparticles (NPs) have gained a lot of attention as carrier systems due to their property of biocompatibility and are being US Food and Drug Administration (FDA) and European Medicines Agency (EMA) drug carriers [174]. Hydrolysis of PLGA leads to metabolite monomers, lactic acid, and glycolic acid, and since these two monomers are endogenous and easily metabolized by the body, the toxicity associated with PLGA is minimal.

PLGA nanoparticles with surface modified with a TNF-α-mimicking eight-amino acid peptide (p8) and encapsulating *L. infantum*-soluble antigen along with monophosphoryl lipid A (MPLA), a TLR4 ligand was assessed against *L. infantum* challenge in BALB/c mice. Results conferred significant protection with nearly complete elimination of parasite along with antigen-specific immune response [175].

Two strategies were used to infer the efficacy of PLGA in terms of protection against CL. In one, immunization was carried out using plasmid DNA encoding *L. infantum chagasi* KMP-11, and in other one, mice were primed with PLGA loaded with recombinant plasmid DNA followed by booster dose of PLGA nanoparticles loaded with recombinant KMP-11. Both the strategies showed significant cellular immune response. However, the decrease in parasitic load at infection site was more prominent in mice immunized with PLGA than with plasmid DNA alone encouraging the use of nanobased delivery systems for *Leishmania* vaccines.

Soluble Leishmania antigen or autoclaved leishmanial antigen was also loaded in PLGA nanoparticles to evaluate the performance of this system both in vitro and in vivo. Contrary to free antigens, both these formulations showed significant potential as evident by higher level of NO production by macrophages. In vivo data suggested increased production of IFN-γ and IL-12 levels and inhibiting IL-4 and IL-10 secretions showing more than 50% protective efficacy in mouse model [176].

These nanoparticles as antigen delivery system were used along with Quillaja saponins (QS) as immunoadjuvant to enhance the immune response of autoclaved

Leishmania major (ALM) against *L. major* challenge. Surprisingly, it was observed that group vaccinated with ALM encapsulated in PLGA showed protection, but in the group wherein QS was also incorporated, no protection was observed, thereby inferring that PLGA have the ability to enhance immune response against Leishmania infection, but it was reversed with QS as an adjuvant [177].

With an aim of improving the immunogenicity of peptide, rationally designed multi-epitope peptide of *Leishmania* cysteine protease A ($CPA_{160-189}$) was co-encapsulated along with MPLA adjuvant in PLGA nanoparticles, and their pro-phylactic efficacy was evaluated against VL. The phenotypic function of DCs and their functional features on exposure to peptide alone and various combinations along with encapsulated peptide and adjuvant were examined using BALB/c bone marrow-derived DCs. It was observed that DCs exposed to PLGA-$CPA_{160-189}$ + MPLA NPs showed signatures of DC maturation. Mice immunized with this combination showed high amounts of IL-2, IFN-γ, and TNF-α and, when challenged with *L. infantum* promastigotes, showed remarkable reduction in para-sitic burden; however, post four months of challenge, the reduced parasitic load in liver and spleen was preserved indicating that vaccine induced partial protec-tion [178].

Similar peptide-based study using PLGA nanoparticles was carried out by designing a chimeric peptide containing HLA-restricted epitopes from three immu-nogenic *L. infantum* proteins (cysteine peptidase A, histone H1, and KMP 11) and their encapsulation in PLGA nanoparticles with or without monophosphoryl lipid A (MPLA) adjuvant or surface modification with an octapeptide targeting the tumor necrosis factor receptor II.

These were tested for their capability to stimulate immunomodulatory functions of DCs. Peptide-based nanovaccine candidates with MPLA incorporation or surface modification stimulated DCs efficiently as evident by prominent IL-12 production, promoting allogeneic T-cell proliferation and intracellular production of IFN-γ by $CD4^+$ and $CD8^+$ T-cell subsets. When HLA A2.1 transgenic mice was immunized with this peptide nanovaccine, it conferred significant protection against *L. infan-tum* infection indicating the protective efficacy of this approach [179].

1.19 ISCOMS and ISCOMATRIX (Immune-Stimulating Complexes)

ISCOMs have a history of 35 years as adjuvant delivery system for various experimental vaccines. These are spherical open cage-like nanoparticles which are prepared spontaneously by mixing of cholesterol, phospholipids and the immune-stimulating saponins under a specific stoichiometry along with vaccine candidate. They have shown strong immunostimulatory property and have been found to enhance the protective potential of various vaccines [180].

ISCOMATRIX have same structure as ISCOM, but these are without antigen and also known as ISCOMATRIX™, a trademark of ISCOTEC AB. Both of these have strong negative charge and exist as spherical, hollow rigid cage of about 40 nm diameter. ISCOMATRIX have an advantage with presence of an in-built adjuvant (Quil A) which is a purified fraction of Quillaja saponaria along with cholesterol and phospholipids which form a cage-like structure [181].

Surface glycoprotein of *Leishmania major* was incorporated in immunostimulating complexes (ISCOMs) and evaluated for its efficacy in Balb/c mice. It was found that two intraperitoneal low doses of this complex showed protective immunity by modulating the immune response toward Th1 type, and lesion size was suppressed after challenge [182].

Different formulations of ISCOMATRIX mixed with soluble Leishmania antigens were tested against *L. major* challenge in BALB/C mice. It was observed that group of mice immunized with ISCOMATRIX DMPC or ISCOMATRIX DSPC showed reduced swelling in footpad and least parasitic burden as compared to other ISCOMATRIX formulations, but it was not significantly different from other vaccinated groups. These groups showed highest level of IFN-γ and IL-4 in the mice splenocytes, thereby indicating the generation of mixed Th1/Th2 response. It was also concluded that efficacy of ISCOMATRIX was not influenced with different phospholipids in their mice model [183].

1.20 Delivery System Using Alginate

Alginate is obtained from cell walls of marine brown algae, and it is an anionic polysaccharide consisting of a chain of (1–4)-linked β-D-mannuronic acid and α-L-guluronic acid. It is a natural, biodegradable polymer with no toxicity. Nanoparticles prepared using alginates are hydrophilic carriers which prolong the release of antigen and also improve the antigenicity of traditional vaccines. Agglomeration of these alginate nanoparticles has not been observed in any major organs which further proves their safety [184, 185].

Alginate microspheres have been used as antigen delivery system and adjuvant for immunization against leishmaniasis. Encapsulation of autoclaved *L. major* along with Quillaja saponin (QS) was carried out. The goal of this study was to prepare and characterize alginate microspheres as an antigen delivery system and adjuvant for immunization against leishmaniasis. Alginate microspheres (ALG) encapsulating autoclaved *L. major* (ALM) and Quillaja saponins (QS) were prepared and tested in BALB/c mice against *L. major* challenge. It was observed that mice immunized with (ALM) ALG + QS showed strongest protection as evident by smaller footpad in immunized mice. Mice immunized individually with (ALM + QS)ALG, ALM, and PBS did not show noticeable protection, whereas (ALM)ALG- and ALM + QS-immunized mice showed an intermediate protection [186].

Another study used alginate microspheres prepared by an emulsification technique as an antigen delivery system along with CpG-ODN as adjuvant to enhance immunoprotective response of autoclaved *Leishmania major* (ALM) vaccine. Immunization groups taken were ALM-encapsulated alginate microspheres [(ALM)

$_{ALG}$], (ALM) $_{ALG}$ + CpG, ALM + CpG, ALM alone, and PBS. It was observed that mice immunized with (ALM + CpG) $_{ALG}$ showed maximum protection as evident by smallest footpad swelling compared to other groups. Other combinations, namely, (ALM + CpG) $_{ALG}$ or ALM + CpG, also showed protective response. Data concluded that alginate microspheres and CpG-ODN adjuvant when used together remarkably enhanced the protective response of ALM [187].

1.21   Immunocircuits-Based Therapeutic Vaccines

Therapeutic vaccines have gained a lot of attention since last decade, especially in case of chronic infections, cancer, and other diseases [188–192]. In case of leishmaniasis, most of population in the endemic disease-affected areas are usually healthy endemic individuals or individuals with asymptomatic infection, which serve as reservoir of parasite and can transform into symptomatic infected individuals over a period of time. Based on this, therapeutic vaccines could be an effective alternative for stimulating the immune system of the patient in these endemic populations for protecting from progression of disease. These therapeutic vaccines have been evaluated in case of leishmaniasis in combination with drugs or adjuvants for their efficacy [193–202].

As discussed above, various strategies and carrier biomaterials have been used to enhance the quality and magnitude of cellular and humoral immune response post immunization for developing new prophylactic and therapeutic vaccines against leishmaniasis, but very few have been successful, with none in clinics against humans. It has been well proven that the kinetic pattern of exposure of antigen used for the vaccine development along with adjuvant to the naive T cells and B cells affects the final immunological response outcome post vaccination. Most of the above-discussed vaccination approaches lack the active control over the progressive pattern of antigen exposure and delivery to lymph nodes. Based on above shortcomings of known strategies, herein we propose the application of synthetic biology for development of nucleic acid-based vaccines where expression dynamics of antigen used in the vaccine along with adjuvant can be externally controlled by the help of a programmed genetic synthetic circuit with mRNA for the first time in leishmaniasis as per our information. Various therapeutic circuits have been developed against cancer, metabolic disorders, allergies, etc. [203–208]. Recently, mRNA-based approaches for vaccine designing have become more popular due to problems associated with DNA vaccines.

These RNA-based synthetic circuit vaccines are usually cost-effective along with a major benefit of external control than protein-based vaccines. Moreover, it is assumed that these programmed vaccine approach might not require a booster dose, which is another hurdle for mass immunization. These synthetic circuits-based nucleic acid vaccines could be optimized for regulated expression of antigens and adjuvants from RNA replicons using some small molecules and with a preprogrammed temporal pattern (Tables 1.1, 1.2 and 1.3).

Table 1.1 Strategies adopted for Leishmania vaccines and few vaccines developed using those strategies

S.no.	Strategy adapted	Rationale behind it	Vaccine candidates	Disease type	Tested in	Final outcome	Clinical trial (if yes, which phase)	References
1.	Introduction of live *Leishmania* parasite (*leishmanization*)	Controlled induction of disease to prevent the consequences of natural infection	*Leishmania major* parasites were used in former USSR, Israel, and Iran.	CL	Was in clinical practice in former USSR, Israel, and Iran and still in Uzbekistan	Many cases of nonhealing lesions	Was in clinical practice	[19–26]
2.	Live attenuated vaccines	These have advantage of partially mimicking the natural course of infection.	Develop attenuated strains in in vitro cultures by selecting for temperature sensitivity, chemical mutagenesis [32], and γ-attenuation [33] or by keeping parasite culture under drug pressure [34].	CL, VL, and MCL	Mice	Mixed. In some studies, it showed protective response, whereas contrary data was also observed.	NA	[38]
		Loss-of-function mutants (knockout)	DHFR-TS (dihydrofolate reductase thymidylate synthase) (first knockout vaccine against Leishmania)	CL	Mice	Induced a potential protection against the virulent parasite	NA	[39]
					Rhesus macaques (Macaca mulatta)	Protective immunity was not observed in monkeys post vaccination.		[40]

	L. major lipophosphoglycan 2 (LPG2−/−)	CL	Mice	Showed protection in mice model	NA	NA
	L. mexicana cysteine proteases+(CPA/CPB−/−)	CL	Mice	Showed protection in mice model	NA	[45]
	L. donovani Centrin (Cen−/−)	VL	Mice and dog	Induces long-lasting protective immunity	NA	[42]
	L. infantum heat shock protein 70 type II (HSP70-II−/−)		Mice	Induced protective response	NA	[46]
	L. infantum silent information regulatory protein 2 (SIR2+/−)		Mice	Induced Th1-type immune response	NA	[44]
	L. donovani protein p 27 (Ldp27−/−)		Mice	Long term protective immunity in BALB/c mice	NA	[47, 48]
Gain-of-function mutants	L. major thymidine kinase (herpes simplex virus)	CL	Mice	Variable levels of protection, from partial to total	NA	[53]
	L. amazonensis δ-aminolevulinate dehydratase, porphobilinogen deaminase knock-in: alad-pbgd+/+)	VL	Hamster and mice	Have shown Th1-type protective response	NA	[36]

(continued)

Table 1.1 (continued)

S.ro.	Strategy adapted	Rationale behind it	Vaccine candidates	Disease type	Tested in	Final outcome	Clinical trial (if yes, which phase)	References
3.	First-generation vaccines	These vaccines elicit a specific memory response without any expected pathology even in immunocompromised individuals.	Leishvaccine (whole-killed promastigotes of *Leishmania amazonensis* and bacillus Calmette-Guérin (BCG))	Canine leishmaniasis	Dog	Successfully to phase I and II clinical trials, wherein it showed good safety and immunogenicity, but it failed to give similar immunoprotective results in randomized phase III clinical trial	Phase I and II clinical trials	[59]
			Alum-precipitated autoclaved *L. major* + BCG	VL and PKDL			NA	[22]
			Psoralen compound amotosalen-treated *L. infantum* and *L. chagasi* + UV radiation	VL and CL	Mice	Induced protective response	NA	[60]
			L. mexicana + BCG	CL	Human	Low levels of leishmanin skin test (LST) and the participants which showed LST conversion had low incidences of leishmaniasis		[61]
			Autoclaved *L. major* (ALM) + BCG	CL	Phase I and II clinical trials within healthy individuals	LST conversion was observed only in 36% of healthy individuals with low levels of IFN-γ production on stimulation with soluble Leishmania antigen.	Phase I and II clinical trials	[62]

#	Category	Description	Disease	Subjects	Result	Status	Ref
		Total or soluble antigens of *L. donovani* used along with MPL-A, BCG, or liposomes as vaccine candidates	VL	Preclinical trials (mice, hamsters, and monkeys)	Showed protection	NA	[43, 64–66]
		97–68 kDa fraction of *L. donovani*-soluble proteins	VL	Hamster	Showed protective Th1-type response		[67, 68]
		Leishmune (purified fraction named as fucose mannose ligand (FML) + a saponin adjuvant which includes QS21 and two deacylated saponins)	VL (canine leishmaniasis)	Dogs	Showed more than 90% efficacy in Brazil	In clinical use	[12, 22]
		CaniLeish® (excreted-secreted proteins (ESP) of *Leishmania infantum* (LiESP) + adjuvanted with QA-21)	VL (canine leishmaniasis)	Dogs	Showed Th1-type immune response	In clinical use	[68]
4.	Second-generation vaccines	More refined products such as recombinant proteins along with adjuvant have advantage of large-scale production.					
		Recombinant proteins: *Membrane antigens:* Hydrophilic acylated surface protein B1 (LdHASPB1), KMP-11, surface protein gp63, etc. along with adjuvant	CL, VL, MCL, and CVL	Mice	Preclinical data showed protective response.	NA	[71, 74–76]
		Soluble antigens: rA2 protein Non-replicative adenovirus vector expressing A2 (rAd5-A2) and boosted with the rA2 protein		Mice and canines	Protective	NA	[77–83]

(continued)

Table 1.1 (continued)

S.no.	Strategy adapted	Rationale behind it	Vaccine candidates	Disease type	Tested in	Final outcome	Clinical trial (if yes, which phase)	References
			Leishmania elongation factor 2 (LeIF-2)		Primate Macaca mulatta	Protective		
			Histone proteins					
			LACK protein + adjuvant		Hamster	Protective	NA	
			LACK- and IL-12-expressing Lactococcus lactis		Hamster, dogs, and monkeys	Partial to complete protection	NA	
					Mice	Protective		
			Cofactor-independent phosphoglycerate mutase		Hamster and mice	Protective	NA	
			Fusion proteins:					
			Protein Q (four fragments of the acidic ribosomal protein Lip2a, Lip2b, P0, and histone 2A (H2A) + BCG/hsp70/ CpG-ODN/pUC18/ pcDNA3/ alum/ Freund's adjuvant as adjuvant)		Dog and mice	Showed protection	NA	[95–98, 100]
			LEISH-F1 (recombinant artificial protein encoded by three genes: L. major homologue of eukaryotic thiol-specific antioxidant (TSA), L. major stress-inducible protein-1 (LmSTI1), and L. braziliensis elongation initiation factor (LeIF) + MPL-SE)		Phase II of clinical trials			

Vaccine/antigen	Disease	Model	Outcome		References
LEISH-F2: Same as LEISH-F1 with removal of N-terminal histidine tag+ MPL-SE		Phase II clinical trials	Showed protection		[88–94]
LEISH-F3 (nucleoside hydrolase (NH) and sterol 24-c-methyltransferase (SMT), derived from *L. donovani* and *L. infantum*, respectively) + GLA-SE		Preclinical studies	Showed protection		
Peptide vaccines:	CL, VL, and MCL				
A single synthetic T-cell epitope (PT3) from histidine zinc-binding region of the metalloprotease gp63		Mice	Provided long-lasting protection		
DCs pulsed with peptide 154-169aa of gp63		Mice	Modulation of the cellular immune response toward a Th1 profile		
Three human HLA-DR1-restricted peptides derived from *L. major* gp63 protein		FVB/N-DR1 transgenic mouse model	High levels of Th1-type immune response		
Epitope-based immunogens, namely, B10 and C01, presented as phage-fused peptides with or without adjuvant		Mice	Partial protection		
Amastigote stage-specific protein: A2, P4, and P8 individually+ *Corynebacterium parvum*	CL and VL	Mice	Th1 cell-mediated immune response (partial to complete protection)	NA	[72]
Hypothetical *Leishmania* amastigote-specific protein (LiHyp1) + saponin		Mice	Showed protection with increased Th1-type immune response		[73]

(continued)

Table 1.1 (continued)

S.no.	Strategy adapted	Rationale behind it	Vaccine candidates	Disease type	Tested in	Final outcome	Clinical trial (if yes, which phase)	References
5.	Third-generation vaccines	Generating an antigen-specific immune response by intramuscular injection of plasmid in animal model brought up the idea of new arm of vaccine research, DNA vaccines.	DNA vaccination with LACK (Leishmania analogue of the receptor kinase C), TSA (thiol-specific-antioxidant) genes alone, or LACK-TSA fusion	CL and VL	Mice	Fusion group showed better efficacy	NA	[102–105]
			Glycosomal phosphoenolpyruvate carboxykinase (PEPCK) DNA vaccine		Mice	Showed significant protection by intradermal route	NA	
			Hemoglobin receptor (HbR) DNA vaccine		Mice and hamster	Significant protection with Th1-type immune response	NA	
			ChAd63-KH (a replication defective simian adenovirus expressing a novel synthetic gene (KH) encoding two Leishmania proteins KMP-11 and HASPB)		Human	Safe and showed increased IFN-γ along with robust CD8+ T-cell response		

Table 1.2 Few biomaterials used for Leishmania vaccine and how they work as adjuvant

S. no.	Formulation	Composition	How they aid vaccine efficacy
1.	Liposomes	Natural amphiphilic phospholipids	(a) Longevity of antigen and targeted delivery (b) Have the ability to channel the proteins and peptide antigens to MHC class I and II pathway of antigen presentation (c) Increase the expression of various chemokine genes such as CCL2 (chemokine (C-C motif) ligand 2), CCL3, and CCL4 by dendritic cells
2.	Virosomes	Spherical, unilamellar vesicles: Reconstituted viral envelopes which contain functional viral envelope glycoproteins: Influenza virus hemagglutinin (HA) and neuraminidase (NA) intercalated in their phospholipid bilayer membrane.	(a) Targeted delivery and are immunogenic themselves as well
3.	Niosomes	Prepared by hydrating synthetic mono- or dialkyl surfactants + cholesterol/ amphiphilic molecules	(a) Can elicit both cell-mediated and humoral immune responses
4.	Cationic solid lipid nanoparticles	Prepared from biocompatible lipids such as triglycerides, fatty acids, free fatty acids, steroids, fatty alcohols, or waxes without use of any solvent by high-pressure homogenization (HPH) method	(a) These positively charged carriers bind with polyanionic DNA, protecting DNA from interacting with other small molecules and its delivery to cells by endocytosis. (b) Moreover, they enhance the stability of the vaccine candidate during transport, since leishmaniasis is widespread in areas lacking cold storage, and this is an important benefit.
5.	PLGA nanoparticles	Prepared by single- or double-emulsion solvent evaporation/extraction, nanoprecipitation, salting out, membrane emulsification, microfluidic technology, and flow focusing	
6.	ISCOMS and ISCOMATRIX	Mixing together cholesterol, phospholipids, and the immune-stimulating saponins under a specific stoichiometry	(a) ISCOMs could prominently enhance the antigen targeting, uptake, and activity of antigen-presenting cells including dendritic and B cells and macrophages resulting in the production of proinflammatory cytokines. (b) ISCOMATRIX adjuvant helps both in antigen delivery and immune stimulation. (c) ISCOMATRIX adjuvant can induce both Th1 and Th2 responses.
7.	Alginate-based microspheres/ nanoparticles	It is an anionic polysaccharide consisting of a chain of (1–4)-linked β-D-mannuronic acid and α-L-guluronic acid.	a) Prolong the release of antigen and also improves the antigenicity of traditional vaccines

Table 1.3 Few examples of various biomaterials used for Leishmania vaccines

S. no.	Type of biomaterial used	Antigen encapsulated	Tested against	Outcome	References
1.	Liposomes				
	Non-phosphatidylcholine (non-PC) liposomes (escheriosomes)	Soluble antigen of L. donovani promastigotes	VL	Stronger protective response than naked antigen	[135]
	Cationic DSPC liposomes	A 63-kDa leishmanial glycoprotein gp63	VL	Decreased parasitic burden with enhancement of antigen-specific IFN-γ response and downregulation of IL-4	[136]
	Cationic liposomes + monophosphoryl lipid A-trehalose dicorynomycolate (MPL-TDM)	Recombinant gp63	VL	Enhanced immune responses that further resulted in high protective levels against VL in the mouse model	[137]
	Egg phosphatidylcholine in lipid bilayer (EPC)	rgp63	CL	Liposomes consisting of EPC induced a Th2 type of immune response, while liposome consisting of DPPC or DSPC induced Th1 type of immune response.	[138]
	Dipalmitoylphosphatidylcholine in lipid bilayer (DPPC)				
	Distearoylphosphatidylcholine in lipid bilayer (DSPC)				
	Liposomes of different sizes including 100, 400, and 1000 nm	rgp63	CL	Larger-size liposomes showed Th1 response, while smaller of 100 nm showed Th2 response.	[139]
	Positively charged liposomes (consisting of egg lecithin/stearylamine/cholesterol)	Leishmanial antigens extracted from the membranes of Leishmania donovani promastigotes	VL	Enhanced the protective efficacy of these antigens	[140]
	Negatively charged liposomes (consisting of egg lecithin/phosphatidic acid/cholesterol)	Leishmania membrane ag	VL	No difference in protection when compared with antigen alone	[141]

Neutral liposomes with dipalmitoylphosphatidylcholine and cholesterol Negatively charged: Neutral liposomes+ dicetyl phosphate (DCP) Positively charged: Neutral liposomes+ dimethyldioctadecylammonium bromide (DDAB)	rgp63	CL	Mice immunized with neutral liposomes showed smaller footpad swelling, significantly lower splenic parasite burden, and the highest IgG2a/IgG1 ratio and IFN-gamma production.	[143]
Liposomes having 1,2-distearoyl-sn-glycero-3-phosphocholine (DSPC) and cholesterol	Leishmania major stress-inducible protein 1 (LmSTI1)	CL	Induced a significant protection against challenge	[144]
Liposomes having lipid phase of distearoylphosphatidylcholine and cholesterol	rLmSTI1 and CpG ODN (lip-rLmSTI1-CpG ODN)	CL	Lip-rLmSTI1-CpG ODN induced remarkable protective response	[145]
Cationic lipid with 1,2-dioleoyl-3-trimethylammonium-propane	Soluble Leishmania antigens (SLA) + CpG ODNs	CL	Showed a significantly smaller footpad swelling, lower spleen parasite burden, higher IgG2a antibody, and lower IL-4 level compared to the control groups	[146]
Cationic liposomes with lipid film of 1,2-distearoyl-sn-glycero-3-phosphocholine, cholesterol, and stearylamine	36 kDa truncated as well as the cloned recombinant EF1-α	VL	Strong resistance to parasitic burden in liver and spleen of BALB/c mice immunized with 36 kDa truncated as well as the cloned recombinant EF1-α-loaded cationic liposomes	[147]

(continued)

Table 1.3 (continued)

S. no.	Type of biomaterial used	Antigen encapsulated	Tested against	Outcome	References
	Liposomes prepared with distearoyl derivative of L-a-phosphatidylcholine (DSPC)	Leishmania donovani membrane antigen	VL	In terms of protection, also DSPC liposomes showed significantly high protection with other two formulations showing no protectivity.	[148]
	Liposomes prepared from dipalmitoyl (DPPC)				
	Liposomes having dimyristoyl (DMPC)				
	Liposomal formulation coated with neoglycolipids containing oligomannose residues (OMLs)	Soluble leishmanial antigen (SLA)	CL	Showed a strong antigen-specific immune response with high production of IFN-γ and IL-2 and lower IL-4 and IL-5	[151, 152]
2.	Virosomes				
	Inactivated *Influenza* virus A/H1N1/California + octaethylene glycol mono (n-dodecyl) ether + phosphatidylcholine	Three different recombinant proteins (LJL143 from *Lutzomyia longipalpis's* saliva as the vector-derived (VD) component and KMP11 and LeishF3+ as parasite-derived (PD) antigens) + GLA-SE	VL	Antigen-specific cellular and humoral responses were higher in immunized versus control groups. The immune responses against the VD protein were reproducibly more robust than those elicited against leishmanial antigens.	[158]
	Virosomal formulations of a synthetic oligosaccharide	A synthetic tetrasaccharide antigen related to lipophosphoglycan was conjugated to a phospholipid and to the influenza virus coat protein hemagglutinin.	VL	Elicited both IgM and IgG anti-glycan antibodies in mice	[160]

3.	Niosomes				
	Positively charged niosomal formulations with the composition of sorbitan esters, cholesterol, and cetyl trimethyl ammonium bromide	Autoclaved Leishmania major (ALM)	CL	Evoked a strong immune response	[162]
	Niosomal formulation	gp63 protein	CL	Immunized mice induced considerable resistance to disease.	[163]
4.	Cationic solid lipid nanoparticles (cSLNs)				
	cSLNs formulated of cetyl palmitate, cholesterol, DOTAP, and tween 80	In vitro for delivery of cysteine proteinases *cpa*, *cpb*, and *cpb*CTE	COS-7 cells	Efficient delivery	[170]
	DOTAP + cetyl palmitate + cholesterol + tween 80	Formulated three pDNAs encoding *L. major* cysteine proteinase type I (cpa), II (cpb), and III (cpc)	CL	SLN-pcDNA-*cpa/b/c* showed higher protection levels with specific Th1 immune response.	[171]
	Cationic solid lipid nanoparticle (cSLN) formulation	A DNA vaccine harboring the *L. donovani* A2 antigen along with *L. infantum* cysteine proteinases [CPA and CPB without its unusual C-terminal extension (CPB^{-CTE})]	VL	Results were comparable to electroporation and showed efficiently protective response.	[172]
	–Same as above–	–Same as above–	Canine visceral leishmaniasis	Increased the efficacy of DNA vaccines	[173]

(continued)

Table 1.3 (continued)

S. no.	Type of biomaterial used	Antigen encapsulated	Tested against	Outcome	References
5.	Poly lactic-co-glycolic acid (PLGA)				
	PLGA nanoparticles with surface modified with a TNF-α-mimicking eight-amino acid peptide (p8)	*L. infantum*-soluble antigen+ MPLA	VL	Significant protection with nearly complete elimination of parasite	[175]
	PLGA nanoparticles	Plasmid DNA-encoding *L. infantum chagasi* KMP-11	VL	Showed significant cellular immune response	
	PLGA nanoparticles	Soluble Leishmania antigen/autoclaved leishmanial antigen	VL	Increased production of IFN-γ and IL-12 levels and inhibiting IL-4 and IL-10 secretions showing more than 50% protective efficacy	[176]
	PLGA nanoparticles	Autoclaved Leishmania major (ALM) + Quillaja saponins (QS)	CL	Group vaccinated with ALM + PLGA showed protection, but in the group wherein QS was also incorporated, no protection was observed.	[177]
	PLGA nanoparticles	Multi-epitope peptide of *Leishmania* cysteine protease A (CPA$_{160-189}$) + MPLA	VL	Showed high amounts of IL-2, IFN-γ and TNF-α along with partial protection	[178]
	PLGA nanoparticles	Chimeric peptides containing HLA-restricted epitopes of cysteine peptidase A, histone H1, and KMP 11 +/− monophosphoryl lipid A (MPLA) adjuvant or surface modification with an octapeptide targeting the tumor necrosis factor receptor II	VL	Significant protection against *L. infantum*	[179]

6.	ISCOMS and ISCOMATRIX (immune-stimulating complexes)				
	ISCOMs (triterpenoids + cholesterol + phosphatidylcholine in the presence of MEGAIO)	gp63	CL	Showed protection by modulating the immune response toward Th1 type	[182]
	ISCOMATRIX formulations with different bilayer compositions with EPC/DMPC/DSPC	Soluble Leishmania antigens (SLA)	CL	ISCOMATRIX DSPC showed the highest elevation of IgG, IgG1, and IgG2a.	[183]
				Mixed Th1/Th2 response that was not protective	
7.	Delivery system using alginate				
	Alginate microspheres	Autoclaved *L. major* (ALM) + Quillaja saponins (QS)	CL	(ALM)ALG + QS showed remarkable protective response.	[186]
		Autoclaved *L. major* (ALM) + CpG-ODN	CL	Remarkably enhanced the protective response	[187]

References

1. World Health Organisation (2017) WHO|Universal health coverage (UHC). WHO fact sheet 2017
2. Centlivre M, Combadière B (2015) New challenges in modern vaccinology. BMC Immunol. https://doi.org/10.1186/s12865-015-0075-2
3. Delany I, Rappuoli R, De Gregorio E (2014) Vaccines for the 21st century. EMBO Mol Med. https://doi.org/10.1002/emmm.201403876
4. Nabel GJ (2013) Designing tomorrow's vaccines. N Engl J Med. https://doi.org/10.1056/NEJMra1204186
5. Griffiths KL, Khader SA (2014) Novel vaccine approaches for protection against intracellular pathogens. Curr Opin Immunol. https://doi.org/10.1016/j.coi.2014.02.003
6. Robinson HL, Amara RR (2005) T cell vaccines for microbial infections. Nat Med. https://doi.org/10.1038/nm1212
7. Servín-Blanco R, Zamora-Alvarado R, Gevorkian G, Manoutcharian K (2016) Antigenic variability: obstacles on the road to vaccines against traditionally difficult targets. Hum Vaccines Immunother. https://doi.org/10.1080/21645515.2016.1191718
8. World Health Organization Leishmaniasis (2015). http://www.who.int/mediacentre/factsheets/fs375/en/. Leishmaniasis Fact Sheet N°375 2015
9. Molyneux DH, Savioli L, Engels D (2017) Neglected tropical diseases: progress towards addressing the chronic pandemic. Lancet. https://doi.org/10.1016/S0140-6736(16)30171-4
10. Alvar J, Yactayo S, Bern C (2006) Leishmaniasis and poverty. Trends Parasitol. https://doi.org/10.1016/j.pt.2006.09.004
11. Cameron MM, Acosta-Serrano A, Bern C, Boelaert M, Den Boer M, Burza S et al (2016) Understanding the transmission dynamics of Leishmania donovani to provide robust evidence for interventions to eliminate visceral leishmaniasis in Bihar, India. The LCNTDR collection: advances in scientific research for NTD control. Parasit Vectors. https://doi.org/10.1186/s13071-016-1309-8
12. Alvar J, Croft SL, Kaye P, Khamesipour A, Sundar S, Reed SG (2013) Case study for a vaccine against leishmaniasis. Vaccine. https://doi.org/10.1016/j.vaccine.2012.11.080
13. Vallur AC, Duthie MS, Reinhart C, Tutterrow Y, Hamano S, Bhaskar KRH et al (2014) Biomarkers for intracellular pathogens: establishing tools as vaccine and therapeutic endpoints for visceral leishmaniasis. Clin Microbiol Infect. https://doi.org/10.1111/1469-0691.12421.
14. Noazin S, Modabber F, Khamesipour A, Smith PG, Moulton LH, Nasseri K et al (2008) First generation leishmaniasis vaccines: a review of field efficacy trials. Vaccine. https://doi.org/10.1016/j.vaccine.2008.09.085
15. Srivastava S, Shankar P, Mishra J, Singh S (2016) Possibilities and challenges for developing a successful vaccine for leishmaniasis. Parasit Vectors 9:1–15. https://doi.org/10.1186/s13071-016-1553-y
16. Halstead SB, Mahalingam S, Marovich MA, Ubol S, Mosser DM (2010) Intrinsic antibody-dependent enhancement of microbial infection in macrophages: disease regulation by immune complexes. Lancet Infect Dis. https://doi.org/10.1016/S1473-3099(10)70166-3
17. Dube A, Rawat K, Yadav N, Joshi S, Ratnapriya S, Sahasrabuddhe A (2016) Management of visceral leishmaniasis with therapeutic vaccines. Vaccine Dev Ther. https://doi.org/10.2147/vdt.s110654
18. Kamhawi S (2017) The yin and yang of leishmaniasis control. PLoS Negl Trop Dis. https://doi.org/10.1371/journal.pntd.0005529
19. Nadim A, Javadian E, Tahvildar Bidruni G, Ghorbani M (1983) Effectiveness of leishmanization in the control of cutaneous leishmaniasis. Bull Soc Pathol Exot Filiales 76(4):377–383
20. Okwor I, Uzonna J (2016) Social and economic burden of human leishmaniasis. Am J Trop Med Hyg. https://doi.org/10.4269/ajtmh.15-0408
21. Kellina OI (1981) Problem and current lines in investigations on the epidemiology of leishmaniasis and its control in the USSR. Bull Soc Pathol Exot Filiales 74(3):306–318

22. Khamesipour A, Rafati S, Davoudi N, Maboudi F, Modabber F (2006) Leishmaniasis vaccine candidates for development: a global overview. Indian J Med Res 123(3):423–438
23. Nadim A, Javadian E, Mohebali M (1997) The experience of leishmanization in the Islamic Republic of Iran. East Mediterr Heal J
24. Hosseini SMH, Hatam GR, Ardehali S (2005) Characterization of leishmania isolated from unhealed lesions caused by leishmanization. East Mediterr Heal J 11(1–2):240–243
25. Khamesipour A, Dowlati Y, Asilian A, Hashemi-Fesharki R, Javadi A, Noazin S et al (2005) Leishmanization: use of an old method for evaluation of candidate vaccines against leishmaniasis. Vaccine. https://doi.org/10.1016/j.vaccine.2005.02.015
26. Khamesipour A, Abbasi A, Firooz A, Amin Mohammadi AM, Eskandari SE, Jaafari MR (2012) Treatment of cutaneous lesion of 20 years' duration caused by leishmanization. Indian J Dermatol. https://doi.org/10.4103/0019-5154.94280
27. Uzonna JE, Wei G, Yurkowski D, Bretscher P (2001) Immune elimination of Leishmania major in mice: implications for immune memory, vaccination, and reactivation disease. J Immunol. https://doi.org/10.4049/jimmunol.167.12.6967
28. Zaph C, Uzonna J, Beverley SM, Scott P (2004) Central memory T cells mediate long-term immunity to Leishmania major in the absence of persistent parasites. Nat Med. https://doi.org/10.1038/nm1108
29. McCall LI, Zhang WW, Ranasinghe S, Matlashewski G (2013) Leishmanization revisited: immunization with a naturally attenuated cutaneous Leishmania donovani isolate from Sri Lanka protects against visceral leishmaniasis. Vaccine. https://doi.org/10.1016/j.vaccine.2012.11.065
30. Sundar S, Singh B (2014) Identifying vaccine targets for anti-leishmanial vaccine development. Expert Rev Vaccines. https://doi.org/10.1586/14760584.2014.894467
31. Mitchell GF, Handman E, Spithill TW (1984) Vaccination against cutaneous Leishmaniasis in mice using nonpathogenic cloned promastigotes of leishmania major and importance of route of injection. Aust J Exp Biol Med Sci. https://doi.org/10.1038/icb.1984.14
32. Kimsey PB, Theodos CM, Mitchen TK, Turco SJ, Titus RG (1993) An avirulent lipophosphoglycan-deficient Leishmania major clone induces CD4+ T cells which protect susceptible BALB/c mice against infection with virulent L. major. Infect Immun 61(12):5205–5213
33. Rivier D, Shah R, Bovay P, Mauel J (1993) Vaccine development against cutaneous leishmaniasis. Subcutaneous administration of radioattenuated parasites protects CBA mice against virulent Leishmania major challenge. Parasite Immunol. https://doi.org/10.1111/j.1365-3024.1993.tb00587.x
34. Daneshvar H, Coombs GH, Hagan P, Phillips RS (2003) Leishmania mexicana and Leishmania major: attenuation of wild-type parasites and vaccination with the attenuated lines. J Infect Dis. https://doi.org/10.1086/374783
35. Das A, Ali N (2012) Vaccine prospects of killed but metabolically active leishmania against visceral leishmaniasis. Expert Rev Vaccines. https://doi.org/10.1586/erv.12.50
36. Jain K, Jain NK (2015) Vaccines for visceral leishmaniasis: a review. J Immunol Methods. https://doi.org/10.1016/j.jim.2015.03.017
37. Saljoughian N, Taheri T, Rafati S (2014) Live vaccination tactics: possible approaches for controlling visceral leishmaniasis. Front Immunol. https://doi.org/10.3389/fimmu.2014.00134
38. Streit JA, Recker TJ, Filho FG, Beverley SM, Wilson ME (2001) Protective immunity against the protozoan Leishmania chagasi is induced by subclinical cutaneous infection with virulent but not avirulent organisms. J Immunol. https://doi.org/10.4049/jimmunol.166.3.1921
39. Titus RG, Gueiros-Filho FJ, De Freitas LAR, Beverley SM (1995) Development of a safe live Leishmania vaccine line by gene replacement. Proc Natl Acad Sci U S A. https://doi.org/10.1073/pnas.92.22.10267
40. Amaral VF, Teva A, Oliveira-Neto MP, Silva AJ, Pereira MS, Cupolillo E et al (2002) Study of the safety, immunogenicity and efficacy of attenuated and killed Leishmania (Leishmania) major vaccines in a rhesus monkey (Macaca mulatta) model of the human disease. Mem Inst Oswaldo Cruz. https://doi.org/10.1590/S0074-02762002000700019

41. Anand S, Madhubala R (2015) Genetically engineered ascorbic acid-deficient live mutants of leishmania donovani induce long lasting protective immunity against visceral leishmaniasis. Sci Rep. https://doi.org/10.1038/srep10706

42. Fiuza JA, Gannavaram S, da Costa Santiago H, Selvapandiyan A, Souza DM, Passos LSA et al (2015) Vaccination using live attenuated Leishmania donovani centrin deleted parasites induces protection in dogs against Leishmania infantum. Vaccine. https://doi.org/10.1016/j.vaccine.2014.11.039

43. Joshi S, Rawat K, Yadav NK, Kumar V, Siddiqi MI, Dube A (2014) Visceral leishmaniasis: advancements in vaccine development via classical and molecular approaches. Front Immunol. https://doi.org/10.3389/fimmu.2014.00380

44. Silvestre R, Cordeiro-Da-Silva A, Santarém N, Vergnes B, Sereno D, Ouaissi A (2007) SIR2-deficient Leishmania infantum induces a defined IFN-γ/IL-10 pattern that correlates with protection. J Immunol. https://doi.org/10.4049/jimmunol.179.5.3161

45. Alexander J, Coombs GH, Mottram JC (1998) Leishmania mexicana cysteine proteinase-deficient mutants have attenuated virulence for mice and potentiate a Th1 response. J Immunol

46. Carrión J, Folgueira C, Soto M, Fresno M, Requena JM (2011) Leishmania infantum HSP70-II null mutant as candidate vaccine against leishmaniasis: a preliminary evaluation. Parasit Vectors. https://doi.org/10.1186/1756-3305-4-150

47. Dey R, Meneses C, Salotra P, Kamhawi S, Nakhasi HL, Duncan R (2010) Characterization of a Leishmania stage-specific mitochondrial membrane protein that enhances the activity of cytochrome c oxidase and its role in virulence. Mol Microbiol. https://doi.org/10.1111/j.1365-2958.2010.07214.x

48. Dey R, Dagur PK, Selvapandiyan A, McCoy JP, Salotra P, Duncan R et al (2013) Live attenuated Leishmania donovani p27 gene knockout parasites are nonpathogenic and elicit long-term protective immunity in BALB/c mice. J Immunol. https://doi.org/10.4049/jimmunol.1202801

49. Elikaee S, Mohebali M, Rezaei S, Eslami H, Khamesipour A, Keshavarz H et al (2018) Development of a new live attenuated Leishmania major p27 gene knockout: safety and immunogenicity evaluation in BALB/c mice. Cell Immunol. https://doi.org/10.1016/j.cellimm.2018.07.002.

50. Elikaee S, Mohebali M, Rezaei S, Eslami H, Khamesipour A, Keshavarz H et al (2019) Leishmania major p27 gene knockout as a novel live attenuated vaccine candidate: protective immunity and efficacy evaluation against cutaneous and visceral leishmaniasis in BALB/c mice. Vaccine. https://doi.org/10.1016/j.vaccine.2019.04.068

51. Davoudi N, Khamesipour A, Mahboudi F, McMaster WR (2014) A dual drug Sensitive L. major induces protection without lesion in C57BL/6 mice. PLoS Negl Trop Dis. https://doi.org/10.1371/journal.pntd.0002785

52. Kumari S, Samani M, Khare P, Misra P, Dutta S, Kolli BK et al (2009) Photodynamic vaccination of hamsters with inducible suicidal mutants of Leishmania amazonensis elicits immunity against visceral leishmaniasis. Eur J Immunol. https://doi.org/10.1002/eji.200838389

53. Raymond F, Boisvert S, Roy G, Ritt JF, Légaré D, Isnard A et al (2012) Genome sequencing of the lizard parasite Leishmania tarentolae reveals loss of genes associated to the intracellular stage of human pathogenic species. Nucleic Acids Res. https://doi.org/10.1093/nar/gkr834

54. Pirdel L, Farajnia S (2017) A non-pathogenic recombinant Leishmania expressing lipophosphoglycan 3 against experimental infection with Leishmania infantum. Scand J Immunol. https://doi.org/10.1111/sji.12557

55. Giunchetti RC, Corrêa-Oliveira R, Martins-Filho OA, Teixeira-Carvalho A, Roatt BM, de Oliveira Aguiar-Soares RD et al (2008) A killed Leishmania vaccine with sand fly saliva extract and saponin adjuvant displays immunogenicity in dogs. Vaccine. https://doi.org/10.1016/j.vaccine.2007.11.057.

56. Petrovsky N, Aguilar JC (2004) Vaccine adjuvants: current state and future trends. Immunol Cell Biol. https://doi.org/10.1111/j.0818-9641.2004.01272.x

57. Raman VS, Duthie MS, Fox CB, Matlashewski G, Reed SG (2012) Adjuvants for Leishmania vaccines: from models to clinical application. Front Immunol. https://doi.org/10.3389/fimmu.2012.00144

58. Teixeira MCA, de Sá Oliveira GG, Santos POM, Bahiense TC, da Silva VMG, Rodrigues MS et al (2011) An experimental protocol for the establishment of dogs with long-term cellular immune reactions to Leishmania antigens. Mem Inst Oswaldo Cruz. https://doi.org/10.1590/S0074-02762011000200011

59. Bruhn KW, Birnbaum R, Haskell J, Vanchinathan V, Greger S, Narayan R et al (2012) Killed but metabolically active Leishmania infantum as a novel whole-cell vaccine for visceral leishmaniasis. Clin Vaccine Immunol. https://doi.org/10.1128/CVI.05660-11

60. Khamesipour A (2014) Therapeutic vaccines for leishmaniasis. Expert Opin Biol Ther. https://doi.org/10.1517/14712598.2014.945415

61. Sharifi I, Aflatoonian MR, Fekri AR, Hakimi Parizi M, Aghaei Afshar A, Khosravi A et al (2015) A comprehensive review of cutaneous leishmaniasis in Kerman province, Southeastern Iran- narrative review article. Iran J Public Health

62. Nagill R, Mahajan R, Sharma M, Kaur S (2009) Induction of cellular and humoral responses by autoclaved and heat-killed antigen of Leishmania donovani in experimental visceral leishmaniasis. Parasitol Int. https://doi.org/10.1016/j.parint.2009.07.008

63. Kumar R, Engwerda C (2014) Vaccines to prevent leishmaniasis. Clin Transl Immunol. https://doi.org/10.1038/cti.2014.4

64. Gholami E, Zahedifard F, Rafati S (2016) Delivery systems for Leishmania vaccine development. Expert Rev Vaccines. https://doi.org/10.1586/14760584.2016.1157478

65. Ravindran R, Bhowmick S, Das A, Ali N (2010) Comparison of BCG, MPL and cationic liposome adjuvant systems in leishmanial antigen vaccine formulations against murine visceral leishmaniasis. BMC Microbiol. https://doi.org/10.1186/1471-2180-10-181

66. Kumari S, Samant M, Khare P, Sundar S, Sinha S, Dube A (2008) Induction of Th1-type cellular responses in cured/exposed Leishmania-infected patients and hamsters against polyproteins of soluble Leishmania donovani promastigotes ranging from 89.9 to 97.1 kDa. Vaccine. https://doi.org/10.1016/j.vaccine.2008.06.102

67. Kumari S, Samant M, Misra P, Khare P, Sisodia B, Shasany AK et al (2008) Th1-stimulatory polyproteins of soluble Leishmania donovani promastigotes ranging from 89.9 to 97.1 kDa offers long-lasting protection against experimental visceral leishmaniasis. Vaccine. https://doi.org/10.1016/j.vaccine.2008.08.021

68. Lemesre JL, Holzmuller P, Gonçalves RB, Bourdoiseau G, Hugnet C, Cavaleyra M et al (2007) Long-lasting protection against canine visceral leishmaniasis using the LiESAp-MDP vaccine in endemic areas of France: double-blind randomised efficacy field trial. Vaccine. https://doi.org/10.1016/j.vaccine.2007.02.083

69. Reed SG, Bertholet S, Coler RN, Friede M (2009) New horizons in adjuvants for vaccine development. Trends Immunol. https://doi.org/10.1016/j.it.2008.09.006

70. Duthie MS, Windish HP, Fox CB, Reed SG (2011) Use of defined TLR ligands as adjuvants within human vaccines. Immunol Rev. https://doi.org/10.1111/j.1600-065X.2010.00978.x

71. Stäger S, Smith DF, Kaye PM (2000) Immunization with a recombinant stage-regulated surface protein from Leishmania donovani induces protection against visceral leishmaniasis. J Immunol. https://doi.org/10.4049/jimmunol.165.12.7064

72. Soong L, Duboise SM, Kima P, McMahon-Pratt D (1995) Leishmania pifanoi amastigote antigens protect mice against cutaneous leishmaniasis. Infect Immun 63(9):3559

73. Martins VT, Chávez-Fumagalli MA, Costa LE, Martins AMCC, Lage PS, Lage DP et al (2013) Antigenicity and protective efficacy of a Leishmania amastigote-specific protein, member of the super-oxygenase family, against visceral leishmaniasis. PLoS Negl Trop Dis. https://doi.org/10.1371/journal.pntd.0002148

74. Basu R, Bhaumik S, Basu JM, Naskar K, De T, Roy S (2005) Kinetoplastid membrane protein-11 DNA vaccination induces complete protection against both pentavalent antimonial-sensitive and -resistant strains of Leishmania donovani that correlates with inducible

nitric oxide synthase activity and IL-4 generation: E. J Immunol. https://doi.org/10.4049/jimmunol.174.11.7160

75. Streit JA, Recker TJ, Donelson JE, Wilson ME (2000) BCG expressing LCR1 of Leishmania chagasi induces protective immunity in susceptible mice. Exp Parasitol. https://doi.org/10.1006/expr.1999.4459
76. Abdelhak S, Louzir H, Timm J, Blel L, Benlasfar Z, Lagranderie M et al (1995) Recombinant BCG expressing the leishmania surface antigen gp63 induces protective immunity against Leishmania major infection in BALB/c mice. Microbiology. https://doi.org/10.1099/13500872-141-7-1585
77. Grimaldi G, Teva A, Porrozzi R, Pinto MA, Marchevsky RS, Rocha MGL et al (2014) Clinical and parasitological protection in a Leishmania infantum-macaque model vaccinated with adenovirus and the recombinant A2 antigen. PLoS Negl Trop Dis. https://doi.org/10.1371/journal.pntd.0002853
78. Kushawaha PK, Gupta R, Sundar S, Sahasrabuddhe AA, Dube A (2011) Elongation Factor-2, a Th1 stimulatory protein of Leishmania donovani, generates strong IFN-γ and IL-12 response in cured Leishmania -infected patients/hamsters and protects hamsters against leishmania challenge. J Immunol. https://doi.org/10.4049/jimmunol.1102081
79. Moreno J, Nieto J, Masina S, Cañavate C, Cruz I, Chicharro C et al (2007) Immunization with H1, HASPB1 and MML leishmania proteins in a vaccine trial against experimental canine leishmaniasis. Vaccine. https://doi.org/10.1016/j.vaccine.2007.05.010
80. Solioz N, Blum-Tirouvanziam U, Jacquet R, Rafati S, Corradin G, Mauël J et al (1999) The protective capacities of histone H1 against experimental murine cutaneous leishmaniasis. Vaccine. https://doi.org/10.1016/S0264-410X(99)00340-0
81. Baharia RK, Tandon R, Sahasrabuddhe AA, Sundar S, Dube A (2014) Nucleosomal histone proteins of L. donovani: a combination of recombinant H2A, H2B, H3 and H4 proteins were highly immunogenic and offered optimum prophylactic efficacy against Leishmania challenge in hamsters. PLoS One. https://doi.org/10.1371/journal.pone.0097911
82. Hugentobler F, Yam KK, Gillard J, Mahbuba R, Olivier M, Cousineau B (2012) Immunization against Leishmania major infection using LACK- and IL-12-expressing lactococcus lactis induces delay in footpad swelling. PLoS One. https://doi.org/10.1371/journal.pone.0030945
83. Tandon R, Chandra S, Baharia RK, Misra P, Das S, Rawat K et al (2018) Molecular, biochemical characterization and assessment of immunogenic potential of cofactor-independent phosphoglycerate mutase against Leishmania donovani: a step towards exploring novel vaccine candidate. Parasitology. https://doi.org/10.1017/S0031182017001160
84. Mizbani A, Taheri T, Zahedifard F, Taslimi Y, Azizi H, Azadmanesh K et al (2009) Recombinant Leishmania tarentolae expressing the A2 virulence gene as a novel candidate vaccine against visceral leishmaniasis. Vaccine. https://doi.org/10.1016/j.vaccine.2009.09.114
85. Shahbazi M, Zahedifard F, Taheri T, Taslimi Y, Jamshidi S, Shirian S et al (2015) Evaluation of live recombinant nonpathogenic leishmania tarentolae expressing cysteine proteinase and A2 genes as a candidate vaccine against experimental canine visceral leishmaniasis. PLoS One. https://doi.org/10.1371/journal.pone.0132794
86. Yam KK, Hugentobler F, Pouliot P, Stern AM, Lalande JD, Matlashewski G et al (2011) Generation and evaluation of A2-expressing Lactococcus lactis live vaccines against Leishmania donovani in BALB/c mice. J Med Microbiol. https://doi.org/10.1099/jmm.0.029959-0
87. Grimaldi G, Teva A, Dos-Santos CB, Santos FN, Pinto IDS, Fux B et al (2017) Field trial of efficacy of the Leish-tec® vaccine against canine leishmaniasis caused by Leishmania infantum in an endemic area with high transmission rates. PLoS One. https://doi.org/10.1371/journal.pone.0185438
88. Spitzer N, Jardim A, Lippert D, Olafson RW (1999) Long-term protection of mice against Leishmania major with a synthetic peptide vaccine. Vaccine. https://doi.org/10.1016/S0264-410X(98)00363-6

89. Tsagozis P, Karagouni E, Dotsika E (2004) Dendritic cells pulsed with peptides of gp63 induce differential protection against experimental cutaneous leishmaniasis. Int J Immunopathol Pharmacol. https://doi.org/10.1177/039463200401700314

90. Rezvan H (2013) Immunogenicity of HLA-DR1 restricted peptides derived from Leishmania major gp63 using FVB/N-DR1 transgenic mouse model. Iran J Parasitol 8(2):273–279

91. Delgado G, Parra-López CA, Vargas LE, Hoya R, Estupiñán M, Guzmán F et al (2003) Characterizing cellular immune response to kinetoplastid membrane protein-11 (KMP-11) during Leishmania (Viannia) panamensis infection using dendritic cells (DCs) as antigen presenting cells (APCs). Parasite Immunol. https://doi.org/10.1046/j.1365-3024.2003.00626.x

92. Herrera-Najera C, Piña-Aguilar R, Xacur-Garcia F, Ramirez-Sierra MJ, Dumonteil E (2009) Mining the leishmania genome for novel antigens and vaccine candidates. Proteomics. https://doi.org/10.1002/pmic.200800533

93. Naouar I, Boussoffara T, Chenik M, Gritli S, Ben Ahmed M, Belhadj Hmida N et al (2016) Prediction of T cell epitopes from Leishmania major potentially excreted/secreted proteins inducing granzyme B production. PLoS One. https://doi.org/10.1371/journal.pone.0147076

94. Costa LE, Chávez-Fumagalli MA, Martins VT, Duarte MC, Lage DP, Lima MIS et al (2015) Phage-fused epitopes from Leishmania infantum used as immunogenic vaccines confer partial protection against Leishmania amazonensis infection. Parasitology. https://doi.org/10.1017/S0031182015000724

95. Gillespie PM, Beaumier CM, Strych U, Hayward T, Hotez PJ, Bottazzi ME (2016) Status of vaccine research and development of vaccines for leishmaniasis. Vaccine. https://doi.org/10.1016/j.vaccine.2015.12.071

96. Chakravarty J, Kumar S, Trivedi S, Rai VK, Singh A, Ashman JA et al (2011) A clinical trial to evaluate the safety and immunogenicity of the LEISH-F1+MPL-SE vaccine for use in the prevention of visceral leishmaniasis. Vaccine. https://doi.org/10.1016/j.vaccine.2011.02.096

97. Llanos-Cuentas A, Calderón W, Cruz M, Ashman JA, Alves FP, Coler RN et al (2010) A clinical trial to evaluate the safety and immunogenicity of the LEISH-F1+MPL-SE vaccine when used in combination with sodium stibogluconate for the treatment of mucosal leishmaniasis. Vaccine. https://doi.org/10.1016/j.vaccine.2010.08.092

98. Christiaansen AF, Dixit UG, Coler RN, Marie Beckmann A, Reed SG, Winokur PL et al (2017) CD11a and CD49d enhance the detection of antigen-specific T cells following human vaccination. Vaccine. https://doi.org/10.1016/j.vaccine.2017.06.013

99. Coler RN, Duthie MS, Hofmeyer KA, Guderian J, Jayashankar L, Vergara J et al (2015) From mouse to man: safety, immunogenicity and efficacy of a candidate leishmaniasis vaccine LEISH-F3+GLA-SE. Clin Transl Immunol. https://doi.org/10.1038/cti.2015.6

100. Parody N, Soto M, Requena JM, Álonso C (2004) Adjuvant guided polarization of the immune humoral response against a protective multicomponent antigenic protein (Q) from Leishmania infantum. A CpG + Q mix protects Balb/c mice from infection. Parasite Immunol. https://doi.org/10.1111/j.0141-9838.2004.00711.x

101. Ferraro B, Morrow MP, Hutnick NA, Shin TH, Lucke CE, Weiner DB (2011) Clinical applications of DNA vaccines: current progress. Clin Infect Dis. https://doi.org/10.1093/cid/cir334

102. Maspi N, Ghaffarifar F, Sharifi Z, Dalimi A, Khademi SZ (2017) DNA vaccination with a plasmid encoding LACK-TSA fusion against leishmania major infection in BALB/c mice. Malays J Pathol 39(3):267–275

103. Louis L, Clark M, Wise MC, Glennie N, Wong A, Broderick K et al (2019) Intradermal synthetic DNA vaccination generates leishmania -specific T cells in the skin and protection against Leishmania major. Infect Immun. https://doi.org/10.1128/iai.00227-19

104. Guha R, Gupta D, Rastogi R, Vikram R, Krishnamurthy G, Bimal S et al (2013) Vaccination with leishmania hemoglobin receptor-encoding DNA protects against visceral leishmaniasis. Sci Transl Med. https://doi.org/10.1126/scitranslmed.3006406

105. Osman M, Mistry A, Keding A, Gabe R, Cook E, Forrester S et al (2017) A third generation vaccine for human visceral leishmaniasis and post kala azar dermal leishmaniasis: first-in-human trial of ChAd63-KH. PLoS Negl Trop Dis. https://doi.org/10.1371/journal.pntd.0005527

106. Ada GL (1991) The ideal vaccine. World J Microbiol Biotechnol. https://doi.org/10.1007/BF00328978
107. Beverley PCL (2002) Immunology of vaccination. Br Med Bull. https://doi.org/10.1093/bmb/62.1.15
108. Deye N, Vincent F, Michel P, Ehrmann S, Da Silva D, Piagnerelli M et al (2016) Changes in cardiac arrest patients' TM temperature management after the 2013 "TTM" trial: results from an international survey. Ann Intensive 6(1). https://doi.org/10.1186/s13613-015-0104-6; Al-Hussaini M, Mustafa S (2016) Adolescents' TM knowledge and awareness of diabetes mellitus in Kuwait. Alexandria J Med 52(1):61–66. https://doi.org/10.1016/j.ajme.2015.04.001; Pollach G, Brunkhorst F, Mipando M, Namboya F, Mndolo S, Luiz T (2016) The "first digit law" – a hypothesis on its possible impact on medicine and development aid. Med Hypotheses 97:102–106. https://doi.org/10.1016/j.mehy.2016.10.021; Asiedu K, Kyei S, Ayobi B, Agyemang FO, Ablordeppey RK (2016) Survey of eye practitioners' preference of diagnostic tests and treatment modalities for dry eye in Ghana. Contact Lens Anterior Eye 39(6):411–415. https://doi.org/10.1016/j.clae.2016.08.001; Barakat KH, Gajewski MM, Tuszynski JA (2012) DNA polymerase beta (pol β) inhibitors: a comprehensive overview. Drug Discov Today 17(15–16):913–920. https://doi.org/10.1016/j.drudis.2012.04.008; Mocan O, Dumitraşcu DL (2016) The broad spectrum of celiac disease and gluten sensitive enteropathy. Clujul Med 89(3):335–342. https://doi.org/10.15386/cjmed-698; et al. Incorporating evidenced based practice into an international mentorship model: a pilot burn nursing experience. J Burn Care Res 2015. https://doi.org/10.1097/BCR.0000000000000251
109. Nascimento IP, Leite LCC (2012) Recombinant vaccines and the development of new vaccine strategies. Braz J Med Biol Res. https://doi.org/10.1590/S0100-879X2012007500142
110. Donnelly JJ, Wahren B, Liu MA (2005) DNA vaccines: progress and challenges. J Immunol. https://doi.org/10.4049/jimmunol.175.2.633
111. Skibinski DAG, Baudner BC, Singh M, O'hagan DT (2011) Combination vaccines. J Glob Infect Dis. https://doi.org/10.4103/0974-777X.77298
112. Huber VC (2014) Influenza vaccines: from whole virus preparations to recombinant protein technology. Expert Rev Vaccines. https://doi.org/10.1586/14760584.2014.852476
113. Sahdev P, Ochyl LJ, Moon JJ (2014) Biomaterials for nanoparticle vaccine delivery systems. Pharm Res. https://doi.org/10.1007/s11095-014-1419-y
114. Shao K, Singha S, Clemente-Casares X, Tsai S, Yang Y, Santamaria P (2015) Nanoparticle-based immunotherapy for cancer. ACS Nano. https://doi.org/10.1021/nn5062029
115. Singh A, Peppas NA (2014) Hydrogels and scaffolds for immunomodulation. Adv Mater. https://doi.org/10.1002/adma.201402105
116. Quinn HL, Kearney MC, Courtenay AJ, McCrudden MT, Donnelly RF (2014) The role of microneedles for drug and vaccine delivery. Expert Opin Drug Deliv. https://doi.org/10.1517/17425247.2014.938635
117. Indermun S, Luttge R, Choonara YE, Kumar P, Du Toit LC, Modi G et al (2014) Current advances in the fabrication of microneedles for transdermal delivery. J Control Release. https://doi.org/10.1016/j.jconrel.2014.04.052.
118. Tsoras AN, Champion JA (2019) Protein and peptide biomaterials for engineered subunit vaccines and immunotherapeutic applications. Annu Rev Chem Biomol Eng. https://doi.org/10.1146/annurev-chembioeng-060718-030347
119. Bookstaver ML, Tsai SJ, Bromberg JS, Jewell CM (2018) Improving vaccine and immunotherapy design using biomaterials. Trends Immunol. https://doi.org/10.1016/j.it.2017.10.002
120. Oelke M, Maus MV, Didiano D, June CH, Mackensen A, Schneck JP (2003) Ex vivo induction and expansion of antigen-specific cytotoxic T cells by HLA-Ig-coated artificial antigen-presenting cells. Nat Med. https://doi.org/10.1038/nm869
121. Sunshine JC, Perica K, Schneck JP, Green JJ (2014) Particle shape dependence of CD8+ T cell activation by artificial antigen presenting cells. Biomaterials. https://doi.org/10.1016/j.biomaterials.2013.09.050

122. Kumar S, Anselmo AC, Banerjee A, Zakrewsky M, Mitragotri S (2015) Shape and size-dependent immune response to antigen-carrying nanoparticles. J Control Release. https://doi.org/10.1016/j.jconrel.2015.09.069.
123. Hardy CL, LeMasurier JS, Mohamud R, Yao J, Xiang SD, Rolland JM et al (2013) Differential uptake of nanoparticles and microparticles by pulmonary APC subsets induces discrete immunological imprints. J Immunol. https://doi.org/10.4049/jimmunol.1203131
124. Shahbazi MA, Fernández TD, Mäkilä EM, Le Guével X, Mayorga C, Kaasalainen MH et al (2014) Surface chemistry dependent immunostimulative potential of porous silicon nanoplatforms. Biomaterials. https://doi.org/10.1016/j.biomaterials.2014.07.050
125. Moyano DF, Goldsmith M, Solfiell DJ, Landesman-Milo D, Miranda OR, Peer D et al (2012) Nanoparticle hydrophobicity dictates immune response. J Am Chem Soc. https://doi.org/10.1021/ja2108905.
126. Watson DS, Endsley AN, Huang L (2012) Design considerations for liposomal vaccines: Influence of formulation parameters on antibody and cell-mediated immune responses to liposome associated antigens. Vaccine. https://doi.org/10.1016/j.vaccine.2012.01.070
127. Sharma A, Sharma US (1997) Liposomes in drug delivery: progress and limitations. Int J Pharm. https://doi.org/10.1016/S0378-5173(97)00135-X.
128. Rao M, Alving CR (2000) Delivery of lipids and liposomal proteins to the cytoplasm and Golgi of antigen-presenting cells. Adv Drug Deliv Rev. https://doi.org/10.1016/S0169-409X(99)00064-2
129. Kersten GFA, Crommelin DJA (1995) Liposomes and ISCOMS as vaccine formulations. BBA Rev Biomembr. https://doi.org/10.1016/0304-4157(95)00002-9
130. Schmidt ST, Foged C, Korsholm KS, Rades T, Christensen D (2016) Liposome-based adjuvants for subunit vaccines: formulation strategies for subunit antigens and immunostimulators. Pharmaceutics. https://doi.org/10.3390/pharmaceutics8010007
131. Carstens MG, Camps MGM, Henriksen-Lacey M, Franken K, Ottenhoff THM, Perrie Y et al (2011) Effect of vesicle size on tissue localization and immunogenicity of liposomal DNA vaccines. Vaccine. https://doi.org/10.1016/j.vaccine.2011.04.081
132. Oussoren C, Zuidema J, Crommelin DJA, Storm G (1997) Lymphatic uptake and biodistribution of liposomes after subcutaneous injection. II. Influence of liposomal size, lipid composition and lipid dose. Biochim Biophys Acta Biomembr. https://doi.org/10.1016/S0005-2736(97)00122-3
133. McLennan DN, Porter CJH, Charman SA (2005) Subcutaneous drug delivery and the role of the lymphatics. Drug Discov Today Technol. https://doi.org/10.1016/j.ddtec.2005.05.006
134. Bachmann MF, Jennings GT (2010) Vaccine delivery: a matter of size, geometry, kinetics and molecular patterns. Nat Rev Immunol. https://doi.org/10.1038/nri2868
135. Sharma SK, Dube A, Nadeem A, Khan S, Saleem I, Garg R et al (2006) Non PC liposome entrapped promastigote antigens elicit parasite specific CD8+ and CD4+ T-cell immune response and protect hamsters against visceral leishmaniasis. Vaccine. https://doi.org/10.1016/j.vaccine.2005.10.025
136. Bhowmick S, Ravindran R, Ali N (2008) Gp63 in stable cationic liposomes confers sustained vaccine immunity to susceptible BALB/c mice infected with Leishmania donovani. Infect Immun. https://doi.org/10.1128/IAI.00611-07
137. Mazumder S, Maji M, Ali N (2011) Potentiating effects of MPL on DSPC bearing cationic liposomes promote recombinant GP63 vaccine efficacy: high immunogenicity and protection. PLoS Negl Trop Dis. https://doi.org/10.1371/journal.pntd.0001429
138. Badiee A, Jaafari MR, Khamesipour A, Samiei A, Soroush D, Kheiri MT et al (2009) Enhancement of immune response and protection in BALB/c mice immunized with liposomal recombinant major surface glycoprotein of Leishmania (rgp63): the role of bilayer composition. Colloids Surfaces B Biointerfaces. https://doi.org/10.1016/j.colsurfb.2009.06.025
139. Badiee A, Khamesipour A, Samiei A, Soroush D, Shargh VH, Kheiri MT et al (2012) The role of liposome size on the type of immune response induced in BALB/c mice against leishmaniasis: rgp63 as a model antigen. Exp Parasitol. https://doi.org/10.1016/j.exppara.2012.09.001

140. Afrin F, Ali N (1997) Adjuvanticity and protective immunity elicited by Leishmania donovani antigens encapsulated in positively charged liposomes. Infect Immun
141. Afrin F, Anam K, Ali N (2000) Induction of partial protection against Leishmania donovani by promastigote antigens in negatively charged liposomes. J Parasitol. https://doi.org/10.2307/3284956
142. Afrin F, Rajesh R, Anam K, Gopinath M, Pal S, Ali N (2002) Characterization of Leishmania donovani antigens encapsulated in liposomes that induce protective immunity in BALB/c mice. Infect Immun. https://doi.org/10.1128/IAI.70.12.6697-6706.2002
143. Badiee A, Jaafari MR, Khamesipour A, Samiei A, Soroush D, Kheiri MT et al (2009) The role of liposome charge on immune response generated in BALB/c mice immunized with recombinant major surface glycoprotein of Leishmania (rgp63). Exp Parasitol. https://doi.org/10.1016/j.exppara.2008.12.015
144. Badiee A, Jaafari MR, Khamesipour A (2007) Leishmania major: immune response in BALB/c mice immunized with stress-inducible protein 1 encapsulated in liposomes. Exp Parasitol. https://doi.org/10.1016/j.exppara.2006.07.002
145. Badiee A, Jaafari MR, Samiei A, Soroush D, Khamesipour A (2008) Coencapsulation of CpG oligodeoxynucleotides with recombinant Leishmania major stress-inducible protein 1 in liposome enhances immune response and protection against leishmaniasis in immunized BALB/c mice. Clin Vaccine Immunol. https://doi.org/10.1128/CVI.00413-07
146. Heravi Shargh V, Jaafari MR, Khamesipour A, Jalali SA, Firouzmand H, Abbasi A et al (2012) Cationic liposomes containing soluble leishmania antigens (SLA) plus CpG ODNs induce protection against murine model of leishmaniasis. Parasitol Res. https://doi.org/10.1007/s00436-011-2806-5
147. Sabur A, Bhowmick S, Chhajer R, Ejazi SA, Didwania N, Asad M et al (2018) Liposomal elongation factor-1α triggers effector CD4 and CD8 T cells for induction of long-lasting protective immunity against visceral leishmaniasis. Front Immunol. https://doi.org/10.3389/fimmu.2018.00018
148. Mazumdar T, Anam K, Ali N (2005) Influence of phospholipid composition on the adjuvanticity and protective efficacy of liposome-encapsulated Leishmania donovani antigens. J Parasitol. https://doi.org/10.1645/ge-356r1
149. Antimisiaris SG, Jayasekera P, Gregoriadis G (1993) Liposomes as vaccine carriers. Incorporation of soluble and particulate antigens in giant vesicles. J Immunol Methods. https://doi.org/10.1016/0022-1759(93)90368-H
150. Moghimi SM, Patel HM (1988) Tissue specific opsonins for phagocytic cells and their different affinity for cholesterol-rich liposomes. FEBS Lett. https://doi.org/10.1016/0014-5793(88)81372-3
151. Kojima N, Ishii M, Kawauchi Y, Takagi H (2013) Oligomannose-coated liposome as a novel adjuvant for the induction of cellular immune responses to control disease status. Biomed Res Int. https://doi.org/10.1155/2013/562924
152. Shimizu Y, Takagi H, Nakayama T, Yamakami K, Tadakuma T, Yokoyama N et al (2007) Intraperitoneal immunization with oligomannose-coated liposome-entrapped soluble leishmanial antigen induces antigen-specific T-helper type immune response in BALB/c mice through uptake by peritoneal macrophages. Parasite Immunol. https://doi.org/10.1111/j.1365-3024.2007.00937.x
153. Almeida JD, Edwards DC, Brand CM, Heath TD (1975) Formation of virosomes from influenza subunits and liposomes. Lancet. https://doi.org/10.1016/S0140-6736(75)92130-3
154. Stegmann T, Morselt HW, Booy FP, van Breemen JF, Scherphof G, Wilschut J (1987) Functional reconstitution of influenza virus envelopes. EMBO J. https://doi.org/10.1002/j.1460-2075.1987.tb02556.x
155. Bron R, Wahlberg JM, Garoff H, Wilschut J (1993) Membrane fusion of Semliki Forest virus in a model system: correlation between fusion kinetics and structural changes in the envelope glycoprotein. EMBO J. https://doi.org/10.1002/j.1460-2075.1993.tb05703.x

156. Bron R, Ortiz A, Dijkstra J, Stegmann T, Wilschut J (1993) [23] Preparation, properties, and applications of reconstituted influenza virus envelopes (virosomes). Methods Enzymol. https://doi.org/10.1016/0076-6879(93)20091-G

157. Gunther-Ausborn S, Schoen P, Bartoldus I, Wilschut J, Stegmann T (2000) Role of hemagglutinin surface density in the initial stages of influenza virus fusion: lack of evidence for cooperativity. J Virol. https://doi.org/10.1128/jvi.74.6.2714-2720.2000

158. Homhuan A, Prakongpan S, Poomvises P, Maas RA, Crommelin DJA, Kersten GFA et al (2004) Virosome and ISCOM vaccines against Newcastle disease: preparation, characterization and immunogenicity. Eur J Pharm Sci. https://doi.org/10.1016/j.ejps.2004.05.005

159. Cecílio P, Pérez-Cabezas B, Fernández L, Moreno J, Carrillo E, Requena JM et al (2017) Preclinical antigenicity studies of an innovative multivalent vaccine for human visceral leishmaniasis. PLoS Negl Trop Dis. https://doi.org/10.1371/journal.pntd.0005951

160. Liu X, Siegrist S, Amacker M, Zurbriggen R, Pluschke G, Seeberger PH (2006) Enhancement of the immunogenicity of synthetic carbohydrates by conjugation to virosomes: a leishmaniasis vaccine candidate. ACS Chem Biol. https://doi.org/10.1021/cb600086b

161. Yoshida H, Lehr CM, Kok W, Junginger HE, Verhoef JC, Bouwstra JA (1992) Niosomes for oral delivery of peptide drugs. J Control Release. https://doi.org/10.1016/0168-3659(92)90016-K

162. Pardakhty A, Shakibaie M, Daneshvar H, Khamesipour A, Mohammadi-Khorsand T, Forootanfar H (2012) Preparation and evaluation of niosomes containing autoclaved Leishmania major: a preliminary study. J Microencapsul. https://doi.org/10.3109/0265204 8.2011.642016

163. Lezama-Dávila CM (1999) Vaccination of C57BL/10 mice against cutaneous leishmaniasis. Use of purified gp63 encapsulated into niosomes surfactants vesicles: a novel approach. Mem Inst Oswaldo Cruz. https://doi.org/10.1590/S0074-02761999000100014

164. Mehnert W, Mäder K (2012) Solid lipid nanoparticles: production, characterization and applications. Adv Drug Deliv Rev. https://doi.org/10.1016/j.addr.2012.09.021.

165. Souto EB, Doktorovova S, Boonme P (2011) Lipid-based colloidal systems (nanoparticles, microemulsions) for drug delivery to the skin: materials and end-product formulations. J Drug Deliv Sci Technol. https://doi.org/10.1016/S1773-2247(11)50005-X

166. Bond ML, Craparo EF (2010) Solid lipid nanoparticles for applications in gene therapy: a review of the state of the art. Expert Opin Drug Deliv. https://doi.org/10.1517/17425240903362410

167. Vighi E, Ruozi B, Montanari M, Battini R, Leo E (2007) Re-dispersible cationic solid lipid nanoparticles (SLNs) freeze-dried without cryoprotectors: characterization and ability to bind the pEGFP-plasmid. Eur J Pharm Biopharm. https://doi.org/10.1016/j.ejpb.2007.02.006

168. Xue HY, Wong HL (2011) Tailoring nanostructured solid-lipid carriers for time-controlled intracellular siRNA kinetics to sustain RNAi-mediated chemosensitization. Biomaterials. https://doi.org/10.1016/j.biomaterials.2010.12.029

169. del Pozo-Rodríguez A, Delgado D, Solinís MÁ, Pedraz JL, Echevarría E, Rodríguez JM et al (2010) Solid lipid nanoparticles as potential tools for gene therapy: in vivo protein expression after intravenous administration. Int J Pharm. https://doi.org/10.1016/j.ijpharm.2009.10.020

170. Doroud D, Vatanara AV, Zahedifard F, Gholami E, Vahabpour R, Najafabadi AR et al (2010) Cationic solid lipid nanoparticles loaded by cysteine proteinase genes as a novel anti-leishmaniasis DNA vaccine delivery system: characterization and in vitro evaluations. J Pharm Pharm Sci. https://doi.org/10.18433/j3r30t

171. Doroud D, Zahedifard F, Vatanara A, Najafabadi AR, Taslimi Y, Vahabpour R et al (2011) Delivery of a cocktail DNA vaccine encoding cysteine proteinases type I, II and III with solid lipid nanoparticles potentiate protective immunity against Leishmania major infection. J Control Release. https://doi.org/10.1016/j.jconrel.2011.04.011

172. Saljoughian N, Zahedifard F, Doroud D, Doustdari F, Vasei M, Papadopoulou B et al (2013) Cationic solid-lipid nanoparticles are as efficient as electroporation in DNA vaccination against visceral leishmaniasis in mice. Parasite Immunol. https://doi.org/10.1111/pim.12042

173. Shahbazi M, Zahedifard F, Saljoughian N, Doroud D, Jamshidi S, Mahdavi N et al (2015) Immunological comparison of DNA vaccination using two delivery systems against canine leishmaniasis. Vet Parasitol. https://doi.org/10.1016/j.vetpar.2015.07.005

174. Danhier F, Ansorena E, Silva JM, Coco R, Le Breton A, Préat V (2012) PLGA-based nanoparticles: an overview of biomedical applications. J Control Release. https://doi.org/10.1016/j.jconrel.2012.01.043

175. Margaroni M, Agallou M, Athanasiou E, Kammona O, Kiparissides C, Gaitanaki C et al (2017) Vaccination with poly(D,L-lactide-co-glycolide) nanoparticles loaded with soluble leishmania antigens and modified with a TNFα-mimicking peptide or monophosphoryl lipid aconfers protection against experimental visceral leishmaniasis. Int J Nanomedicine. https://doi.org/10.2147/IJN.S141069

176. Abamor E, Allahverdiyev A, Tosyali O, Bagirova M, Acar T, Mustafaeva Z et al (2019) Evaluation of in vitro and in vivo immunostimulatory activities of poly (lactic-co-glycolic acid) nanoparticles loaded with soluble and autoclaved Leishmania infantum antigens: a novel vaccine candidate against visceral leishmaniasis. Asian Pac J Trop Med. https://doi.org/10.4103/1995-7645.262564

177. Tafaghodi M, Eskandari M, Kharazizadeh M, Khamesipour A, Jaafari MR (2010) Immunization against leishmaniasis by PLGA nanospheres loaded with an experimental autoclaved Leishmania major (ALM) and Quillaja saponins. Trop Biomed

178. Agallou M, Margaroni M, Athanasiou E, Toubanaki DK, Kontonikola K, Karidi K et al (2017) Identification of BALB/c immune markers correlated with a partial protection to Leishmania infantum after vaccination with a rationally designed multi-epitope cysteine protease a peptide-based nanovaccine. PLoS Negl Trop Dis. https://doi.org/10.1371/journal.pntd.0005311

179. Athanasiou E, Agallou M, Tastsoglou S, Kammona O, Hatzigeorgiou A, Kiparissides C et al (2017) A poly(lactic-co-glycolic) acid nanovaccine based on chimeric peptides from different Leishmania infantum proteins induces dendritic cells maturation and promotes peptide-specific IFNγ-producing CD8+ T cells essential for the protection against experiment. Front Immunol. https://doi.org/10.3389/fimmu.2017.00684.

180. Lövgren Bengtsson K, Morein B, Osterhaus AD (2011) ISCOM technology-based matrix M™ adjuvant: success in future vaccines relies on formulation. Expert Rev Vaccines. https://doi.org/10.1586/erv.11.25

181. Morelli AB, Becher D, Koernig S, Silva A, Drane D, Maraskovsky E (2012) ISCOMATRIX: a novel adjuvant for use in prophylactic and therapeutic vaccines against infectious diseases. J Med Microbiol. https://doi.org/10.1099/jmm.0.040857-0

182. Papadopoulou G, Karagouni E, Dotsika E (1998) ISCOMs vaccine against experimental leishmaniasis. Vaccine. https://doi.org/10.1016/S0264-410X(97)00308-3

183. Mehravaran A, Jaafari MR, Jalali SA, Khamesipour A, Ranjbar R, Hojatizade M et al (2016) The role of ISCOMATRIX bilayer composition to induce a cell mediated immunity and protection against leishmaniasis in BALB/c mice. Iran J Basic Med Sci. https://doi.org/10.22038/ijbms.2016.6542.

184. Draget KI, Taylor C (2011) Chemical, physical and biological properties of alginates and their biomedical implications. Food Hydrocoll. https://doi.org/10.1016/j.foodhyd.2009.10.007

185. Downs EC, Robertson NE, Riss TL, Plunkett ML (1992) Calcium alginate beads as a slow-release system for delivering angiogenic molecules in vivo and in vitro. J Cell Physiol. https://doi.org/10.1002/jcp.1041520225

186. Tafaghodi M, Eskandari M, Khamesipour A, Jaafaric MR (2016) Immunization against cutaneous Leishmaniasis by alginate microspheres loaded with autoclaved Leishmania major (ALM) and Quillaja saponins. Iran J Pharm Res. https://doi.org/10.22037/ijpr.2016.1832.

187. Tafaghodi M, Eskandari M, Khamesipour A, Jaafari MR (2011) Alginate microspheres encapsulated with autoclaved Leishmania major (ALM) and CpG-ODN induced partial protection and enhanced immune response against murine model of leishmaniasis. Exp Parasitol. https://doi.org/10.1016/j.exppara.2011.07.007

188. Beaumier CM, Gillespie PM, Strych U, Hayward T, Hotez PJ, Bottazzi ME (2016) Status of vaccine research and development of vaccines for Chagas disease. Vaccine. https://doi.org/10.1016/j.vaccine.2016.03.074

189. Ye Z, Li Z, Jin H, Qian Q (2016) Therapeutic cancer vaccines. Adv Exp Med Biol. https://doi.org/10.1007/978-94-017-7555-7_3

190. Bachmann MF, Dyer MR (2004) Therapeutic vaccination for chronic diseases: a new class of drugs in sight. Nat Rev Drug Discov. https://doi.org/10.1038/nrd1284

191. Gröschel MI, Prabowo SA, Cardona PJ, Stanford JL, Van der Werf TS (2014) Therapeutic vaccines for tuberculosis-a systematic review. Vaccine. https://doi.org/10.1016/j.vaccine.2014.03.047

192. Kim TJ, Jin HT, Hur SY, Yang HG, Seo YB, Hong SR et al (2014) Clearance of persistent HPV infection and cervical lesion by therapeutic DNA vaccine in CIN3 patients. Nat Commun. https://doi.org/10.1038/ncomms6317

193. Mukhopadhyay S, Bhattacharyya S, Majhi R, De T, Naskar K, Majumdar S et al (2000) Use of an attenuated leishmanial parasite as an immunoprophylactic and immunotherapeutic agent against murine visceral leishmaniasis. Clin Diagn Lab Immunol. https://doi.org/10.1128/CDLI.7.2.233-240.2000

194. Datta S, Roy S, Manna M (2015) Therapy with radio-attenuated vaccine in experimental murine visceral leishmaniasis showed enhanced T cell and inducible nitric oxide synthase levels, suppressed tumor growth factor-beta production with higher expression of some signaling molecules. Braz J Infect Dis. https://doi.org/10.1016/j.bjid.2014.10.009

195. Borja-Cabrera GP, Mendes AC, Paraguai De Souza E, Okada LYH, Trivellato FADA, Kawasaki JKA et al (2004) Effective immunotherapy against canine visceral leishmaniasis with the FML-vaccine. Vaccine. https://doi.org/10.1016/j.vaccine.2003.11.039

196. Musa AM, Khalil EAG, Mahgoub FAE, Elgawi SHH, Modabber F, Elkadaru AEMY et al (2008) Immunochemotherapy of persistent post-kala-azar dermal leishmaniasis: a novel approach to treatment. Trans R Soc Trop Med Hyg. https://doi.org/10.1016/j.trstmh.2007.08.006

197. Ghosh M, Pal C, Ray M, Maitra S, Mandal L, Bandyopadhyay S (2003) Dendritic cell-based immunotherapy combined with antimony-based chemotherapy cures established murine visceral leishmaniasis. J Immunol. https://doi.org/10.4049/jimmunol.170.11.5625

198. Miret J, Nascimento E, Sampaio W, França JC, Fujiwara RT, Vale A et al (2008) Evaluation of an immunochemotherapeutic protocol constituted of N-methyl meglumine antimoniate (Glucantime®) and the recombinant Leish-110f® + MPL-SE® vaccine to treat canine visceral leishmaniasis. Vaccine. https://doi.org/10.1016/j.vaccine.2008.01.026

199. Trigo J, Abbehusen M, Netto EM, Nakatani M, Pedral-Sampaio G, de Jesus RS et al (2010) Treatment of canine visceral leishmaniasis by the vaccine Leish-111f + MPL-SE. Vaccine. https://doi.org/10.1016/j.vaccine.2010.02.089

200. Seifert K, Juhls C, Salguero FJ, Croft SL (2015) Sequential chemoimmunotherapy of experimental visceral leishmaniasis using a single low dose of liposomal amphotericin B and a novel DNA vaccine candidate. Antimicrob Agents Chemother. https://doi.org/10.1128/AAC.00273-15

201. Bhowmick S, Ravindran R, Ali N (2007) Leishmanial antigens in liposomes promote protective immunity and provide immunotherapy against visceral leishmaniasis via polarized Th1 response. Vaccine. https://doi.org/10.1016/j.vaccine.2007.05.042

202. Maroof A, Brown N, Smith B, Hodgkinson MR, Maxwell A, Losch FO et al (2012) Therapeutic vaccination with recombinant adenovirus reduces splenic parasite burden in experimental visceral leishmaniasis. J Infect Dis. https://doi.org/10.1093/infdis/jir842

203. Xie Z, Wroblewska L, Prochazka L, Weiss R, Benenson Y (2011) Multi-input RNAi-based logic circuit for identification of specific cancer cells. Science 80. https://doi.org/10.1126/science.1205527

204. Kemmer C, Gitzinger M, Daoud-El Baba M, Djonov V, Stelling J, Fussenegger M (2010) Self-sufficient control of urate homeostasis in mice by a synthetic circuit. Nat Biotechnol. https://doi.org/10.1038/nbt.1617

205. Ye H, Daoud-El Baba M, Peng RW, Fussenegger M (2011) A synthetic optogenetic transcription device enhances blood-glucose homeostasis in mice. Science 80. https://doi.org/10.1126/science.1203535

206. Ye H, Charpin-El Hamri G, Zwicky K, Christen M, Folcher M, Fussenegger M (2013) Pharmaceutically controlled designer circuit for the treatment of the metabolic syndrome. Proc Natl Acad Sci U S A. https://doi.org/10.1073/pnas.1216801110
207. Rössger K, El Hamri GC, Fussenegger M (2013) Reward-based hypertension control by a synthetic brain-dopamine interface. Proc Natl Acad Sci U S A. https://doi.org/10.1073/pnas.1312414110
208. Ausländer D, Eggerschwiler B, Kemmer C, Geering B, Ausländer S, Fussenegger M (2014) A designer cell-based histamine-specific human allergy profiler. Nat Commun. https://doi.org/10.1038/ncomms5408

Systems Immunology Approach in Understanding the Association of Allergy and Cancer

2

Sreyashi Majumdar and Sudipto Saha

Abstract

Epidemiological studies on allergy/asthma and cancer suggest that there exists association between these two types of immunological diseases. Atopic allergy can promote protection from certain types of cancer such as colorectal and esophageal cancers, whereas it may also serve as a risk factor for cancers like lung cancer. There are key immune cells like Tregs and macrophages that play a crucial role in immunoregulation of both the diseases. Besides, PD-1, PD-L1/L2, CTLA4, IgE, Type 2 cytokines regulate allergic manifestations and malignant conditions in the human system. In this chapter, the association of atopic allergy with different types of cancer, and the key immune cells and important molecules associated with both the diseases have been highlighted. In the end, the future perspectives of the field of allergo-oncology and possible therapeutic approaches to modulate the immune systems have been described.

Keywords

Allergy · Cancer stem cells · Tregs · PD-L1 · IgE · Th2 cytokine

2.1 Introduction

Atopy or allergy refers to allergic hypersensitivity, characterized by heightened immune responses [1]. A plethora of immune cells, namely, mast cells, dendritic cells, macrophages, B cells, CD4+ T cells and a variety of mast cell mediators, immunoglobulin E (IgE) and Th2 cytokines, lie central to the onset and progression of allergic diseases. Cancer, on the other hand, is characterized by abnormal

S. Majumdar · S. Saha (✉)

Division of Bioinformatics, Bose Institute, Kolkata, India

e-mail: ssaha4@jcbose.ac.in

© Springer Nature Singapore Pte Ltd. 2020

S. Singh (ed.), *Systems and Synthetic Immunology*,

https://doi.org/10.1007/978-981-15-3350-1_2

53

uncontrolled cell growth and proliferation. Evasion of the immune system is crucial for cancer progression and metastasis. Several cells in the tumor microenvironment (TME), namely, cancer-associated fibroblasts (CAFs), stromal cells, cancer stem cells (CSCs), M2 macrophages, regulatory T cells (Tregs), myeloid-derived suppressor cells (MDSCs) cross-talk amongst themselves through cytokines, signaling molecules, and ECM-modifying agents to generate a heterogeneous network that mediates and maintains immunosuppressive milieu in TME [2–4]. This contributes to tumor progression and resistance to cancer therapy. Both allergy and cancer are associated with dysregulated immune response [5].

Several epidemiological studies since 1985 have shown allergy to influence the occurrence of cancer by acting either in protective manner or as a risk factor [6, 7]. The relationship between allergies and malignancies varies from organ to organ. An inverse association was observed between allergy and malignancies of colon, rectum, pancreas, and esophagus, while a positive association was noted for lung cancer, bladder cancer, and prostate cancer. Four immunological hypotheses have been architected to explain the impact of allergy on cancer, namely, antigenic stimulation, inappropriate Th2 immune skewing, immunosurveillance, and prophylaxis [8]. Several immune cells and molecules have been implicated to express differentially and play pivotal roles in regulating the immune system in allergic condition and malignancies [5]. Present-day allergo-oncology research primarily focuses on revealing the roles of these molecules in asthma and cancer pathogenesis for the development of novel therapeutics. Mechanistic insight into key immune cells and molecules operating to mediate such complex association (Fig. 2.1) is crucial for regulating allergy and cancer via reprogramming of altered cellular function and rewiring of key networks.

Here, first the concept of allergy and immunological mechanisms driving allergic reactions has been discussed. Thereafter, cancer and mechanisms of immune evasion during disease progression have been explained. Next, the complex relationship between allergies and cancer susceptibility have been addressed with special reference to: i) immunological hypotheses explaining such associations, ii) case studies showing the influence of allergic manifestations on cancers at different sites, and iii) the key molecules and cells underlying immune tolerance in allergy and cancer. Finally, the chapter deals with the developments, challenges, and future perspectives of allergo-oncology with special emphasis on mechanistic understanding of such association and rewiring of pathways/networks in immune cells for efficient control of the diseases.

2.2   Allergy and Allergenicity

Allergy arises due to hyperactivity of the immune system [9]. The hypersensitive immune system elicits inappropriate and exaggerated immune response to typically harmless substances coming in contact with or entering the body, thereby leading to the occurrence of allergic reaction. Any substance, protein or non-protein, that can trigger allergic reaction is referred to as an allergen. The most commonly

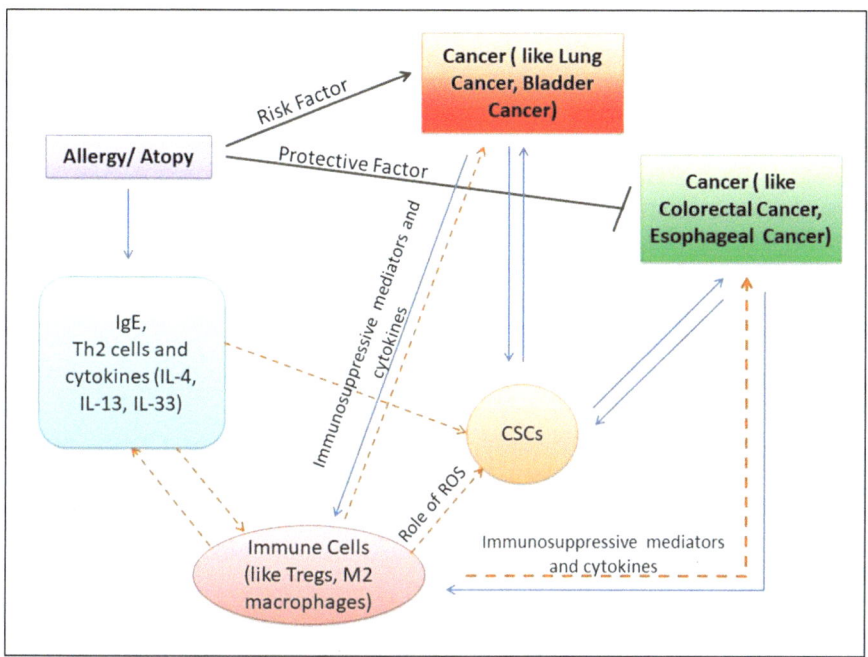

Fig. 2.1 **Complex, heterogeneous interactive network mediating association between allergy and cancer**. Mechanistic insight of these pathways is vital for rewiring of key networks for developing novel therapeutics. The solid black lines denote the pattern of association between allergy and different types of cancer. The solid blue lines denote the cellular and molecular components involved in allergy and cancer. The dotted orange lines denote indirect interaction

recognized allergens include pollens, animal dander, house dust mites, molds, insect stings, food allergens (like milk, peanuts, eggs, crustaceans, mustard, sesame, lupins, glutens, nuts, fish, soybeans, celery, molluscs), and drug allergens. The property or potential of an allergen to induce sensitization and allergic reactions is known as its allergenic potential or allergenicity [10]. Allergic manifestations can be localized or systemic. The common allergic conditions include hives or urticaria, atopic dermatitis, atopic eczema, hay fever or allergic rhinitis, allergic conjunctivitis, asthma, and anaphylactic shock. The incidences of the different types of allergic diseases have increased over decades. These allergic diseases are often associated with co-morbidity and also contribute to high economic burden and substantial morbidity [11].

2.2.1   Immunological Reactions Leading to Allergy

Exposure to allergen triggers a series of immune reactions leading to allergic manifestation. At first, these allergens encounter the antigen-presenting cells (APCs) at/near the site of exposure. Upon recognition of the allergen, the APCs (namely, dendritic cells, B cells, and macrophages) undergo activation and promote the differentiation of naive T cells into Th2 cells (CD4+ T cells). Activated allergen-specific CD4+ T cells release Th2 cytokines, namely, IL-4, IL-5, IL-6, IL-9, IL-10, IL-13, and mediate B-cell differentiation and the formation of Immunoglobulin E (IgE) by B cells [9]. These IgE molecules bind to their specific Fc receptors present on innate immune cells, namely, mast cells and granulocytes-like basophils and eosinophils and mediate cross-linking of the Fc receptors upon allergen binding. Cross-linkage of Fc receptors leads to cascade of signaling events within the mast cells and/or granulocytes, leading to degranulation and release of vaso-active mediators (like histamines, proteases, heparin, leukotrienes, prostaglandins), chemokines, and cytokines, which in turn acts on a plethora of cells like smooth muscle cells, mucous glands, epithelial cells, stromal and muscle cells, small blood vessels, nerve endings, and eosinophils. This causes inflammation, tissue damage and remodeling, and acute changes in functionality and thus results in allergic manifestations.

2.3 Cancer and Onco-immunology

Cancer is a chronic disease marked by uncontrollable division of abnormal cells and is one of the major causes of death all over the world [12]. Cancer can be primary, staying localized at the site of origin, or may be metastatic, spreading to other sites within the body [13]. Mutations like activation of oncogenes, silencing of tumor suppressor genes and DNA repair genes, chromosomal aberration like translocation, posttranslational modification like glycosylation, and epigenetic changes like alteration in methylation status trigger the process of oncogenesis [4, 13]. The pathogenesis of cancer is complicated and largely varies with the site of origin [13].

2.3.1 Tumor Microenvironment

The tumor microenvironment (TME) consists of the following: (i) a diverse variety of cells like cancer-associated fibroblasts (CAFs), stromal cells, cancer cells, cancer stem cells (CSCs), myeloid-derived suppressor cells (MDSCs), blood vessels, immune cells like Tregs, M2 macrophages, tumor-infiltrating lymphocytes (TILs), tumor-associated macrophages (TAMs), and neutrophils (TANs); (ii) different signaling molecules; (iii) cytokines; and (iv) several extracellular matrix (ECM) remodeling agents [3, 14–17]. Apart from diversity in cell type, there exists considerable degree of heterogeneity among each cell type which adds to the complexity of cancer pathogenesis [18]. The cross-talk between different cancerous cells with immune cells in TME lies central to the process of cancer development and progression. The tumor microenvironment mediates immune tolerance and largely dictates the responsiveness to cancer therapy [19, 20].

2.3.2   Onco-immunology: Role of Immune System in Cancer Pathogenesis and Progression

The immune system plays a crucial role in cancer progression via the process of tumor immunoediting [21]. This process comprises three phases, namely, elimination, equilibrium, and escape. In the elimination phase, members of the innate and adaptive immune system, namely, natural killer cells (NK cells), cytotoxic T cells (CD8+ T cells), natural killer T cells (NKT cells), and γδT cells, recognize and eliminate the cancer cells by perforin secretion, complement-dependent cytotoxicity (CDC), or antibody-dependent cellular cytotoxicity (ADCC) [13, 22, 23]. During the second phase, an equilibrium exists between immunity-mediated elimination of cancer cells and cancer progression [24]. In the last phase, that is, the escape phase, tumor cells efficiently escape anti-cancer immune responses mainly by decreased immune recognition and by establishing an immunosuppressive tumor microenvironment. Decreased MHC-I expression and reduction of co-stimulatory molecule primarily contributes to reduced immune recognition [24]. Cancer stem cells (CSCs) and immune cells like MDSCs, M2 macrophages, and Tregs play a vital role in inducing and maintaining immune-suppressive environment. CSCs are a small subpopulation of cancer cells endowed with the property of self-renewal, differentiation, tumor initiation, and propagation. These cells can escape immune surveillance and therapeutic effectiveness and mediate relapse [4, 15]. CSCs are regulated by fibroblasts via release of CCL2 [25]. These cells secrete cytokines like TGF-β, IL-10, VEGF which drives T cell population from effector T cells to Tregs. CSCs also release factors like MIF, STAT3, and VEGF that polarize TAM toward M2 phenotype in the TME. Immunosuppressive factors like COX-2 and IDO-1 released by cancer cells further deteriorates the situation. IL-10 and TGF-β released by TAMs blocks effector T cell activity and dendritic cell maturation. M2 macrophages also secrete EGF and MMP9, which, in turn, promotes cancer proliferation and angiogenesis [2]. Another immune cell population that play crucial role in mediating and restoring immunosuppression in TME are the regulatory T cells (Tregs). The immunosuppressive action of Tregs is dependent primarily on the expression of transcription factor, Forkhead Box P3 (FOXP3) [26]. Tregs (CD4+ CD25+ FOXP3+ T cells) induce immunosuppression in TME by i) contact-dependent method involving immune checkpoint molecules like PD1, PD-L1, LAG-3, CTLA4, CD39/73, FOXP3 and ii) contact-independent mechanism via mediators like TGF-β, IL-35, IL-10, STAT3, VEGF, PGE2, adenosine, galectin-1 [2, 3, 26]. Other immune cells like Bregs, MDSCs, and TANs also aid in mediating immune suppression [2, 3]. The cross-talk of the different cancer cells and immune cells mediated by cytokines, signaling molecules, creates an interactive network between immune and cancer cells that further enhances the immunosuppressive milieu in the tumor microenvironment.

2.3.3 Pathways Leading to Escape from Host Immune System

Several signaling pathways have been implicated for mediating immune suppression in the tumor microenvironment. The Wnt/β catenin pathway is an intrinsic oncogene pathway that prevents anti-tumor T cell activity within the tumor microenvironment (TME) [27]. The STAT3 pathway activated by regulatory T cells (Tregs) mediates immune suppression through activation of M2 macrophages. This pathway also promotes cancer survival and angiogenesis [3]. TGF-β signaling in TME promotes Tregs and TANs and mediates FOXP3 expression in Tregs, thereby restoring immune tolerance in the cancerous cell milieu [3, 28, 29]. Another crucial pathway is the PI3K/PTEN/AKT pathway. This pathway is associated with the recruitment of TAMs via production of mediators like VEGF, IL-6, IL-8. It also activates hypoxia-inducible factor 1α (HIF-1α) via Mammalian target of rapamycin complex 1 (MTORC1) and induces epithelial mesenchymal transition (EMT) and metastasis. Release of CXCL12 by cancer-associated fibroblasts and subsequent CXCL12 signaling polarizes macrophages toward M2 phenotype. Other ancillary pathways, namely, STAT5 pathway, NF-κβ pathway, COX2/PGE2 pathway, and aberrant p53 signaling pathway, also aid in establishing immune-suppressive ambience in tumor microenvironment [3, 30]. These pathways are also associated with poor response to various cancer therapies. Remodeling of these pathways and restoring aberrant networks might help in restoring anti-tumor immune response, leading to control of tumor progression and enhancement of therapeutic efficacy.

2.4 Role of Allergy in Cancer Susceptibility

Epidemiological studies conducted before and after 1985 denoted potential association between allergic diseases and cancer susceptibility [6, 31]. The association between allergic diseases and cancer is complex and may be organ specific [32, 33]. Many studies have reported positive association between allergic condition and cancer [34]. On the contrary, several other studies have reported an inverse association between allergic manifestations and cancer malignancies [34]. Several hypotheses have been proposed to explain such associations between allergies and cancer in the light of immunology (Fig. 2.2).

The antigenic stimulation and inappropriate Th2 immune skewing hypotheses account for the positive association between allergy and cancer. In 1988, McWhorter proposed the "Antigenic Stimulation" hypothesis, also called the "Chronic Inflammation" hypothesis [8, 35]. Allergic diseases cause chronic inflammation, and allergenic stimulation leads to activation of neutrophils and phagocytes, which generates reactive oxygen species and free radicals [36]. This reduces antioxidant levels and increases the likelihood of inducing mutation in tumor suppressor genes, causing genetic damage of actively dividing stem cells and inducing modifications in proteins involved in DNA repair and apoptosis, thereby promoting malignant

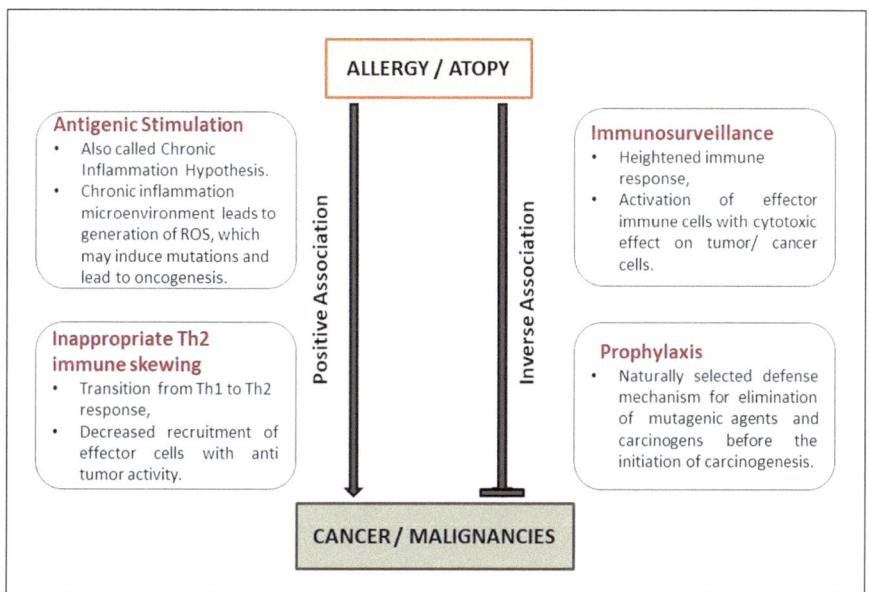

Fig. 2.2 **Hypotheses driving positive and inverse association between allergy and malignancy. Four major hypotheses have been proposed**. Antigenic stimulation and inappropriate Th2 immune skewing provide the immunological explanation for positive association between occurrence of allergy and cancer. Immunosurveillance and prophylaxis explain the immune mechanism that reduces incidences of cancer in allergic condition

transformation of cells and cancer development [8, 37, 38]. The "Inappropriate Th2 immune skewing" hypothesis further substantiates the positive association between allergy and cancer. According to this hypothesis, during allergic reaction, there is a shift from Th1 response to Th2 response, leading to decrease in Th1 cytokines (namely, IFN-γ, IL-2, IL-3, GM-CSF, and TNF-α) and reduced recruitment of effector cells with tumor-eradicating features (like M1 macrophages and cytotoxic T lymphocytes), allowing the tumor cells to grow and spread [8].

Immunosurveillance and prophylaxis explain the inverse association between allergy and cancer. In 1957, the "Immunosurveillance" hypothesis, stated by Burnet, proposed that allergy arises due to exaggerated immune response. Overstimulation of immune cells upon exposure to allergen leads to further production of IgE and activation of effector cells that might exert cytotoxicity on tumor cells and prevent oncogenesis [8, 39]. On the other hand, the "Prophylaxis" hypothesis, introduced by Profet in 1991, is based on Darwin's principle of evolution by natural selection [40]. The immune mediators and physical manifestations of allergic reactions might be naturally selected for efficient elimination or destruction of mutagenic toxins and environmental carcinogens, thereby conferring protection against cancer [8].

Table 2.1 Overview of association between allergic conditions and different types of cancer

System/organ affected	Cancer	Association with allergic condition/history of atopy
Blood	Leukemia	Inverse association [8, 42, 43]
	Lymphoma	Inverse association [41]
	Myeloma	Inverse association [44]
		Positive association [34]
	Non-Hodgkin lymphoma	Inverse association [31, 34, 45]
Brain	Head and neck cancer	Inverse association [31]
	Glioma	Inverse association [46]
		Positive association [47]
	Meningioma	Inverse association [48, 49]
Breast	Breast cancer	Inverse association [55]
		Positive association [54, 56]
Gastrointestinal tract	Colorectal cancer	Inverse association [57–61]
		Positive association [62, 63]
	Pancreatic cancer	Inverse association [59, 64–66]
		Positive association [68]
		No association [67, 69]
	Esophageal cancer	Inverse association [58, 70]
	Stomach cancer	Inverse association [44, 62]
	Liver cancer	Inverse association [85]
	Gastric cancer	Inverse association [62, 71]
	Small bowel cancer	Inverse association [56]
Gynecological organs	Gynecological malignancies	Inverse association [56, 62, 72]
	Uterine body cancer	Inverse association [44]
Larynx	Cancer of larynx	Inverse association [34, 62]
Lung	Lung cancer	Positive association [8, 31, 73–75]
Oral cavity	Oral cancer	Inverse association [34]
Prostate gland	Prostate cancer	Positive association [76]
Skin	Skin cancer	Positive association [31, 44, 76, 77]
		Inverse association [56, 78, 79]
		No association [76]
Thyroid gland	Thyroid cancer	Inverse association [34, 82, 84]
Urinary tract	Bladder and urothelial cancer	Positive association [62, 80, 81]
Others	Kaposi sarcoma	Positive association [83]

This table enlists positive or inverse association existing between malignancies and allergies

2.4.1 Association Between Allergies and Different Types of Cancer

Several epidemiological studies were conducted to assess the pattern of influence of allergic history on different types of cancers (Table 2.1). The protective effects of atopy were observed for some malignancies like colorectal cancer, esophageal cancer, and glioma, whereas atopy was a risk factor for cancer like lung cancer, prostate cancer, and bladder cancer. The susceptibility of different cancers in allergic condition as observed in epidemiological studies is stated underneath.

Leukemia, Lymphoma, and Myeloma Atopic conditions were found to confer protection against hematological malignancies in most cases. History of allergic condition and increased IgE levels were found to be associated with reduced risks of B-cell lymphoma [41]. Inverse association was also observed between history of allergic condition and both adult and childhood acute lymphoid leukemia (ALL) [8]. Cohort studies conducted on general population of UK revealed inverse association between atopy and chronic lymphoid leukemia (CLL) [42]. Decreased risk of acute myeloid leukemia (AML) was reported with the history of any form of allergy [43]. Inconsistent results have been obtained in case of myeloma. Case control study highlighted the fact that allergic condition was associated with an increased risk for multiple myeloma [34]. However, other studies were suggestive of an inverse association between allergy and myeloma [44].

Non-Hodgkin Lymphoma (NHL) An inverse association was observed between allergic condition and NHL in statistically significant studies [31]. Allergy and asthma were associated with significant reduction in childhood NHL [45]. HIV-positive homosexual males allergic to pollen, grass, hay, and leaves were protective to NHL. Non-medication allergies in HIV-negative homosexual men reduced the risk of development of NHL [34].

Brain Tumors Allergic conditions were mostly inversely associated with head and neck cancer (HNC) [31]. Reduced risk of glioma was noted in patients with a history of allergy (including asthma, eczema, and hay fever) [46]. There was a greater reduction in the risk of glioma with the increasing number of allergies [47]. Decreased risk of meningioma was observed in patients having eczema and allergies [48, 49]. Serum IgE level was found to be reduced in patients with glioma and meningioma [50, 51]. Such reductions could be attributed to immune-modulating properties of chemotherapy [52]. An increase in serum IgE was found to be associated with better survivality in glioma [53].

Breast Cancer Inconsistent results were obtained for association between allergy and breast cancer. History of atopy served as a risk factor for premenstrual breast cancer [54]. Again, another study revealed reduced risk of breast cancer in women older than 35 years, with allergic predisposition [55]. In a nationwide cohort study conducted in Taiwan, allergic rhinitis was found to be significantly positively associated with incidences of breast cancer [56].

Colorectal Cancer An inverse association was observed among allergies and colorectal cancer in both case-control and cohort-based studies. In the meta-analysis of prospective studies, allergic individuals were found to have reduced colorectal cancer risk and mortality [57]. An allergic history also resulted in decreased risk for colon cancer or rectal cancer [58]. The relative risk of developing colorectal cancer was 20% lower in patients with asthma and hay fever [59]. Similar trends were also noted in a prospective study conducted for female candidates where asthma, hay fever, eczema, and other allergic conditions conferred protection against oncogenesis [60]. Even individuals with drug allergy were less prone to developing colorectal cancer [61]. However, a single study showed asthmatics to be at a greater risk for colorectal cancer and few studies yielded no concrete conclusion [62, 63].

Pancreatic Cancer History of allergy (namely, hay fever, mold allergy, allergy to animal dander and stings, and other allergies) was associated with reduced incidence of pancreatic cancer and greater survivality in pancreatic cancer patients [64–66]. Reduced pancreatic mortality was associated with incidences of hay fever [59]. No association was noted between drug allergy and pancreatic cancer [67]. Only a retrospective cohort study revealed positive association of asthma and pancreatic cancer [68]. A multiethnic prospective study, however, revealed no association between atopic allergic conditions, antihistamine usage, and incidence of pancreatic cancer [69].

Other Gastrointestinal Cancers Allergic history had inverse association with esophageal cancer and vice-versa [58, 70]. Asthmatics were found to be at a reduced risk for developing gastric cancer while individuals with allergic rhinitis had reduced instances of small bowel cancer [56, 62, 71]. Both the cases are indicative of protective role of allergy.

Gynecological Malignancies Asthmatic and allergic females were at a reduced risk for developing endometrial, cervical, and ovarian cancer, suggesting a protective role of allergy in these malignancies [56, 62, 72].

Lung Cancer Atopic diseases elevate the risk for the development of lung cancer [31]. Both retrospective and prospective studies denote a greater risk of lung cancer in asthmatics. Chronic asthmatic inflammation and tissue remodeling might possibly contribute to such oncogenic predisposition [8]. Such positive association between asthma and lung cancer was observed even among non-smokers [73]. Prospective studies have also highlighted higher death from lung cancer in patients with bronchial asthma [74]. Recent studies have reported chronic asthmatic inflammation to be associated with polymorphisms in pro-inflammatory genes (like IL-1β,

IL-6 and IL1RN). Such genetic polymorphisms might serve to elevate the risk of bronchial carcinoma [75].

Prostate Cancer History of asthma and allergic sensitization to pollen and house dust mites was associated with a greater risk of developing prostate cancer [76].

Skin Cancer Asthma was found to confer protection against melanoma, but hay fever and eczema, on the other hand, separately served as risk factors for malignant melanoma [44, 76, 77]. However, no association between atopy and melanoma was observed in prospective study [76]. Reduced risk of non-melanoma skin cancer was noted in case of allergic rhinitis and eczema [56, 78]. Asthma exhibited inverse association with basal cell carcinoma [79]. Allergy and atopic condition were found to have positive association with squamous cell carcinoma [31].

Bladder Cancer Asthmatics, especially males, were found to be at a greater risk for developing bladder cancer and urothelial cancer [62, 80, 81].

Other Types of Cancer Asthma was found to render protection against stomach cancer, uterine body cancer, cancer of larynx, oral cancer, and thyroid cancer but was a risk factor for Kaposi sarcoma [44, 62, 82, 83]. History of allergy was linked with lower risk of thyroid cancer, oral cancer, and cancer of larynx [34, 84]. Drug allergies rendered protection against liver cancer [85].

2.4.2 Immune Players Involved in Allergy and Cancer

Immunological balance lies central to maintaining homeostasis and healthy condition. In case of allergy, the immune system overreacts, while in case of malignancies, active suppression of immune system is observed. An immune phenomenon called immune tolerance lies central to the development of allergy and cancer [86]. Immune tolerance is the mechanism by which the immune system is rendered unresponsive to self-antigens and potentially harmless antigens [87]. Suppression of immune tolerance is associated with allergic manifestation, and enhanced immune tolerance is related to oncogenesis. The cellular and molecular players involved in maintaining immune tolerance are inhibited in allergies and upregulated in malignancies [86].

2.4.2.1 Key Cellular Players Mediating Immune Tolerance in Allergy and Cancer

Immune cells exhibit differential behavioral, interaction, and secretion pattern to ensure immune hyperactivity under allergic condition and enhanced immune

Table 2.2 List of cells with roles in immune tolerance in allergy and cancer

Immune cells	Role in allergy	Role in cancer
Dendritic cells	Induce cascade of immune signaling, leading to allergy [5]	Induce the conversion of T cells into Tregs and promote tumor progression [88]
Eosinophils	Play a major role in allergic and atopic reaction	Exert anti-tumorigenic effect on solid tumors [89]; support metastasis [90]
Epithelial cells	Epithelial barrier disruption allows allergen entry; secrete TSLP and IL-33 and promote Th2 branch of immune responses [5, 91]	Lead to innate immune suppression and contribute to cancer progression [92]
Macrophages	M2a macrophages support allergic manifestation [5]; M1 macrophages promote airway remodeling in atopic asthma [93].	M1 macrophages promote increased survivality of cancer patients; M2b and M2c play a vital role in tumor progression and poor prognosis [5]
Mast cells	Mediate allergic reaction [9]	Promote tumor growth [5, 87]
T regulatory cells (Tregs)	Suppress allergic reaction [5]	Associated with tumor progression [5]

This table enlists the different immune cells and their probable roles in allergic response and tumorigenesis
Partially adapted from [5]

tolerance under malignant condition [5]. The key immune cells with pivotal roles in allergo-oncology are enlisted in Table 2.2.

The differential activity pattern under allergic and malignant condition is not just restricted to immune cells. Such diversification of immune roles in allergies and cancer is also visible at the molecular level.

2.4.2.2 Immunoregulatory Molecules in Allergy and Malignancy

Vital immune proteins and molecules are differentially expressed in allergy and tumor. These molecules often execute contrasting roles under allergic and malignant condition. Two molecules that have drawn primary attention are programmed cell death-1 (PD-1) and its ligand 1/2 (PD-L1/L2) and cytotoxic T lymphocyte–associated protein 4 (CTLA4). These are often referred to as checkpoint molecules [5]. Other key molecules involved include IgE, IgG4, mast cell mediators, cytokines, and lipocalins.

PD-1, PD-L1, and PD-L2: These proteins have been long implicated for their role in cancer. PD-1 induces immunotolerance and restores immunosuppressive microenvironment in cancer by inhibiting the activation of T cells, restricting cytotoxic T cell proliferation and preventing cancer cell lysis [94]. Expression of PD-L1/L2 and CTLA4 on tumor cells further exaggerates this effect and ensures cancer progression [95]. Up regulation of PD-L2 has been found in malignancies like Hodgkin Lymphoma [96]. Recently, the roles of PD-L1/L2 have been extensively studied in allergic diseases and atopic asthma. These proteins play a vital but opposing role in guiding the polarization of T cells [97]. In allergic diseases, blocking of PD-L2 resulted in enhanced eosinophil infiltration. Deficiency of

PD-L2 has been linked to increased severity in atopic asthma. On the other hand, PD-L1 deficiency is associated with elevated Th1 cytokine response and reduced inflammation in allergic condition [97]. This is indicative of PD-L2 as a protective factor and PD-L1 as a risk factor in allergy.

CTLA4: This protein plays a vital role in dictating T cell activation and differentiation [98]. It has been targeted clinically for cancer. CTLA4 molecules expressed on T cells present in tumor microenvironment inhibit further T cell activation and proliferation, thereby restoring immune tolerance [99]. This ensures tumor growth and progression without much hindrance from host immune system. Role of CTLA4 has also been highlighted in allergy. Blockade of CTLA4 triggers Th2 cytokine response, elevates eosinophil-mediated inflammation, and increases allergic sensitization [100].

IgE and IgG repertoires: IgE is the major antibody-mediating allergic reaction. IgE and IgG have also been implicated in cancer as well. Monoclonality of IgE and IgG is observed in myeloma. Reemergence of small sub-clones is noted along with the dominant clone in case of B-cell leukemia [5].

IgG4 antibodies: IgG4 are anti-inflammatory antibodies with protective role in allergy [5]. IgG4 elevate in cancer and correlate with tumor progression [5, 101, 102].

Mast cell mediators: Mast cells secrete vasoactive mediators (histamines, leukotrienes, prostaglandins) and cytokines that act on smooth muscle cells, sensory nerve endings, blood vessels, and mucous cells to mediate inflammation and allergic reaction [9]. Mast cell mediators have a controversial role in oncogenesis. They promote tumorigenesis by secreting histamine, NGF, IL-8, and restricting T-cell responses. Mediators like heparin, TGFβ, VEGF promote neovascularization [103]. On the other hand, TNFα, IFNγ, PAR-1/2 induce cellular disruption and apoptosis in tumor cells. IL-8, TNFα inhibit carcinogenesis by attracting inflammatory leukocytes. These mediators inhibit metastasis via chondroitin sulfate. Mast cells release amphiregulin and contribute to immunosupression in tumor [5].

Cytokines: TGFβ and IL-10 play a vital role in the establishment and perpetuation of immune tolerance. TGFβ and IL-10 modulate immunosuppressive microenvironment in cancer and connect with different stages of oncogenesis [5, 104]. IL-17A shows upregulation in allergic disease like asthma. The role of IL-17A is controversial in cancer, acting both as a tumor suppressor and a promoter [86].

Lipocalins (LCNs): Lipocalins play an important role in iron sequestration. Their expression and subsequently the serum iron level are decreased in case of allergic reaction [5]. However, this protein exhibits an opposite expression profile in case of malignancies. LCN2 is overexpressed in various types of cancer [105]. Elevated iron level also increases the risk for cancer development [106]. LCN2 also forms complex with matrix metallopeptidase-9 (MMP-9), a prognostic factor in different cancers [107].

These molecules can be targeted differentially for controlling allergy and cancer. Present-day scientific research focuses largely on unraveling the mechanisms driving such differential molecular expression and cellular behavior. A better and detailed understanding of the role of these molecules and immune cells shall provide efficient means of targeting these molecules, thus opening up new gateways for the development of novel therapeutic approaches for allergy and cancer.

2.5 Developments, Challenges, and Future Perspectives in Allergo-oncology

Recent research in allergo-oncology has enabled better understanding of Th2 immune response, contrasting roles of immune cells in tumor microenvironment and allergic condition, and pivotal roles of immune molecules in regulating immunity in allergies and malignancies. However, there still lies a considerable gap in translating the current knowledge and research into clinical practice for developing novel therapeutic strategies [108]. One of the major challenges lies in the availability of suitable animal models for further studies in immune tolerance and assessing efficacy of targeting key cellular and molecular players [5]. Present knowledge on immune response in allergic and malignant milieu may be combined efficiently to generate optimal in vivo animal models for allergo and onco-immunological studies [5]. Humanized mice and canines are emerging as desirable animal models for allergo-oncology-related studies for spontaneous development of allergies and malignancies [109, 110]. Another issue arises due to inconsistency in the association pattern among allergy and cancer in epidemiological studies. Methodological limitations (associated with case-control and cohort study, retrospective and prospective study); screening biasness and lack of proper consideration of confounding factors (like smoking, alcohol consumption, obesity, socioeconomic status) during data analysis often affect the significance of such epidemiological studies. Prospective cohort studies and meta-analyses, properly adjusted for confounding variables, might serve to set off study-related issues and allow better understanding of the association between allergy and cancer [31]. Mechanistic study of such associations using co-culture systems, multi-omics, and systems immunology–based approaches shall enable identification of key pathways and networks involved in these associations. Reprogramming and rewiring of key interactions in these regulatory networks and pathways of specific immune cells can be employed for designing novel therapeutics. Besides, impact of allergen, allergenic peptide harboring IgE-binding epitopes, IgE, or Th2 cytokines on cancer stem cell population can be investigated further for exploring allergy-mediated control of cancer. Such novel therapeutic strategies can be applied along with regular cancer therapy for better efficacy [111]. Reprogramming of immune cells and remodeling of pathways may also aid in controlling allergic manifestations. Thus, the association between allergy and cancer might be exploited in a constructive manner to control the occurrence and progression of both the disease conditions.

References

1. Han H, Roan F, Ziegler SF (2017) The atopic march: current insights into skin barrier dysfunction and epithelial cell-derived cytokines. Immunol Rev 278(1):116–130
2. Gonzalez H, Hagerling C, Werb Z (2018) Roles of the immune system in cancer: from tumor initiation to metastatic progression. Genes Dev 32(19–20):1267–1284
3. Najafi M, Farhood B, Mortezaee K (2019) Contribution of regulatory T cells to cancer: a review. J Cell Physiol 234(6):7983–7993
4. Codony-Servat J, Rosell R (2015) Cancer stem cells and immunoresistance: clinical implications and solutions. Transl Lung Cancer Res 4(6):689–703
5. Jensen-Jarolim E, Bax HJ, Bianchini R, Crescioli S, Daniels-Wells TR, Dombrowicz D et al (2018) AllergoOncology: opposite outcomes of immune tolerance in allergy and Cancer. Allergy 73(2):328–340
6. Wang H, Diepgen TL (2005) Is atopy a protective or a risk factor for cancer? A review of epidemiological studies. Allergy 60(9):1098–1111
7. Vena JE, Bona JR, Byers TE, Middleton E Jr, Swanson MK, Graham S (1985) Allergy-related diseases and cancer: an inverse association. Am J Epidemiol 122(1):66–74
8. Josephs DH, Spicer JF, Corrigan CJ, Gould HJ, Karagiannis SN (2013) Epidemiological associations of allergy. IgE Cancer Clin Exp Allergy 43(10):1110–1123
9. Nauta AJ, Engels F, Knippels LM, Garssen J, Nijkamp FP, Redegeld FA (2008) Mechanisms of allergy and asthma. Eur J Pharmacol 585(2–3):354–360
10. Verhoeckx KCM, Vissers YM, Baumert JL, Faludi R, Feys M, Flanagan S et al (2015) Food processing and allergenicity. Food Chem Toxicol 80:223–240
11. D'Arcy M, Rivera DR, Grothen A, Engels EA (2019) Allergies and the subsequent risk of cancer among elderly adults in the United States. Cancer Epidemiol Biomark Prev 28(4):741–750
12. Ferlay J, Soerjomataram I, Dikshit R, Eser S, Mathers C, Rebelo M et al (2015) Cancer incidence and mortality worldwide: sources, methods and major patterns in GLOBOCAN 2012. Int J Cancer 136(5):E359–E386
13. Pandya PH, Murray ME, Pollok KE, Renbarger JL (2016) The immune system in cancer pathogenesis: potential therapeutic approaches. J Immunol Res 2016:4273943
14. Kato T, Noma K, Ohara T, Kashima H, Katsura Y, Sato H et al (2018) Cancer-associated fibroblasts affect Intratumoral CD8(+) and FoxP3(+) T cells via IL6 in the tumor microenvironment. Clin Cancer Res 24(19):4820–4833
15. Toledo-Guzman ME, Bigoni-Ordonez GD, Ibanez Hernandez M, Ortiz-Sanchez E (2018) Cancer stem cell impact on clinical oncology. World J Stem Cells 10(12):183–195
16. Kim J, Bae JS (2016) Tumor-associated macrophages and neutrophils in tumor microenvironment. Mediat Inflamm 2016:6058147
17. Ciszewski WM, Sobierajska K, Wawro ME, Klopocka W, Chefczynska N, Muzyczuk A et al (2017) The ILK-MMP9-MRTF axis is crucial for EndMT differentiation of endothelial cells in a tumor microenvironment. Biochim Biophys Acta, Mol Cell Res 1864(12):2283–2296
18. Prasetyanti PR, Medema JP (2017) Intra-tumor heterogeneity from a Cancer stem cell perspective. Mol Cancer 16(1):41
19. Wu T, Dai Y (2017) Tumor microenvironment and therapeutic response. Cancer Lett 387:61–68
20. Ostrand-Rosenberg S (2016) Tolerance and immune suppression in the tumor microenvironment. Cell Immunol 299:23–29
21. Selvarajoo K (2014) Advances in systems immunology and Cancer. Front Physiol 5:249
22. Vesely MD, Kershaw MH, Schreiber RD, Smyth MJ (2011) Natural innate and adaptive immunity to cancer. Annu Rev Immunol 29:235–271
23. Schreiber RD, Old LJ, Smyth MJ (2011) Cancer immunoediting: integrating immunity's roles in cancer suppression and promotion. Science 331(6024):1565–1570

24. Mittal D, Gubin MM, Schreiber RD, Smyth MJ (2014) New insights into cancer immunoediting and its three component phases – elimination, equilibrium and escape. Curr Opin Immunol 27:16–25
25. Tsuyada A, Chow A, Wu J, Somlo G, Chu P, Loera S et al (2012) CCL2 mediates cross-talk between Cancer cells and stromal fibroblasts that regulates breast cancer stem cells. Cancer Res 72(11):2768–2779
26. Ward-Hartstonge KA, Kemp RA (2017) Regulatory T-cell heterogeneity and the cancer immune response. Clin Transl Immunol 6(9):e154
27. Spranger S, Bao R, Gajewski TF (2015) Melanoma-intrinsic beta-catenin signalling prevents anti-tumour immunity. Nature 523(7559):231–235
28. Albini A, Bruno A, Noonan DM, Mortara L (2018) Contribution to tumor angiogenesis from innate immune cells within the tumor microenvironment: implications for immunotherapy. Front Immunol 9:527
29. Zhao P, Wang Y, Kang X, Wu A, Yin W, Tang Y et al (2018) Dual-targeting biomimetic delivery for anti-glioma activity via remodeling the tumor microenvironment and directing macrophage-mediated immunotherapy. Chem Sci 9(10):2674–2689
30. Spranger S, Gajewski TF (2016) Tumor-intrinsic oncogene pathways mediating immune avoidance. Oncoimmunology 5(3):e1086862
31. Cui Y, Hill AW (2016) Atopy and specific Cancer sites: a review of epidemiological studies. Clin Rev Allergy Immunol 51(3):338–352
32. Olson SH, Hsu M, Satagopan JM, Maisonneuve P, Silverman DT, Lucenteforte E et al (2013) Allergies and risk of pancreatic cancer: a pooled analysis from the pancreatic cancer case-control consortium. Am J Epidemiol 178(5):691–700
33. Hemminki K, Forsti A, Fallah M, Sundquist J, Sundquist K, Ji J (2014) Risk of cancer in patients with medically diagnosed hay fever or allergic rhinitis. Int J Cancer 135(10):2397–2403
34. Merrill RM, Isakson RT, Beck RE (2007) The association between allergies and cancer: what is currently known? Ann Allergy Asthma Immunol 99(2):102–116; quiz 17–9, 50
35. McWhorter WP (1988) Allergy and risk of cancer. A prospective study using NHANESI followup data. Cancer 62(2):451–455
36. Kelly FJ (1999) Gluthathione: in defence of the lung. Food Chem Toxicol 37(9–10):963–966
37. Boffetta P, Ye W, Boman G, Nyren (2002) Lung cancer risk in a population-based cohort of patients hospitalized for asthma in Sweden. Eur Respir J 19(1):127–133
38. Sherman PW, Holland E, Sherman JS (2008) Allergies: their role in cancer prevention. Q Rev Biol 83(4):339–362
39. Burnet M (1957) Cancer: a biological approach. III. Viruses associated with neoplastic conditions. IV. Practical applications. Br Med J 1(5023):841–847
40. Profet M (1991) The function of allergy: immunological defense against toxins. Q Rev Biol 66(1):23–62
41. Ellison-Loschmann L, Benavente Y, Douwes J, Buendia E, Font R, Alvaro T et al (2007) Immunoglobulin E levels and risk of lymphoma in a case-control study in Spain. Cancer Epidemiol Biomark Prev 16(7):1492–1498
42. Helby J, Bojesen SE, Nielsen SF, Nordestgaard BG (2015) IgE and risk of cancer in 37 747 individuals from the general population. Ann Oncol 26(8):1784–1790
43. Severson RK, Davis S, Thomas DB, Stevens RG, Heuser L, Sever LE (1989) Acute myelocytic leukemia and prior allergies. J Clin Epidemiol 42(10):995–1001
44. Kallen B, Gunnarskog J, Conradson TB (1993) Cancer risk in asthmatic subjects selected from hospital discharge registry. Eur Respir J 6(5):694–697
45. Dikalioti SK, Chang ET, Dessypris N, Papadopoulou C, Skenderis N, Pourtsidis A et al (2012) Allergy-associated symptoms in relation to childhood non-Hodgkin's as contrasted to Hodgkin's lymphomas: a case-control study in Greece and meta-analysis. Eur J Cancer 48(12):1860–1866
46. Wigertz A, Lonn S, Schwartzbaum J, Hall P, Auvinen A, Christensen HC et al (2007) Allergic conditions and brain tumor risk. Am J Epidemiol 166(8):941–950

47. Wiemels JL, Wiencke JK, Sison JD, Miike R, McMillan A, Wrensch M (2002) History of allergies among adults with glioma and controls. Int J Cancer 98(4):609–615
48. Claus EB, Calvocoressi L, Bondy ML, Schildkraut JM, Wiemels JL, Wrensch M (2011) Family and personal medical history and risk of meningioma. J Neurosurg 115(6):1072–1077
49. Turner MC, Krewski D, Armstrong BK, Chetrit A, Giles GG, Hours M et al (2013) Allergy and brain tumors in the INTERPHONE study: pooled results from Australia, Canada, France, Israel, and New Zealand. Cancer Causes Control 24(5):949–960
50. Wiemels JL, Wiencke JK, Patoka J, Moghadassi M, Chew T, McMillan A et al (2004) Reduced immunoglobulin E and allergy among adults with glioma compared with controls. Cancer Res 64(22):8468–8473
51. Wiemels JL, Wrensch M, Sison JD, Zhou M, Bondy M, Calvocoressi L et al (2011) Reduced allergy and immunoglobulin E among adults with intracranial meningioma compared to controls. Int J Cancer 129(8):1932–1939
52. Wiemels JL, Wilson D, Patil C, Patoka J, McCoy L, Rice T et al (2009) IgE, allergy, and risk of glioma: update from the San Francisco Bay area adult Glioma study in the temozolomide era. Int J Cancer 125(3):680–687
53. Wrensch M, Wiencke JK, Wiemels J, Miike R, Patoka J, Moghadassi M et al (2006) Serum IgE, tumor epidermal growth factor receptor expression, and inherited polymorphisms associated with glioma survival. Cancer Res 66(8):4531–4541
54. Eriksson NE, Holmen A, Hogstedt B, Mikoczy Z, Hagmar L (1995) A prospective study of cancer incidence in a cohort examined for allergy. Allergy 50(9):718–722
55. Hedderson MM, Malone KE, Daling JR, White E (2003) Allergy and risk of breast cancer among young women (United States). Cancer Causes Control 14(7):619–626
56. Hwang CY, Chen YJ, Lin MW, Chen TJ, Chu SY, Chen CC et al (2012) Cancer risk in patients with allergic rhinitis, asthma and atopic dermatitis: a nationwide cohort study in Taiwan. Int J Cancer 130(5):1160–1167
57. Ma W, Yang J, Li P, Lu X, Cai J (2017) Association between allergic conditions and colorectal Cancer risk/mortality: a meta-analysis of prospective studies. Sci Rep 7(1):5589
58. Negri E, Bosetti C, La Vecchia C, Levi F, Tomei F, Franceschi S (1999) Allergy and other selected diseases and risk of colorectal cancer. Eur J Cancer 35(13):1838–1841
59. Turner MC, Chen Y, Krewski D, Ghadirian P, Thun MJ, Calle EE (2005) Cancer mortality among US men and women with asthma and hay fever. Am J Epidemiol 162(3):212–221
60. Prizment AE, Folsom AR, Cerhan JR, Flood A, Ross JA, Anderson KE (2007) History of allergy and reduced incidence of colorectal cancer, Iowa Women's Health Study. Cancer Epidemiol Biomark Prev 16(11):2357–2362
61. La Vecchia C, D'Avanzo B, Negri E, Franceschi S (1991) History of selected diseases and the risk of colorectal cancer. Eur J Cancer 27(5):582–586
62. Vesterinen E, Pukkala E, Timonen T, Aromaa A (1993) Cancer incidence among 78,000 asthmatic patients. Int J Epidemiol 22(6):976–982
63. Wang H, Rothenbacher D, Low M, Stegmaier C, Brenner H, Diepgen TL (2006) Atopic diseases, immunoglobulin E and risk of cancer of the prostate, breast, lung and colorectum. Int J Cancer 119(3):695–701
64. Gandini S, Lowenfels AB, Jaffee EM, Armstrong TD, Maisonneuve P (2005) Allergies and the risk of pancreatic cancer: a meta-analysis with review of epidemiology and biological mechanisms. Cancer Epidemiol Biomark Prev 14(8):1908–1916
65. Olson SH, Chou JF, Ludwig E, O'Reilly E, Allen PJ, Jarnagin WR et al (2010) Allergies, obesity, other risk factors and survival from pancreatic cancer. Int J Cancer 127(10):2412–2419
66. Silverman DT, Schiffman M, Everhart J, Goldstein A, Lillemoe KD, Swanson GM et al (1999) Diabetes mellitus, other medical conditions and familial history of cancer as risk factors for pancreatic cancer. Br J Cancer 80(11):1830–1837
67. La Vecchia C, Negri E, D'Avanzo B, Ferraroni M, Gramenzi A, Savoldelli R et al (1990) Medical history, diet and pancreatic cancer. Oncology 47(6):463–466
68. Ji J, Shu X, Li X, Sundquist K, Sundquist J, Hemminki K (2009) Cancer risk in hospitalised asthma patients. Br J Cancer 100(5):829–833

69. Huang BZ, Le Marchand L, Haiman CA, Monroe KR, Wilkens LR, Zhang ZF et al (2018) Atopic allergic conditions and pancreatic cancer risk: results from the Multiethnic Cohort Study. Int J Cancer 142(10):2019–2027

70. Dai Q, Zheng W, Ji BT, Shu XO, Jin F, Cheng HX et al (1997) Prior immunity-related medical conditions and oesophageal cancer risk: a population-based case-control study in Shanghai. Eur J Cancer Prev 6(2):152–157

71. El-Zein M, Parent ME, Ka K, Siemiatycki J, St-Pierre Y, Rousseau MC (2010) History of asthma or eczema and cancer risk among men: a population-based case-control study in Montreal, Quebec. Can Ann Allergy Asthma Immunol 104(5):378–384

72. Johnson LG, Schwartz SM, Malkki M, Du Q, Petersdorf EW, Galloway DA et al (2011) Risk of cervical cancer associated with allergies and polymorphisms in genes in the chromosome 5 cytokine cluster. Cancer Epidemiol Biomark Prev 20(1):199–207

73. Gonzalez-Perez A, Fernandez-Vidaurre C, Rueda A, Rivero E, Garcia Rodriguez LA (2006) Cancer incidence in a general population of asthma patients. Pharmacoepidemiol Drug Saf 15(2):131–138

74. Huovinen E, Kaprio J, Vesterinen E, Koskenvuo M (1997) Mortality of adults with asthma: a prospective cohort study. Thorax 52(1):49–54

75. Lim WY, Chen Y, Ali SM, Chuah KL, Eng P, Leong SS et al (2011) Polymorphisms in inflammatory pathway genes, host factors and lung cancer risk in Chinese female never-smokers. Carcinogenesis 32(4):522–529

76. Talbot-Smith A, Fritschi L, Divitini ML, Mallon DF, Knuiman MW (2003) Allergy, atopy, and cancer: a prospective study of the 1981 Busselton cohort. Am J Epidemiol 157(7):606–612

77. Arana A, Wentworth CE, Fernandez-Vidaurre C, Schlienger RG, Conde E, Arellano FM (2010) Incidence of cancer in the general population and in patients with or without atopic dermatitis in the U.K. Br J Dermatol 163(5):1036–1043

78. Ming ME, Levy R, Hoffstad O, Filip J, Abrams BB, Fernandez C et al (2004) The lack of a relationship between atopic dermatitis and nonmelanoma skin cancers. J Am Acad Dermatol 50(3):357–362

79. Pelucchi C, Naldi L, Di Landro A, La Vecchia C (2008) Oncology Study Group of Italian Group for Epidemiologic Research in D. Anthropometric measures, medical history and risk of basal cell carcinoma in an Italian case-control study. Dermatology 216(3):271–276

80. Kim WJ, Lee HL, Lee SC, Kim YT, Kim H (2000) Polymorphisms of N-acetyltransferase 2, glutathione S-transferase mu and theta genes as risk factors of bladder cancer in relation to asthma and tuberculosis. J Urol 164(1):209–213

81. Steineck G, Adolfsson J, Scher HI, Whitmore WF Jr (1995) Distinguishing prognostic and treatment-predictive information for localized prostate cancer. Urology 45(4):610–615

82. Hallquist A, Hardell L, Degerman A, Boquist L (1994) Thyroid cancer: reproductive factors, previous diseases, drug intake, family history and diet. A case-control study. Eur J Cancer Prev 3(6):481–488

83. Goedert JJ, Vitale F, Lauria C, Serraino D, Tamburini M, Montella M et al (2002) Risk factors for classical Kaposi's sarcoma. J Natl Cancer Inst 94(22):1712–1718

84. Negri E, Ron E, Franceschi S, La Vecchia C, Preston-Martin S, Kolonel L et al (2002) Risk factors for medullary thyroid carcinoma: a pooled analysis. Cancer Causes Control 13(4):365–372

85. La Vecchia C, Negri E, D'Avanzo B, Boyle P, Franceschi S (1990) Medical history and primary liver cancer. Cancer Res 50(19):6274–6277

86. Andreev K, Graser A, Maier A, Mousset S, Finotto S (2012) Therapeutical measures to control airway tolerance in asthma and lung cancer. Front Immunol 3:216

87. Akbari O, DeKruyff RH, Umetsu DT (2001) Pulmonary dendritic cells producing IL-10 mediate tolerance induced by respiratory exposure to antigen. Nat Immunol 2(8):725–731

88. Raker VK, Domogalla MP, Steinbrink K (2015) Tolerogenic dendritic cells for regulatory T cell induction in man. Front Immunol 6:569

89. Gatault S, Delbeke M, Driss V, Sarazin A, Dendooven A, Kahn JE et al (2015) IL-18 is involved in eosinophil-mediated tumoricidal activity against a colon carcinoma cell line by upregulating LFA-1 and ICAM-1. J Immunol 195(5):2483–2492

90. Zaynagetdinov R, Sherrill TP, Gleaves LA, McLoed AG, Saxon JA, Habermann AC et al (2015) Interleukin-5 facilitates lung metastasis by modulating the immune microenvironment. Cancer Res 75(8):1624–1634

91. Majumdar S, Ghosh A, Saha S (2018) Modulating interleukins and their receptors interactions with small chemicals using in Silico approach for asthma. Curr Top Med Chem 18(13):1123–1134

92. Jiang L, Shen Y, Guo D, Yang D, Liu J, Fei X et al (2016) EpCAM-dependent extracellular vesicles from intestinal epithelial cells maintain intestinal tract immune balance. Nat Commun 7:13045

93. Khanduja KL, Kaushik G, Khanduja S, Pathak CM, Laldinpuii J, Behera D (2011) Corticosteroids affect nitric oxide generation, total free radicals production, and nitric oxide synthase activity in monocytes of asthmatic patients. Mol Cell Biochem 346(1–2):31–37

94. Probst HC, McCoy K, Okazaki T, Honjo T, van den Broek M (2005) Resting dendritic cells induce peripheral CD8+ T cell tolerance through PD-1 and CTLA-4. Nat Immunol 6(3):280–286

95. Robainas M, Otano R, Bueno S, Ait-Oudhia S (2017) Understanding the role of PD-L1/PD1 pathway blockade and autophagy in cancer therapy. Onco Targets Ther 10:1803–1807

96. Rosenwald A, Wright G, Leroy K, Yu X, Gaulard P, Gascoyne RD et al (2003) Molecular diagnosis of primary mediastinal B cell lymphoma identifies a clinically favorable subgroup of diffuse large B cell lymphoma related to Hodgkin lymphoma. J Exp Med 198(6):851–862

97. Singh AK, Stock P, Akbari O (2011) Role of PD-L1 and PD-L2 in allergic diseases and asthma. Allergy 66(2):155–162

98. Munthe-Kaas MC, Carlsen KH, Helms PJ, Gerritsen J, Whyte M, Feijen M et al (2004) CTLA-4 polymorphisms in allergy and asthma and the TH1/TH2 paradigm. J Allergy Clin Immunol 114(2):280–287

99. Bertrand A, Kostine M, Barnetche T, Truchetet ME, Schaeverbeke T (2015) Immune related adverse events associated with anti-CTLA-4 antibodies: systematic review and meta-analysis. BMC Med 13:211

100. Hellings PW, Vandenberghe P, Kasran A, Coorevits L, Overbergh L, Mathieu C et al (2002) Blockade of CTLA-4 enhances allergic sensitization and eosinophilic airway inflammation in genetically predisposed mice. Eur J Immunol 32(2):585–594

101. Saul L, Ilieva KM, Bax HJ, Karagiannis P, Correa I, Rodriguez-Hernandez I et al (2016) IgG subclass switching and clonal expansion in cutaneous melanoma and normal skin. Sci Rep 6:29736

102. Harshyne LA, Nasca BJ, Kenyon LC, Andrews DW, Hooper DC (2016) Serum exosomes and cytokines promote a T-helper cell type 2 environment in the peripheral blood of glioblastoma patients. Neuro-Oncology 18(2):206–215

103. Ribatti D (2016) Mast cells as therapeutic target in cancer. Eur J Pharmacol 778:152–157

104. Akdis M, Akdis CA (2014) Mechanisms of allergen-specific immunotherapy: multiple suppressor factors at work in immune tolerance to allergens. J Allergy Clin Immunol 133(3):621–631

105. Lippi G, Meschi T, Nouvenne A, Mattiuzzi C, Borghi L (2014) Neutrophil gelatinase-associated lipocalin in cancer. Adv Clin Chem 64:179–219

106. Wen CP, Lee JH, Tai YP, Wen C, Wu SB, Tsai MK et al (2014) High serum iron is associated with increased cancer risk. Cancer Res 74(22):6589–6597

107. Li H, Zhang K, Liu LH, Ouyang Y, Bu J, Guo HB et al (2014) A systematic review of matrix metalloproteinase 9 as a biomarker of survival in patients with osteosarcoma. Tumour Biol 35(6):5487–5491

108. Maio M, Coukos G, Ferrone S, Fox BA, Fridman WH, Garcia PL et al (2019) Addressing current challenges and future directions in immuno-oncology: expert perspectives from the 2017 NIBIT Foundation Think Tank, Siena, Italy. Cancer Immunol Immunother 68(1):1–9

109. Walsh NC, Kenney LL, Jangalwe S, Aryee KE, Greiner DL, Brehm MA et al (2017) Humanized mouse models of clinical disease. Annu Rev Pathol 12:187–215
110. Jensen-Jarolim E, Einhorn L, Herrmann I, Thalhammer JG, Panakova L (2015) Pollen allergies in humans and their dogs, cats and horses: differences and similarities. Clin Transl Allergy 5:15
111. Jensen-Jarolim E, Achatz G, Turner MC, Karagiannis S, Legrand F, Capron M et al (2008) AllergoOncology: the role of IgE-mediated allergy in cancer. Allergy 63(10):1255–1266

Genome Engineering Tools in Immunotherapy

Rashmi Dahiya, Taj Mohammad, and Md. Imtaiyaz Hassan

Abstract

Immunotherapy is a breakthrough in the potential treatment of cancer as well as preventing future relapses by stimulating the immune system in the recognition and killing of cancer cells. Numerous strategies are ongoing in the clinical laboratories for the advancements in the immunotherapy approaches which include therapeutic engineered T lymphocytes, vector-based (noncellular) cancer vaccines, dendritic cell vaccines, and immune checkpoint blockade. Regardless of their capacity, continuous research is required to recognize the failures of cancer response toward strong immunotherapy treatment as well as to envisage the therapeutic combinatorial strategies appropriate for patient-specific ways. Fundamental to these challenges underlie the technological methods for rapid and thorough characterization of tumors-immune microenvironments, immune response monitoring of patients, predictive tools to screen potential and sensitive therapies, tumor regression, and tumor dissemination throughout and after the therapy. The emerging field of immune engineering addresses these challenges and contributed the tools and approaches to facilitate the clinical transformation of immunotherapy. Customized and programmable site-specific nucleases have already revolutionized our ability to interrogate genomic functions and introduce genetic manipulations in diseases which are intractable with traditional therapies for potential clinical applications. In this chapter, we highlight the developments, recent technological advances, and applications of these tools in the diagnosis, treatment, and cancer monitoring, as well as the ongoing challenges in their uses as a platform technology in the context of immunotherapy.

R. Dahiya · T. Mohammad · Md. I. Hassan (✉)
Centre for Interdisciplinary Research in Basic Sciences, Jamia Millia Islamia, New Delhi, India
e-mail: mihassan@jmi.ac.in

© Springer Nature Singapore Pte Ltd. 2020
S. Singh (ed.), *Systems and Synthetic Immunology*,
https://doi.org/10.1007/978-981-15-3350-1_3

Keywords
Adaptive immunity · Adoptive immunotherapy · Gene transfer · Chimeric antigen receptor · Transcription activator-like effector

3.1 Introduction

For the past decades, the development of whole-genome sequencing methods for the implementation of large genomic annotation projects has challenged the scientific community to deliver the genomic revolution from basic science into personalized medicine in translational research. The conversion of enormous data from a plethora of genome sequence information into clinically relevant knowledge is of utmost requirement. There is a need of efficient and reliable methods to determine the influence of genotype on phenotype changes. To this end, the inactivation of targeted genes via homologous recombination is a very powerful approach capable of providing conclusive information for the evaluation of gene functions [1]. However, this approach is hampered by numerous limiting factors which include the low efficiency of insertions of engineered constructs at correct chromosomal locations, the enormous time needed for selection/screening procedures, and the potential mutagenic effects caused by the adverse expression [2].

Manipulation in the eukaryotic genomes is extremely difficult and error-prone due to the presence of billions of DNA bases. The discovery of genetic targeting by homologous recombination (HR) is considered as the breakthrough in genome modifications, which led to the integration of exogenous repair DNA templates containing sequence homology to the donor site [3]. HR-directed genome targeting has enabled the construction of knock-in and knockout animal models via germline manipulation of competent stem cells leading to dramatic advancement of biological research areas. Although gene targeting mediated by HR generates highly accurate alterations, the anticipated recombination events arise extremely infrequently with an average of 1 in 10^6–10^9 cells, offering a massive challenge for its large-scale applications in gene-editing experiments [4].

The discovery that targeted induced DNA double-strand breaks (DSBs) could stimulate the cellular repair machinery is foundational and exceptionally significant to the field of gene editing [5]. DNA breaks are classically repaired by two major pathways: homology-directed repair (HDR) and nonhomologous end joining (NHEJ) [6]. HDR repair mechanism employs a donor DNA sequence homologous to the genomic site flanking the DSB, which can introduce novel genetic information at the DSB sites. In NHEJ repair pathway, the DNA ends at the DSB sites are ligated back together, incorporating small insertions or deletions; therefore, NHEJ is error-prone and lead to gene disruption. HDR repair pathways are highly precise and potentially used for large gene replacements, integration of selective markers, deletions, and base mutations. RNAi approach for the knockdown of targeted genes has provided a very rapid, high-throughput, and inexpensive alternative to homologous recombination [7]. However, gene knockdown offered by RNAi is partial and results in temporary inhibition of gene functions. Moreover, RNAi has unpredictable off-target effects, and the results obtained from RNAi inhibitions vary between laboratories and experiments. These drawbacks of RNAi approach have restricted the identification of direct link between phenotype and genotype and resulted in its limited applications.

Several approaches have emerged which enabled investigators to manipulate genomic regions in various cell types and diverse range of organisms. One such approach is commonly termed as "genome editing" mainly based on the manipulative functions of engineered nucleases. For high-throughput and precise genome editing, a series of programmable nuclease-based genetic tools have been developed enabling specific targeting and effective modification of numerous eukaryotic and mammalian species. Along with these nucleases are fused the sequence-specific DNA-binding domains which together function as a potential DNA cleavage module [8, 9]. The fused products of DNA binding modules with nucleases generate chimeric nucleases which enable precise, high-throughput, and efficient genomic modifications by the inductions of DSBs and lead to the activation of DNA repair pathways including HDR and NHEJ [10–12]. The flexibility of these newer approach is generated by the versatility of the DNA-binding modules derived from various protein structural motifs like zinc finger, transcription activator-like effector (TALE), and Cas9 (Fig. 3.1). Out of the existing generation of editing tools, CRISPR-Cas (clustered regulatory interspaced short palindromic repeats/CRISPR associated) represents the most rapidly developing class driven by RNA-guided endonucleases originating from the microbial adaptive immune system. CRISPR technology can be effortlessly targeted by a short RNA guide to practically any organism of choice achieving targeted perturbation of endogenous genomic regions.

The combination of this high flexibility and simple experimentations has catapulted these genome editing techniques toward the forefront of genetic engineering. The recent advances in genome editing technologies have markedly improved our capability to make accurate genomic changes in the eukaryotic genomes. Here, we discuss the current advances in site-specific nucleases and review their potential applications for precise genome editing and functional analysis within cells and model organisms. We will also discuss the therapeutic potential of these advanced technologies and examine their projections with the major focus on the development and applications of CRISPR/Cas endonucleases along with future challenges and avenues for innovation.

Our current understanding of genome editing procedures to engineer cells and redirecting them to precise targets, bestowing the immune system with tremendous functions along with safety features, and uniting them with additional targeted immune therapies is discussed in this chapter. We exemplify how monitoring of the immune system and potential biomarkers can govern the effects and destiny of cell therapies in clinical settings. We finally conclude with a brief discussion of the genetic and molecular elements essential for the establishment of new pillars of clinical treatments constructed around personalized cell therapies. This chapter provides an overview of existing progress in the development of targeted genome editing and will also discuss the current state of ACT for the treatment of human cancer, as well as approaches and the underlying principles of effective treatments pointing toward further advances in these methodologies.

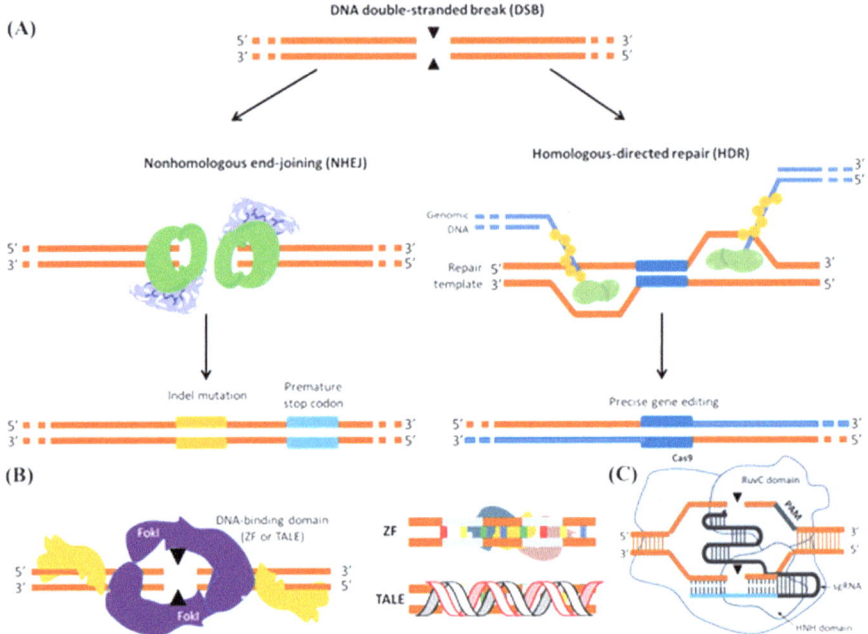

Fig. 3.1 The approaches of genome editing exploit endogenous DNA repair pathways. (**a**) The DNA double-strand breaks (DSBs) are classically repaired by either the error-prone nonhomologous end joining (NHEJ) or homology-directed repair (HDR). In NHEJ, Ku heterodimers function as a molecular scaffold for other repair proteins after binding to DSB ends. End resection within the complementary strands and microhomology-mediated misaligned repair eventually generate indels and lead to frameshift mutations and gene knockout. Alternatively, in HDR, Rad51 proteins bind DSB ends and recruit accessory factors directing homologous recombination with an exogenous repair template. (**b**) Modular domains like zinc finger (ZFs) and transcription activator-like effectors (TALEs) are naturally occurring DNA-binding proteins recognizing 3 and 1 bp of DNA, respectively. These domains can be assembled and fused to the FokI endonuclease to construct programmable site-specific nucleases targeting specific sequences. (**c**) CRISPR adaptive immune system consists of Cas9 nuclease that can be targeted to specific DNA sequences guided by its cognate guide RNA (black) through direct base pairing with target DNA. Protospacer adjacent motif (PAM, green) directs the Cas9-mediated DSBs

3.2   Programmable Nucleases as Tools for Efficient and Precise Genome Editing

3.2.1   Genome Editing with Site-Specific Nucleases

Homologous recombination has proven highly successful in the inactivation of gene by additions or deletions of specific genomic regions; however, two limitations dramatically constrain the utilization of recent genome engineering technologies: the low frequency of homologous recombination in mammalian cells as well as the model organisms and the high percentage of non-targeted genomic integration of the vector DNA. Subsequent discovery elucidating that DSB induction substantially increases the HDR frequency by previously unbelievable extent has provided the emergence of targeted nucleases of choice for the improved efficiency of HDR-mediated genetic modifications. Integration of multiple transgenes can be efficiently triggered at the location of choice with the help of a donor plasmid DNA consisting of homology sequence for the desired genomic site along with the site-specific engineered nuclease [13]. Homologous sequences less than 50 bp as linear donor sequences [14] as well as the ssDNA oligonucleotides [15] could functionally induce the targeted mutations, insertions, or deletions in virtually any DNA sequence. Moreover, in addition to facilitating DSB-mediated HDR, engineered site-specific nuclease also permits the quick development of cell lines as well as whole organisms having null phenotypes which are mediated by the nonhomologous end joining (NHEJ) repair of DSB that potentially introduce small deletions or insertions at the target genomic sites resulting in functional knockout of gene generated by frameshift mutations [16]. Additionally, engineered nuclease also facilitates site-specific deletions within large chromosomal regions [17]. Moreover, these approaches also function to induce large chromosomal translocations at specified genomic loci [18] as well as chromosomal duplications and inversions [19] as reported within the human genome. Finally, NHEJ-mediated ligation (ObLiGaRe, Obligate Ligation-Gated Recombination) facilitated by the synchronization of nuclease-driven site-specific cleavage of donor DNA along with the specified chromosomal region enabled the introduction of large transgenes (14 kb) into several endogenous loci [20]. Site-specific integration of genetic sequence significantly controls the positional effects which enabled the structural-functional relationships of numerous genetic analyses in a native chromosomal environment. Zinc finger nucleases (ZFNs) and transcription activator-like effector (TALENs) have been extended to target specific gene in human embryonic stem cells (ESCs) and induced pluripotent stem cells (iPSCs) [21, 22]. Both zinc finger nucleases and TALENS have proven unprecedented in the gene functions study, disease modelling by alteration of gene mimicking known and uncharacterized genotypes (Fig. 3.2). These approaches encouraged their employment in the modelling of a wide range of genetic disorders. Moreover, these approaches have also been reported in the functional elucidation of noncoding DNA and RNA regulating the bulk genome including the use of multiplexed approaches for the identification of unknown regulatory sites within the choice of genes [23, 24].

3.2.2 Cys$_2$-His$_2$ Zinc Finger Nucleases

For successful genetic engineering, an endonuclease must possess an astonishing combination of abilities which includes the precise recognition of lengthy target sequences uniquely occurring in a eukaryotic genome along with satisfactory flexibility which allows their retargeting to user-defined genomic regions. The architecture of ZFN meets the above qualities by linking the DNA-binding domain of zinc finger proteins (ZFPs) with the nuclease domain of the FokI restriction enzyme. ZFNs combine the favorable assets of both apparatuses: the flexibility and

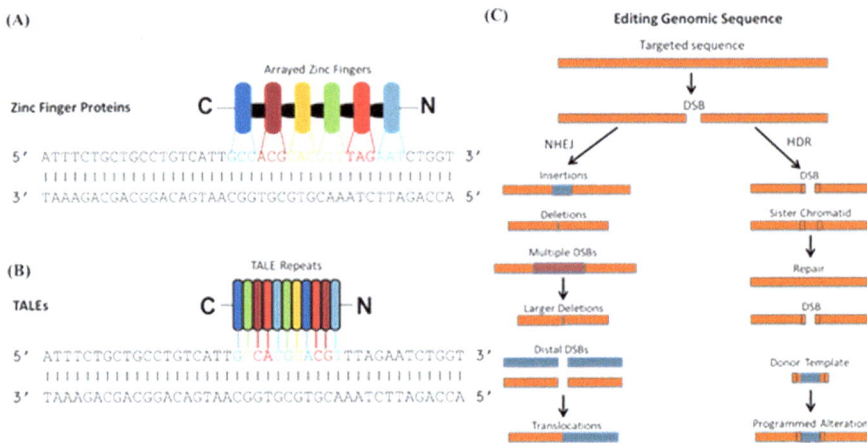

Fig. 3.2 Engineering platforms for editing genome sequence. Individual ZFs (**a**) and TALEs (**b**) uniquely recognizing triplets or single base pairs, respectively, can be engineered in arrays to target specific sequences. (**c**) ZFs and TALEs fused to targeted nucleases or Cas9-gRNA complex can potentially cleave genomic sequences to generate double-strand breaks (DSBs). The eventual resolution of DSBs through NHEJ or HDR can lead to various genomic alterations

specificity of DNA-binding ZFPs which retain functional modularity and a robust cleavage activity which only occurs in the presence of a specific DNA-binding incident. Consequently, both the DNA-binding and catalytic domains can be easily manipulated in isolated platforms further simplifying the improved retargeting efforts. The Cys_2-His_2 zinc finger domain characterizes as the frequently encoded motif in the human genome and represented as the most common and versatile class of DNA-binding eukaryotic transcription factors. The ZFP region within a ZFN facilitates its binding to a distinct base sequence. A specific zinc finger motif comprises of ~30 amino acid residues in a well-conserved $\beta\beta\alpha$ arrangement [25]. This region consists of a tandem array of Cys_2-His_2 fingers [26, 27]. The specific amino acids present on the α-helical surface made of Cys_2-His_2 fingers are known to typically contact 3 bp of DNA in its major groove, with variable selectivity. The modular architecture of ZFPs has recognized them as a striking framework to devise custom DNA-binding proteins.

The key to the successful application of ZFPs in highly specific and precise DNA recognition is the development of arrays which consists of more than three zinc finger domains. In previous studies, modified ZFNs employed three zinc fingers to bind DNA target made up of 9 base pairs, which could enable the highly active ZFN dimers to bind 18 bp of DNA at potential cleavage sites. Recent studies have facilitated the addition of more fingers (up to 6 per ZFNs) for the specifications of long and rare cleavage targets. The discovery of the conserved linker sequence eased by the structure-based studies has enabled the construction of synthetic ZFPs which can recognize a DNA sequence of 9–18 bp in length [25, 28]. Interestingly, a DNA sequence formed with 18 bp confers specificity within 68 billion bp of DNA; hence, the advanced ZFNs allow the targeting of specific sequences within the human genome [29, 30]. The recent design has optimized for ZFPs constructions which could potentially recognize highly specific contiguous DNA regions within complex genomes. Later, numerous methods for the construction of ZFPs with exclusive user-chosen DNA-binding specificity were established. The initial strategy was first emerged from the observations of ZFP-DNA co-crystal structure. The interactions of zinc fingers with DNA region suggested a functional autonomy [31, 32]. The initial approach was termed as "modular assembly" in which candidate ZFPs for a user-chosen target DNA sequences were generated by the identification of fingers for each constituent triplet and then linking them into a complex multi-finger peptide which facilitates its binding to the corresponding DNA sequence. This approach also comprises the use of a library of finger modules which are generated by rational designs or the preselection of combinatorial libraries [25, 33]. By this method, ZF domains have been designed for the recognition of all the potential 64 nucleotide triplets, which then can be linked together in a tandem fashion against any sequence which contains any combination of DNA triplets. An alternative approach relies on the selection-based method, termed as oligomerized pool engineering (OPEN), that can be used for the selection of new arrays of zinc fingers from randomized libraries which include a context-dependent interaction among adjoining fingers [34].

The key success of ZFNs underlies in the crucial function of the FokI domain which possesses numerous characteristics which facilitate the targeted cleavage within highly complex genomes. Interestingly, FokI dimerization is essential for the DNA cleavage at target loci [35]. However, the interaction between FokI dimers is very weak, and cleavage requires the independent binding of two ZFNs in an adjacent fashion. Moreover, the binding events of ZFNs must occur in the precise orientations along with appropriate space to allow the FokI dimer formation [35]. Two independent and adjacent ZFNs-binding events permit the precise targeting of longer and unique recognition sequences (from 18 to 36 bp). The requirement of dimerized nuclease has encouraged the development of ZFN variants with improved specificity which can cleave only as a heterodimer pair and eliminates the undesirable homodimers [36, 37]. ZFP-FokI linkers have been further modified to develop ZFN dimers with novel spacing between two-monomer binding events. The catalytic domain variants of FokI were also reported to show boosted cleavage activities [38].

3.2.3   Gene Disruption by ZFNs in Model Organisms

In *Drosophila melanogaster*, ZFNs can be delivered to the early fly embryos via mRNA injection, and it has been shown that ~10% of the progeny adult flies were mutated for the gene of interest [39]. Different alleles of the same gene can be targeted by the ZFNs, and homozygotes of each mutated allele would completely lack the protein expression. Engineered ZFNs encoded by specific mRNA when injected into zebrafish embryos were also used to generate desired genetic lesions in ~50% of progeny [40, 41].

For gene disruption in rats, engineered zinc finger nucleases (ZFNs) were designed against two independent endogenous rat genes, IgM and Rab38, along with an integrated reporter and demonstrated that microinjection of mRNA-encoding ZFNs to rat embryo leads 25–100% disruption of target locus in progeny [42]. Moreover, a faithful and proficient transmission of mutated alleles was observed through the progeny [42]. In an independent study, severe combined immune deficiency (SCID) rat was also generated using the similar strategy [43]. In *Arabidopsis thaliana,* the transgenesis of a stable and inducible expression cassette of engineered ZFN allows gene disruption [44].

3.2.4   Gene Disruption in Mammalian Cells

Customized and programmable nucleases have been immensely used for the disruption of genes in mammalian cells. Conventional targeting of genomic sites coupled with the strategies of positive and negative selection is a powerful approach for gene knockouts. Moreover, the use of engineered adeno-associated viruses (AAVs) has permitted its application in transformed and primary human cells [21, 22]. Gene knockouts with ZFNs preclude the necessity for selection based on drugs which further extend its application toward any potential cell type and organisms given the availability of transient delivery of either DNA or mRNA. Interestingly, knockouts by ZFNs resulted in 1–50% of all cells. The first report of gene knockout in mammalian Chinese hamster ovary (CHO) cells by ZFNs has demonstrated in the disruption of dihydrofolate reductase (*Dhfr*) gene [16]. Transient transfection of ZFNs targeting *Dhfr* gene encoded by a plasmid DNA resulted in ~15% (2 clones out of 60) frequencies of biallelic disruption in the cell populations as observed with genotyping and measurably lacked DHFR protein expression. Later, in CHO and K562 cells, ZFNs were reported to target locus-specific DNA regions and to construct double [45] and triple [46] gene knockouts. Moreover, the engineered ZFNs-driven knockout approach has proven highly successful in a range of cell types which also includes the human ES cells and CD4[+] T cells.

Numerous approaches have been developed which combine several methods to utilize ZFNs in a context-dependent selection using longer assembled arrays. For many years, ZFPs represented the only technology and approach available to construct conventional sequence-specific DNA-binding modules. Broadly, the ZFN approach facilitates the targeting of virtually any genomic sequence. Moreover, to bypass the constructions of ZFNs and to evade their validation altogether, thousands of engineered zinc fingers modular proteins are commercially available by a joint venture of Sangamo Biosciences (Richmond, CA, USA) and Sigma-Aldrich (St. Louis, MO, USA), and a propriety platform (CompoZr) has been developed in partnership which allows investigators to understand the genotype to phenotype changes.

3.2.5 TALEs

TALEs are natural proteins secreted by the plant pathogenic bacteria Xanthomonas. TALE proteins consist of a sequence of 33–35 amino acid repeat domains which functions as DNA-binding modules recognizing a single DNA base pair. The discovery of modular TALE proteins recognizing a DNA recognition code has led to the development of an alternative approach for the construction of engineered and programmable DNA-binding modules [20, 21]. The position of two hyper-variable residues within TALE proteins determines their specificity toward the target sequence. These hyper-variable residues are known as repeat-variable di-residues (RVDs) [22, 23]. As discussed above for the ZFPs, TALE repeats are also linked together for the recognition of any contiguous DNA sequences. However, unlike ZFPs, the linkage between repeats cannot be reengineered for the construction of longer TALE arrays capable of targeting single genome sites. After approximately two decades of groundbreaking research on ZFPs, several effector domains including nucleases [24–26], site-specific recombinases [28], and transcriptional activators [26, 27] were developed to fuse to TALEs for targeted genome engineering. Investigators face an advanced technical challenge in the cloning of TALE repeat arrays due to the presence of extensive indistinguishable repeat sequences; however, the recognition of a single base by TALE repeats-DNA binding enables extensive flexibility in their designing as compared to triplet-confined ZFPs. The other limitation is that TALE recognition of DNA sequence should start with a T base. Numerous methods were developed to overcome this challenge which enabled the quick assembly of customized arrays of TALE repeats.

A high-throughput method for the construction of TALENs has been reported in several studies. These strategies include rapid PCR-based molecular cloning approach termed as "Golden Gate" to assemble multiple DNA fragments [47]; solid-surface-based rapid, high-throughput, and cost-effective method for large-scale TALENs assemblies such as fast ligation-based automatable solid-phase high-throughput (FLASH) system; and iterative capped assembly (ICA) for the synthesis of TALENs of variable length of target DNA site and demonstrated their abilities to trigger gene editing by a donor oligonucleotide in human cells [48, 49] and ligation-independent cloning techniques [50]. Several large systematic studies utilizing various assembly methods have indicated that TALE repeats can be combined to recognize virtually any user-defined sequence. Indeed, the TALE repeats can be assembled easily as evident in the previous report suggesting the construction of a TALENs library to target 18,740 protein-coding genes from human [51]. These technological accomplishments will encourage future ambitious endeavors and facilitate new studies. Moreover, custom-designed TALE arrays are also available commercially through Transposagen Biopharmaceuticals (Lexington, KY, USA), Cellectis Bioresearch (Paris, France), and Life Technologies (Grand Island, NY, USA).

3.2.6 Improving the Performance of Site-Specific Nucleases

Complex and large genomes consist of multiple copies of highly homologous or identical sequences potentially leading to several off-target activity and toxicity toward target cells. Therefore, customizable nucleases must demonstrate stringent target specificity toward proposed DNA targets to carry out relevant genetic analysis and further clinical application. To overcome this challenge, both structural [52, 53] and selection-driven [54, 55] methods have been employed for the generation of highly specific and improved heterodimers of ZFN and TALEN. The cleavage specificity of ZFNs and TALENs has been optimized for enhanced specificity along with reduced toxicity. Moreover, a directed evolutionary approach has been utilized for the generation of a hyper-activated variant of FokI termed as Sharkey (cleavage domain of FokI). Sharkey displays highly significant compatibility with several ZFN architectures [54] and represents >15-fold enhanced cleavage activity as compared to traditional ZFNs [55]. Furthermore, various evidence suggested that 4–6 ZF domains for individual ZFN could significantly boost its activity and target specificity [13, 55–57]. Further procedures to improve the ZFN activities include brief hypothermic culture environments to enhance the levels of nuclease expression [58], co-transformation of DNA end-processing enzymes along with targeted nucleases [59], and co-delivery of vectors expressing fluorescent surrogate reporter allowing the propagation of ZFN- and TALEN-transformed cells [60]. The target specificity of ZFNs has been further enhanced with the advancement of ZF nickases, which facilitates the induction of DNA nicks stimulating HDR for DNA repair without the activation of error-promoting NHEJ pathway [64]. The nickase approach led to minimal off-target mutational effects as compared to conventional DSB-mediated genome edition; however, unlike traditional ZNFs, the frequency of HDR through ZFNickase is comparatively low. Lastly, the traditional delivery methods of ZFNs using DNA or RNA are restricted to certain cell types and are also linked to unwanted side effects which include minimal efficiency, mutagenic insertions, and high toxicity. To overcome these limitations, purified ZFNs proteins have been delivered directly into the cells as an alternative process. This approach leads to minimal off-target effects and does not result in insertional mutagenesis [52]. This type of platform for targeted delivery might represent optimal strategy but suffer with other challenges due to extensive design strategy and high cost of production.

3.3 Brief History of CRISPR-Cas

The story of CRISPR started in the year 1987 when Nakata and colleagues reported a set of 29 nucleotide repeats present downstream of *iap* gene involved in isozyme conversion of alkaline phosphatase in E. coli [53]. These 29 nt repeats were curiously interspaced by five intervening 32 nt nonrepetitive sequences. With the advent of genome sequencing by the next decade, additional interspaced repeat elements were reported from bacterial and archaeal genomes which were eventually classified as a unique family of clustered repeat elements present in 90% of archaea and > 40% of sequenced bacterial genomes [54]. These initial findings stimulated much interest in microbial repeats. In 2002, the acronym CRISPR was coined to describe microbial genomic loci consisting of an interspaced repeat array [55–57]. Interestingly, CRISPR loci were found to be transcribed [58]. Later, several well-conserved clustered elements were identified typically adjacent to CRISPR and were named as CRISPR-associated genes (cas) [55]. Cas genes serve as a basis for the classification of CRISPR systems (types I–III) [59, 60]. For the recognition and destruction of the target site, types I and III CRISPR loci consist of multiple Cas proteins and forms independent complexes with crRNA (type I forms CASCADE and type III forms Cmr or Csm RAMP complex) [61, 62]. Type II system consists of a smaller number of Cas proteins. In 2005, sequence analysis of the spacers separating the CRISPR suggested that they are originated extra chromosomally and are associated with phage genomes [63–65]. Moreover, viruses are unable to infect archaeal cells carrying spacers representing their own genomes [63]. Together, CRISPR arrays were speculated to serve as defense mechanisms against bacteriophage infection [63, 64]. Later, the RNAi-like mechanism underlying the spacers functioning as small-guide RNAs and directing Cas enzymes for degradation of viral DNA was uncovered [65, 66].

3.4 Genome Editing Using CRISPR-Cas9 in Eukaryotic Cells

A dual-RNA hybrid composed of crRNA and tracrRNA together with Cas9 are the three essential components of type II CRISPR nuclease system along with endogenous RNase III, required for processing the CRISPR array transcript into mature crRNAs [67, 68]. Biochemical characterizations of Cas9 purified from *Streptococcus thermophilus* or *Streptococcus pyogenes* showed that it can be guided by crRNAs for degradation of target DNA in vitro [69, 70]. Cas9-mediated degradation requires the presence of a protospacer adjacent motif (PAM) immediately downstream of the target site. A single-guide RNA (sgRNA) could be generated by the fusion of crRNA and tracrRNA, which then potentially facilitates DNA cleavage by Cas9 [69]. crRNA or sgRNA contains a 20 nt guide sequence which directly matches the target sequence. Till date, Cas9 from *Streptococcus pyogenes* (SpCas9) is broadly used for genome editing in a variety of cell types and species that include human cells, mouse, monkey, drosophila, yeast, bacteria, zebrafish, and so on [71]. Targeting through SpCas9 can be achieved with either a pair of crRNA and tracrRNA [72] or a chimeric sgRNA [72–74]. Human genome editing using the engineered dual-guide RNA system along with SpCas9 showed higher levels of NHEJ-induced indels compared to the engineered sgRNA scaffold. Moreover, an extension of the 30 tracrRNA sequence generates additional stem loops hairpin structures which enhance the stability of the sgRNA critical for effective in vivo sgRNA-mediated genome editing through Cas9-sgRNA-DNA ternary complex formation [72, 75, 76]. CRISPR-Cas9 system has an inherent ability of efficiently cleaving multiple target sites in parallel by conversion of pre-crRNA transcript containing many different spacers into specific guide RNAs duplexes (crRNA-tracrRNA) [67, 68]. This unique aspect of the CRISPR system is harnessed to enable scalable multiplex genome perturbations. Indeed, co-expression of SpCas9 together with CRISPR array consisting of spacers targeting multiple distinct genes [72], or numerous sgRNAs [73, 77], resulted in efficient multiplex genome editing in mammalian cells.

3.5 Functional Screening of Genomes

The ability of CRISPR-Cas9 to edit many genomic targets in parallel with high efficiency and precision enabled the identification of genes of interest using unbiased genome-wide functional screens. Lentiviral delivery of sgRNAs together with Cas9 directed against genes could potentially perturb thousands of genomic regions in parallel. Many reports have demonstrated the ability of CRISPR-Cas in robust positive and negative selection screens in human cells by the introduction of loss-of-function mutations of a distinct gene in each cell [78, 79]. Previously, RNAi was employed for genome-wide loss-of-function screens; however, this approach is limited to transcribed genes, has many extensive off-target effects, and leads to only partial knockdown. Contrastingly, Cas9-sgRNA screens can be designed for targeting nearly any DNA sequence and are reported to provide increased screening sensitivity with no off-target effects [78].

Approximately 76% of the human genome is transcribed into RNAs while less than 2% encodes for proteins. The human genome generates a plethora of long noncoding RNAs (lncRNAs), many of which are shown to be functional. lncRNAs consist of at least 200 nucleotides in length and represent a major subset of the human transcriptome [80]. Functional lncRNAs were first identified through a specially designed high-throughput CRISPR approach which employed paired gRNAs (pgRNAs) for generating genomic deletions. Multiplexed gRNA libraries facilitate the dissection of large genomic regions through perturbation of noncoding genetic elements. CRISPR approach also led to the dissection of large, uncharacterized genomic regions which were previously implicated in GWAS studies as functional zones.

Previously, high-throughput screening strategy for the deletion of genomic segments for the identification of functional long noncoding RNAs (lncRNAs) has identified 51 lncRNAs that can potentially regulate the growth of human cancer cells either positively or negatively [81]. This approach is based on a pgRNA library specific for 671 human lncRNAs having a total of 12,472 gRNA pairs constructed with lentiviral paired guide RNA. CRISPR-Cas9-mediated validation of these 9 out of 51 lncRNA hits using deletion, functional rescue, and gene expression profiling confirmed their cellular functions. Moreover, the systematic activation or disruption of additional regulatory elements like general promoters, distant enhancers, and various other regions of genes facilitated their functional elucidation. Researchers all over the world have admired over the remarkable ease and versatility of CRISPR as a gene-editing tool; however, "killing" the catalytic activity of its nuclease Cas9 came out as equally significant in the functional characterization of genomes. Mutation D10A in the RuvC domain and H840A in the HNH domain of the nuclease domains of Cas9 generated a nuclease-deficient dCas9 (termed as dCas9 null mutant) [82]. Interestingly, when guided by sgRNA, this dead or inactive version of Cas9 can still precisely bind DNA; however, it is unable to cleave its target site. The specific binding of dCas9 potentially interferes with the transcriptional status of the target site despite altering its sequence.

Moreover, dCas9 binding also resulted in the reversible transcriptional activation or silencing of the target gene. Therefore, the functional tethering of dCas9 to diverse effector domains facilitated the genomic screens outside the loss-of-function phenotypes. Transcriptional activation through dCas9 allowed the screening of gain-of-function phenotypes. To attain high expression levels with a single sgRNA, multiple transcriptional activators are recruited to TSS by CRISPRa methods. dCas9 fused with transcriptional activator domains of multiple proteins, e.g., VP64, HSF1, p65, and GCN4, which were then recruited to multiple arrays in synergistic system [83–85]. Moreover, fusion of dCas9 with epigenetic modifiers was used to study the posttranslational modification effects on the cellular differentiation as well as various disease pathologies.

3.6 Personalized Immunotherapy: Adoptive Cell Therapy (ACT) in Human Cancer

Adoptive Immunotherapy

Adoptive immunotherapy or adoptive cell therapy (ACT) is a highly personalized therapy that involves administration of immune cells with direct anticancer and antiviral activities. ACT involves the infusion of lymphocytes considered as a promising approach for the treatment of cancer and certain chronic viral infections. Adoptive T-cell therapy employs the power of T cells which can recognize and kill target cells. Hence, it is not surprising that most ACT investigations have targeted various cancer as well as chronic viruses. The application of the principles of synthetic biology to enhance T-cell function has resulted in substantial increases in clinical efficacy. The primary challenge to the field is to identify tumor-specific targets to avoid off-tumor, on-target toxicity. Given recent advances in efficacy in numerous pilot trials, the next steps in clinical development will require multicenter trials to establish adoptive immunotherapy as a mainstream technology. Compared to other cancer immunotherapy approaches, ACT has numerous advantages which rely on the vigorous in vivo development of tumor reacting T cells coupled with their functions required to facilitate cancer regression. Enormous amounts of tumor-reacting lymphocytes (up to 10^{11}) can be easily propagated in vitro for their effector properties and can be selected for recognition of tumor with high avidity to mediate cancer regression in ACT (Fig. 3.3). In vitro growth and activation render antitumor T cells for their release from the inhibitory factors present in vivo. Most notably, ACT delivers a favorable environment for T-cell propagation supporting improved immunity against tumor as well as enables the manipulation of the tumor-reacting cell before their transfer. ACT is considered as a living treatment since the T cells are proliferated in vivo for the maintenance of their antitumor effector properties before administration in the host and are revived within the host organism.

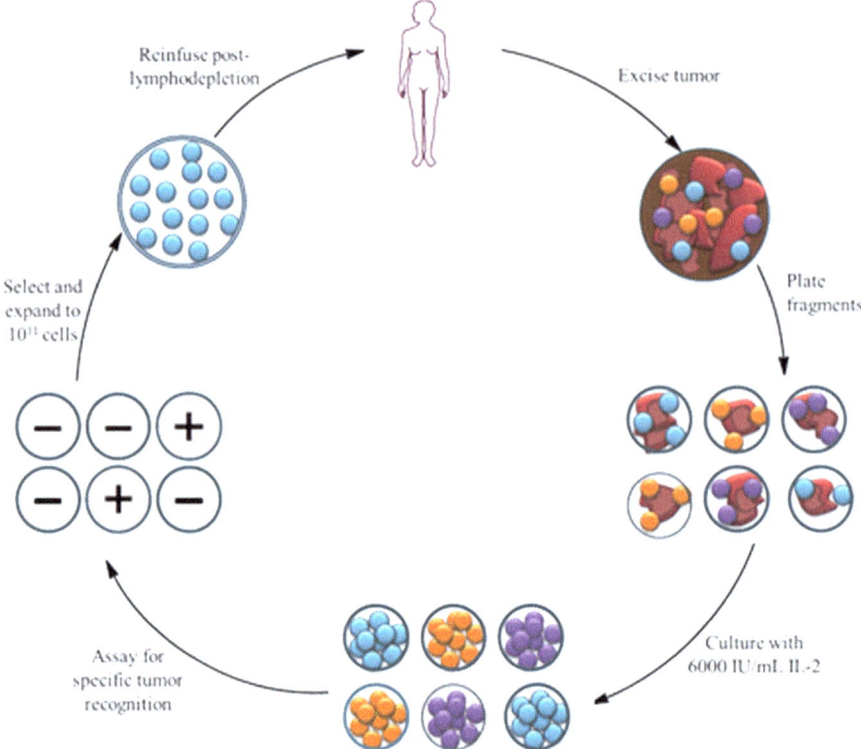

Fig. 3.3 The adoptive cell transfer (ACT) scheme for naturally occurring autologous tumor-infiltrated lymphocytes (TILs). The melanoma specimen is resected from the patient and either digested into a single-cell suspension or divided into multiple fragments and independently grown in the presence of IL-2. TILs grow extensively and destroy tumor cells within 2–3 weeks. Pure cultures of lymphocytes are generated which can be tested for reactivity in coculture experiments. Independent cultures then expanded in the presence of IL-2, irradiated feeder lymphocytes, and OKT3. Up to 10^{11} TILs can be obtained by 5–6 weeks after the tumor resection and then infused into cancer patients

A major off-putting issue for the successful ACT treatment in humans is the stringent identification of immune cells that could potentially recognize and target antigens selectively represented on the cancer cells and are absent on essential normal tissues. Two downstream mechanism can be used in successful ACT: one is the use of natural host cells which exhibit reactivity on tumor-specific cells, and the other is the use of genetically engineered host cells possessing T-cell receptors (TCRs) or chimeric antigen receptors (CARs) against tumor. With these approaches, ACT has facilitated regressions and cure in a diverse cancer histolopathologies, primarily including melanoma, lymphoma, cervical cancer, leukemia, neuroblastoma, and bile duct cancer.

3.7   ACT: A Brief History

Until the 1960s, scarce information about T-lymphocytes functions was available. Later that time, it was reported that lymphocytes mediate the rejection of allografts in animal models. Initial attempts to treat the murine models transplanted with tumors were restricted due to the continuous failures of expansion and manipulation of T cells in culture conditions. Previously, ACT used the transfer of tumor-immunized T lymphocytes from syngeneic mice which resulted in minimal growth inhibition of established tumors [86]. The identification of interleukin-2 in 1976 as a potent T-cell growth factor has facilitated the use of ACT by providing a platform to grow T cells in culture conditions which is majorly affecting their effector properties [87]. Indeed, intravenously injected T cells proliferated in IL-2 presence effectively inhibited subcutaneously grown FBL3 lymphomas [88]. Moreover, it was demonstrated that IL-2 administration in high doses potentially inhibited tumor progression in mice [89]. In fact, IL-2 administration following the T-lymphocytes transfer potentially augmented their therapeutic functions [90]. Moreover, early preclinical studies also suggested the importance of lymphodepletion by radiation or chemotherapy before the ACT and showed a substantial increase in the T-cells reactivity against cancer [91]. In metastatic melanoma patients, it has been demonstrated that IL-2 administration leads to complete tumor regressions [92]. These studies provided an impetus for the identification of specific T lymphocytes and their related antigens intricate in cancer immunotherapy.

Adoptive T-cell therapy represents a highly promising and earliest form of immunotherapy which employs patient's tumor-infiltrating lymphocytes (TILs: T cells isolated from tumor). T cells have the inherent capability to localize and traffic to the cancerous site; however, the identification of TILs at tumor site and their isolation in sufficient amounts from patient are challenging underscoring their potential [93, 94]. These extracted TILs are allowed to expand ex vivo and transfuse back into the patient as an anticancer therapy. Indeed, stromal region of transplantable and growing tumors represents a concentrated source of tumor-infiltrating T lymphocytes (TILs), which can efficiently recognize tumors in vitro. CD8$^+$ and CD4$^+$ T-cells mixtures constitute the general TILs populations isolated from tumors with few contaminating cells in mature cultures. However, the ability of pure cultures of T lymphocytes facilitating human cancer regression has provided the direct evidences that T cells played a vital role in cancer immunotherapy. Previous reports from tumor models generated in mice have demonstrated that these TILs proliferated in the presence of IL-2 facilitated the liver and lung tumors regression [95]. Later, propagated TILs isolated from resected melanomas recognized specific autologous tumors, and these autologous TILs could also lead to complete regression of metastatic melanoma [96, 97]. A research on the exomic mutation rates on >3000 tumor-normal pairs discovered that the non-synonymous mutations frequency varied more than 1000-fold across different tumor types [98]. However, T cells cannot recognize all expressed mutations. Therefore, small peptides (~9 residues) presented on the cell surface of MHC class 1 and MHC class 2. Methods have developed to eliminate the need for predicted peptide binding to MHC and facilitate the screening of all candidate peptides on all MHC loci in a single test (Fig. 3.4).

Although T-lymphocyte cultures can be propagated from distinct tumor types, melanoma represents the only type of cancer from which TILs with specific antitumor recognition can be isolated. The responses mediated with the administered TILs are short-lived, and T cells are rarely identified in circulation days after administration. In 2002, it was reported that TIL transfer followed by the administration of lymphodepletion usually with nonmyeloablative chemotherapy could enhance cancer regression, along with the continuous repopulation of cancer-directed lymphocytes within the host [99]. ACT application has been critically improved with the demonstration that up to 80% of antitumor CD8+ T cells are majorly represented in the circulation after months of infusion. Studies with melanoma provide the stimulus for wide application of ACT against multiple cancer treatments which include

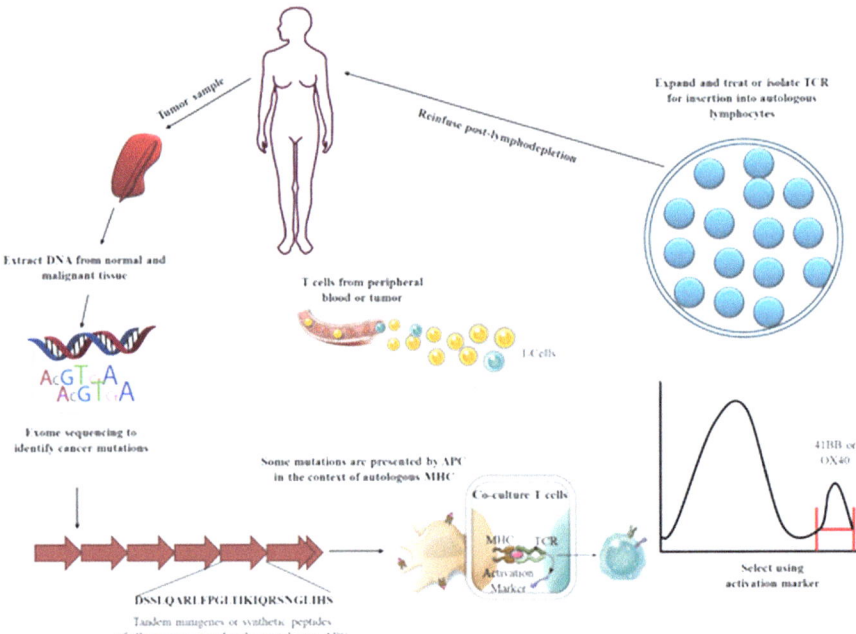

Fig. 3.4 A treatment outline for the tumor-specific mutations recognized by patient T cells. A comparison of the exomic sequences of the patient's tumor and normal cells to identify tumor-specific mutations is performed, which is further used for the synthesis of minigenes encoding mutated residues or peptides flanked by 10 to 12 non-mutated residues. These can be expressed by the patient's autologous APCs, for the processing and presentations by patient's MHC. Coculture of the patient's T cells with these APCs can identify all mutations by MHC class I and class II molecules. Activation markers are expressed by T cells, such as 41BB (CD8+ T cells) and OX40 (CD4+ T cells) after the recognition of their cognate target mutagenic antigens. Activation markers expressing T cells are then purified by flow cytometry eventually leading to their expansion and reinfusion into patients

the genetic manipulation of T lymphocytes with potential to express diverse antitumor receptors. The first demonstration of cancer immunotherapy through genetic modifications came from the mice model [100] followed by humans which showed that the transfusion of T lymphocytes genetically modified with a retroviral system-encoding T-cell receptor (TCR) which could recognize melanoma-melanocyte antigen (MART-1) potentially mediates melanoma regression in patients [101].

Almost a decade back, it was reported that administration of genetically engineered T lymphocytes expressing a chimeric antigen receptor (CAR) targeting CD19 (B-cell antigen) mediates significant regression in the patients of B-cell lymphoma [102]. With the first use of two unspecific immunomodulating agents, interferon and interleukin-2 (IL-2), there was an intensive emphasis on other immunological approaches. Antibody against CTLA-4 protein of cytotoxic T lymphocytes has been approved by FDA for the treatment of advanced metastatic stage disease by the generic name of ipilimumab [103]. All the above findings toward the administration of either unmodified autologous or syngeneic lymphocytes or genetically engineered antitumor T cells provide the impetus for the developmental advancements of ACT against human cancer.

3.8   Genetic Engineering and Cellular Immunotherapy: A Potent Combination Against Tumors

To expand the scope of ACT to distinct cancer types, several approaches were developed for the introduction of engineered antitumor receptors into unmodified T cells which could further be used for therapy (Fig. 3.5). The limitations of TILs have accelerated the energy of scientific communities for redirecting the specificity of T lymphocytes for cancer cells despite relying on the T-cells isolation with intrinsic tumor-aiming abilities. Toward this end, T lymphocytes from a cancer patient can be genetically edited with genes-encoding receptors that could target the tumor-specific antigens and subsequently will "teach" the T cells to bind and ultimately kill cancer cells [104]. Typically, CD8T cells are isolated from the cancer patient and propagated ex vivo with genetic modification, rendering them to express the receptor, and then transfused back into the patient. Two distinct versions of receptors have been used for this purpose. The target specificity of T cells can be readdressed through the genomic integration of receptor genes encoding for either conventional T-cell receptors (TCRs) or genetically manipulated chimeric antigen receptors (CARs). TCR can be genetically engineered to detect and bind cancer-specific epitopes [105, 106], while CAR consisted of a tumor-specific antigen having single-chain variable fragment (scFv) which is fused to the signaling domains of T-cell receptor that could trigger the activation and proliferation of T lymphocytes [107, 108]. The strategy of CARs has undergone various genetic engineering approaches through the addition of diverse T-cell signaling domains which could potentially drive T-lymphocytes proliferation and activation and could lead to the therapeutic variations between these diverse designs.

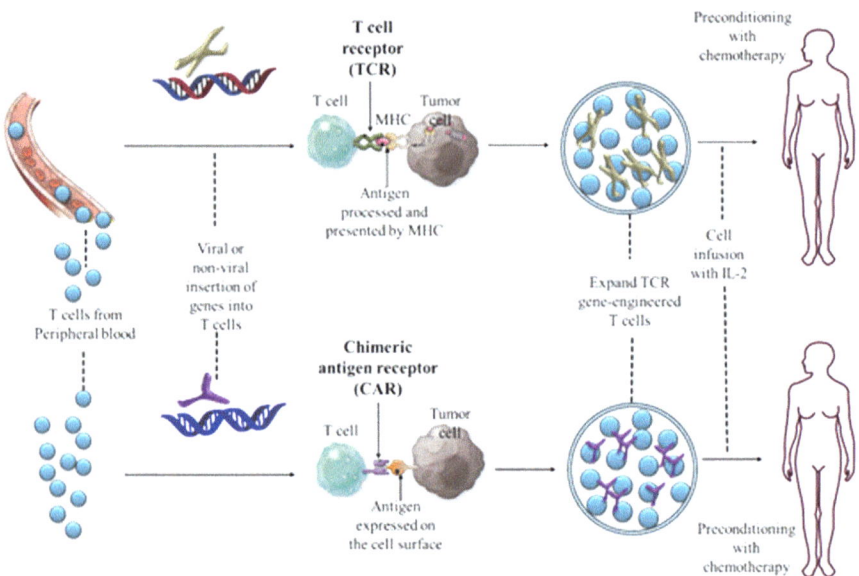

Fig. 3.5 Genetic manipulation of peripheral lymphocytes by the introduction of antitumor receptors into normal T lymphocytes for therapeutic intervention. Techniques are being developed to expand the ACT reach toward other cancers. Top panel depicts the expansion and infusion of autologous T lymphocytes into the patient following the integration of a conserved T-cell receptor (TCR). The bottom panel represents the insertion of a chimeric antigen receptor (CAR) into a patient's T cell, followed by the expansion and their reinfusion into the patient's body. TCRs and CARs have different structures and recognition. TCRs consist of one alpha and one beta chain and recognize antigens processed and presented by MHC molecules. CARs are artificially constructed receptors by linking the variable segments of heavy and light chains of the antibody with intracellular signaling chains (e.g., CD3-zeta, CD28, 41BB). CARs are non-MHC-restricted; however, they recognize the antigens presented on tumor cell surface

TCRs and CARs are differentiated from each other by the type and method of their recognition of cancer antigens. Engineered TCRs expressed on CD8+ T lymphocytes could recognize the protein antigens expressed and derived from the antigen-presenting cells (APCs) and presented on the APC surface by the major histocompatibility complex-1 (MHC-1). However, CARs directly bind to antigens or markers expressed at the surface of the cancer cell. Both TCR- and CAR-based therapeutic interventions have been established in clinical trials and showed promising results. Clinical trial associated with the treatment of 20 metastatic melanoma patients using TCRs specifically targeting MART-1 (melanoma antigen recognized by T cells 1) demonstrated that 33% (Clinical Trials: NCT00509288, NCT00509496) of the patients had objective responses (OR) [106]. CARs therapy was introduced by Gross et al. in 1989 [109] by linking the variable regions of heavy and light chains of the antibody with the intracellular proteins such as CD3-zeta, often including co-stimulatory signaling domains such as TCRzeta/CD28 [110] or CD137 for complete activation of T lymphocytes [111, 112]. One of the advantages of CARs is that they can be easily introduced with high efficiency into T cells using viral vectors and potentially offer the recognition of non-MHC-restricted cell surface components. In adults, relapsed or refractory B-cell acute lymphoblastic leukemia (B-ALL) has a median survival of <6 months and shows a poor prognosis. CD19-specific CAR therapy has emerged as a treatment of B-ALL and demonstrated up to 90% complete response rates [113–116].

TCRs and CARs are extremely encouraging and already considered as breakthroughs in the war against cancer; however, toxicities have been reported with both forms of genetically engineered T-cell therapy associated with clinical trials. Therefore, the précised selectivity between cancer and normal vital organs is a particularly significant safety question that has arose with both TCRs and CARs [117]. Another significant question confronting the usage of genetically engineered T cells in the adoptive cancer immunotherapy involves the choice of the idyllic human T-cell subpopulation which can be used for genetic integration, as well as the selection of suitable antigenic targets for the modified TCRs or CARs. Identifying target epitopes and antigens for TCR and CAR therapies is severely limited by the possible expression of these epitopes on noncancerous and normal cells which could lead to autoimmune responses against healthy cells. Autoimmune toxicity has been demonstrated in TCR therapy in the case of MART-1 with "on-target, off-tumor" effects [106]. Furthermore, one colon cancer patient in a clinical trial (NCT00924287) treated with an ERBB2 (human epidermal growth factor receptor 2)-specific CAR-bearing T lymphocytes died after responsiveness toward low levels of ERBB2 in the vital organs [118]. An excessively robust, life-threatening T-cell response posits another key safety alarm for the potential use of engineered T lymphocytes. Clinical trials using CARs therapy for the treatment of leukemia outburst with the massive release of enormous amounts of cytokines [113], which led to cytokine release syndrome (CRS), including severe symptoms of, for instance, high fever, hypoxia, and hypotension [113]. With this line, immunosuppressive steroids and antibodies have been used to treat CRS which can temper the immune system responses [116].

To determine the extreme burden of CAR-bearing T cells which can be provided to cancer patient with minimal severity of CRS, a clinical trial was conducted [116]. Altogether, despite the severe adverse side effects observed with the engineered T cells, the highly promising outcomes of ACT in clinical trials have generated hope for cancer patients. Preclinical studies conducted in mice have demonstrated that responses against tumors are best observed when T lymphocytes in their early phases of differentiation, for instance, naive or in CNS, are employed for transduction [119]. This observation was further supported by the studies performed in monkeys which showed an enhanced persistence of T lymphocytes from CNS as compared to effector memory cells [103]. Based on the differentiation states, CD8$^+$ T cells are majorly categorized into discrete memory subsets and follow a pathway of progressive differentiation from naive T cells into effector memory T-cell populations [120]. Paradoxically, CD8$^+$ T cells, in the course of their development, lose antitumor functionality along with their ability to lyse the target cells and the production of interferon-γ cytokine which are considered important in antitumor efficacy [121]. These findings are clinically relevant, since, interestingly, there is an inverse correlation of the differentiation states of CD8+ T cells and their capacity for proliferation and persistence [121–124]. Conversely, a statistically significant positive correlation is observed with young or naive T cells with high efficacy in ACT clinical trials [125].

Moreover, CD8+ T cells are capable of clonal repopulation with a stem cell-like state with T-memory stem cells expressing a gene expression program which enabled them to widely differentiate and proliferate into distinct T-cell populations [126]. A considerable amount of the existing research for adoptive cancer immunotherapy has focused on CD8+ T cells. CD4+ T cells could also promote tumor rejection competently. The notion that CD4+ T cells play more straightforward functions in tumor eradication has been validated in humans [127]. The antitumor immune response played by CD4+ T cells is crucially reliant on their polarization, which is determined by the key transcription factors expression. Evidence suggested that CD4+ cells can efficiently destroy tumor cells and that adoptively transferred T-helper 17 cells can promote long-lived antitumor immunity [122].

3.9   Adoptive Immunotherapy for Viruses

Genetically modified T-lymphocytes CARs were first clinically used to treat HIV infection. CARs comprised of the fusion protein (CD4z CAR) generated from extracellular HIV envelope receptor protein (CD4 transmembrane regions) and T-cell receptor (TCR)-ζ signaling molecule. The transduction of these genetically modified T cells potentially lyses the cells expressing HIV envelope proteins. Clinical studies conducted during 1998 and 2005 on active viremic patients have demonstrated the generation of CD4z CAR fusion expression in autologous CD4+ and CD8+ T lymphocytes using retroviral vector [128] as well as in chronic HIV-1 patients [129]. These studies established the feasible and safe transfusion of virus-directed T cells and showed significant effects on viremia by the trafficking of T cells to mucosal reservoirs of infection. Data collected from these trials led to a long-term follow-up analytical studies a decade later and demonstrated the efficacy and safety of retroviral-mediated genetically modified human CAR T cells in their long-term persistence, with an estimated half-life >16 years [130]. These initial research approaches have also revealed that as compared to hematopoietic stem cells (HSCs), T cells are less vulnerable for insertional mutagenesis caused by retroviral-mediated genetic editing.

Almost a decade back, the notable story of Berlin patient came out as the first demonstration of a complete cure of an HIV-infected patient after the transplantation of allogeneic HSCs for AML (acute myelogenous leukemia) [131]. It was later identified that the allogeneic donor for HSCs was genetically homozygous for CCR5 Δ32 mutation, conferring resistance against HIV infection. The significant discovery from this study has challenged the scientific community for the development of cell therapy approaches functions in the absence of allogeneic donors or advanced myeloablative chemotherapy. Later, gene therapy strategies were developed to genetically downregulate the expression of CCR5, either through lentiviral vectors encoding shRNA against CCR5 [132] or gene-editing strategies for the disruption of CCR5 using ZFNs [133]. Genetically engineered autologous T lymphocytes were then reinfused for their reconstitution in patients infected with HIV. Moreover, thorough monitoring and control of HIV infection with possible interpretation of T-cell therapeutic effects on viremia with highly active antiretroviral therapy (HAART), along with cautiously designed and scheduled trials, have changed the entire therapeutic pathways for viral infections.

Allogeneic bone marrow transplantation in patients suffering with hematologic malignancy is highly susceptible for severe chronic viral illnesses mainly from the recurrence of human herpesvirus, Epstein-Barr virus (EBV), cytomegalovirus (CMV), and primary adenoviral infection. These transplants-associated viremia also causes severe and acute illnesses in immunocompromised patients. Pharmacologic interventions against these viremic infections are available with limited efficacy and show substantial side effects when administered recursively. To overcome these issues, transplantation centers have developed the strategies for donor lymphocyte infusion (DLI) against virus infections [134, 135]. However, with the limitations of healthy allogeneic donors for virus-directed T cells, "third-party" T-cell banks have been developed which selectively span the most common HLA alleles isolated from a panel of donors [136, 137]. Scientific group has pioneered the administration of either donor-derived or third-party-derived specific T lymphocytes simultaneously directed against many viruses as lymphocyte infusions. Most importantly, the occurrence of graft-versus-host disease (GvHD) was found to be either partial or bearable in DLI studies. These versions of adoptive immunotherapy are in clinically advanced stages, with many publications of phase II and III, multicenter trials.

3.10 Concluding Remarks

Recent advancements in genome engineering approaches based on programmable and site-specific endonucleases have enabled the systematic examination of functions of mammalian genomes as well as their targeted modifications. Using these approaches, DNA sequences and their functional outputs within the endogenous genome can be easily modulated or edited in virtually any organism type. Engineered nucleases-mediated perturbation of genome is simple and highly scalable which empowered the researchers to establish the causal linkages between genotypes and phenotypes and to elucidate the functional organization of the genome at the systems level. It can be referred as analogous to the search function in word processors, by which nuclease can be guided within complex genomic locations. Here, we have described the applications and development of engineered nuclease for numerous research purposes and their translational applications while highlighting the challenges as well as their future directions.

The various fundamental barriers have now been overcome for the application of engineered nuclease as a platform for designing DNA editing modules with novel specificities. Moreover, several commercial methods are available to produce large-scale novel engineered nucleases against investigator-specified and chosen genetic loci. Programmable nucleases have shown to permit bona fide reverse genetics not only in diverse model organisms but also enabled editing of human cell genetics. Furthermore, the current generation of these approaches exploits two evolutionarily conserved pathways: DNA-protein interactions and pathways of DNA repair offering successful visions to a wide-ranging experimental and applied setting. Nucleases-driven genome editing like ZFNs, TALENs, and CRISPR/Cas offers sophisticated capabilities to understand gene function studies directly and modulate the unhealthy and disease-driving genes.

Adoptive cell therapy (ACT) represents a highly promising strategy against multiple cancers. The clinical consequences of such therapy are closely linked to the capability of effector cells (T lymphocytes) to infuse, engraft, expand, proliferate, and specifically recognize and kill cancer cells within patients. Specific identification, targeting, and killing of cancer cells and not the essential normal tissues pose a tremendous challenge and considered as the major factor limiting the successful use of ACT in humans. Development of cells that can target antigens selectively expressed on cancer and not on essential normal tissues is therefore the prime requirement. ACT is currently undergoing a dramatic period of extensive growth and enthusiasm following encouraging data regarding the clinical efficacy of its administration. ACT uses either natural host cells exhibiting antitumor activity or genetically engineered host cells which express antitumor T-cell receptors (TCRs) or chimeric antigen receptors (CARs). Using these approaches, ACT has led to dramatic regressions in a variety of cancer pathologies, including melanoma, lymphoma, leukemia, cervical cancer, bile duct cancer, and neuroblastoma. Significant results obtained from ACT administration have expanded their reach to the treatment of common epithelial cancers. For instance, ACT using naturally occurring tumor-reactive lymphocytes has mediated durable and complete regressions in melanoma patients, probably by targeting somatic mutations exclusive to each cancer.

Moreover, genetically engineered lymphocytes expressing conventional T-cell receptors or chimeric antigen receptors (CARs) have further extended the successful application of ACT for other cancers treatment. ACT directed against viruses are under critical investigation for the treatment of chronic viral infections as well as for viruses that cause morbidity and mortality in immunocompromised settings such as transplantation of bone marrow as seen with HIV infections. Additionally, cell therapies are taking a prominent role in both hematologic malignancies and solid tumors. Here, we reviewed and discussed the history, current state of ACT, and rationale of immunotherapy for the treatment of diseases and advances in understanding the principles of effective T-cell transfer that point toward impactful clinical results. We also shed light on the strategies and methods in developing effective, appropriate, reliable, and scalable culture systems of ACT driven by programmable nucleases. We hope that more significant and driving innovative applications will expand from basic biology to applied biotechnology and medicine.

References

1. Capecchi MR (2005) Gene targeting in mice: functional analysis of the mammalian genome for the twenty-first century. Nat Rev Genet 6(6):507–512. https://doi.org/10.1038/nrg1619
2. Saha SK, Saikot FK, Rahman MS, Jamal MAHM, Rahman SMK, Islam SMR, Kim KH (2019) Programmable molecular scissors: applications of a new tool for genome editing in biotech. Mol Ther Nucleic Acids 14:212–238. https://doi.org/10.1016/j.omtn.2018.11.016
3. Maeder ML, Gersbach CA (2016) Genome-editing technologies for gene and cell therapy. Mol Ther 24(3):430–446. https://doi.org/10.1038/mt.2016.10
4. Chari R, Church GM (2017) Beyond editing to writing large genomes. Nat Rev Genet 18(12):749–760. https://doi.org/10.1038/nrg.2017.59
5. Rouet P, Smih F, Jasin M (1994) Expression of a site-specific endonuclease stimulates homologous recombination in mammalian cells. Proc Natl Acad Sci U S A 91(13):6064–6068
6. Ceccaldi R, Rondinelli B, D'Andrea AD (2016) Repair pathway choices and consequences at the double-strand break. Trends Cell Biol 26(1):52–64. https://doi.org/10.1016/j.tcb.2015.07.009
7. McManus MT, Sharp PA (2002) Gene silencing in mammals by small interfering RNAs. Nat Rev Genet 3(10):737–747. https://doi.org/10.1038/nrg908
8. Carroll D (2011) Genome engineering with zinc-finger nucleases. Genetics 188(4):773–782. https://doi.org/10.1534/genetics.111.131433
9. Urnov FD, Rebar EJ, Holmes MC, Zhang HS, Gregory PD (2010) Genome editing with engineered zinc finger nucleases. Nat Rev Genet 11(9):636–646. https://doi.org/10.1038/nrg2842
10. Wyman C, Kanaar R (2006) DNA double-strand break repair: All's well that ends well. Annu Rev Genet 40:363–383. https://doi.org/10.1146/annurev.genet.40.110405.090451
11. Hicks WM, Yamaguchi M, Haber JE (2011) Real-time analysis of double-strand DNA break repair by homologous recombination. Proc Natl Acad Sci 108(8):3108–3115. https://doi.org/10.1073/pnas.1019660108
12. Povirk LF (2012) Processing of damaged DNA ends for double-strand break repair in mammalian cells. ISRN Mol Biol. https://doi.org/10.5402/2012/345805
13. Moehle EA, Rock JM, Lee Y-L, Jouvenot Y, DeKelver RC, Gregory PD, Urnov FD, Holmes MC (2007) Targeted gene addition into a specified location in the human genome using designed zinc finger nucleases. Proc Natl Acad Sci 104(9):3055–3060. https://doi.org/10.1073/pnas.0611478104
14. Orlando SJ, Santiago Y, DeKelver RC, Freyvert Y, Boydston EA, Moehle EA, Choi VM, Gopalan SM, Lou JF, Li J et al (2010) Zinc-finger nuclease-driven targeted integration into mammalian genomes using donors with limited chromosomal homology. Nucleic Acids Res 38(15):e152. https://doi.org/10.1093/nar/gkq512
15. Chen F, Pruett-Miller SM, Huang Y, Gjoka M, Duda K, Taunton J, Collingwood TN, Frodin M, Davis GD (2011) High-frequency genome editing using ss DNA oligonucleotides with zinc-finger nucleases. Nat Methods 8(9):753–755. https://doi.org/10.1038/nmeth.1653
16. Santiago Y, Chan E, Liu P-Q, Orlando S, Zhang L, Urnov FD, Holmes MC, Guschin D, Waite A, Miller JC et al (2008) Targeted gene knockout in mammalian cells by using engineered zinc-finger nucleases. Proc Natl Acad Sci 105(15):5809–5814. https://doi.org/10.1073/pnas.0800940105
17. Lee HJ, Kim E, Kim JS (2010) Targeted chromosomal deletions in human cells using zinc finger nucleases. Genome Res 20(1):81–89. https://doi.org/10.1101/gr.099747.109
18. Brunet E, Simsek D, Tomishima M, DeKelver R, Choi VM, Gregory P, Urnov F, Weinstock DM, Jasin M (2009) Chromosomal translocations induced at specified loci in human stem cells. Proc Natl Acad Sci 106(26):10620–10625. https://doi.org/10.1073/pnas.0902076106
19. Lee HJ, Kweon J, Kim E, Kim S, Kim JS (2012) Targeted chromosomal duplications and inversions in the human genome using zinc finger nucleases. Genome Res 22(3):539–548. https://doi.org/10.1101/gr.129635.111

20. Maresca M, Lin VG, Guo N, Yang Y (2013) Obligate ligation-gated recombination (ObLiGaRe): custom-designed nuclease-mediated targeted integration through nonhomologous end joining. Genome Res 23(3):539–546. https://doi.org/10.1101/gr.145441.112
21. Hockemeyer D, Soldner F, Beard C, Gao Q, Mitalipova M, Dekelver RC, Katibah GE, Amora R, Boydston EA, Zeitler B et al (2009) Efficient targeting of expressed and silent genes in human ESCs and iPSCs using zinc-finger nucleases. Nat Biotechnol 27(9):851–857. https://doi.org/10.1038/nbt.1562
22. Hockemeyer D, Wang H, Kiani S, Lai CS, Gao Q, Cassady JP, Cost GJ, Zhang L, Santiago Y, Miller JC et al (2011) Genetic engineering of human pluripotent cells using TALE nucleases. Nat Biotechnol 29(8):731–734. https://doi.org/10.1038/nbt.1927
23. Gutschner T, Baas M, Diederichs S (2011) Noncoding RNA gene silencing through genomic integration of RNA destabilizing elements using zinc finger nucleases. Genome Res 21(11):1944–1954. https://doi.org/10.1101/gr.122358.111
24. Sanyal A, Lajoie BR, Jain G, Dekker J (2012) The long-range interaction landscape of gene promoters. Nature 489(7414):109–113. https://doi.org/10.1038/nature11279
25. Beerli RR, Barbas CF (2002) Engineering polydactyl zinc-finger transcription factors. Nat Biotechnol 20(2):135–141. https://doi.org/10.1038/nbt0202-135
26. Miller J, McLachlan AD, Klug A (1985) Repetitive zinc-binding domains in the protein transcription factor IIIA from Xenopus oocytes. EMBO J 4:1609–1614
27. Wolfe SA, Nekludova L, Pabo CO (2000) DNA recognition by Cys2His2 zinc finger proteins. Annu Rev Biophys Biomol Struct 25(1):22–29
28. Liu Q, Segal DJ, Ghiara JB, Barbas CF (2002) Design of polydactyl zinc-finger proteins for unique addressing within complex genomes. Proc Natl Acad Sci 94:5525. https://doi.org/10.1073/pnas.94.11.5525
29. Beerli RR, Segal DJ, Dreier B, Barbas CF III, Barbas CF (1998) Toward controlling gene expression at will: specific regulation of the erbB-2/HER-2 promoter by using polydactyl zinc finger proteins constructed from modular building blocks. Proc Natl Acad Sci U S A 95(25):14628–14633
30. Beerli RR, Dreier B, Barbas CF (2002) Positive and negative regulation of endogenous genes by designed transcription factors. Proc Natl Acad Sci. https://doi.org/10.1073/pnas.040552697
31. Segal DJ, Beerli RR, Blancafort P, Dreier B, Effertz K, Huber A, Koksch B, Lund CV, Magnenat L, Valente D et al (2003) Evaluation of a modular strategy for the construction of novel polydactyl zinc finger DNA-binding proteins. Biochemistry 42(7):2137–2148. https://doi.org/10.1021/bi026806o
32. Pavletich NP, Pabo CO (1991) Zinc finger-DNA recognition: crystal structure of a Zif268-DNA complex at 2.1 Å. Science (80-.) 252(5007):809–817. https://doi.org/10.1126/science.2028256
33. Segal DJ, Dreier B, Beerli RR, Barbas CF (2002) Toward controlling gene expression at will: selection and design of zinc finger domains recognizing each of the 5'-GNN-3' DNA target sequences. Proc Natl Acad Sci 96(6):2758–2763. https://doi.org/10.1073/pnas.96.6.2758
34. Maeder ML, Thibodeau-Beganny S, Osiak A, Wright DA, Anthony RM, Eichtinger M, Jiang T, Foley JE, Winfrey RJ, Townsend JA et al (2008) Rapid 'open-source' engineering of customized zinc-finger nucleases for highly efficient gene modification. Mol Cell 31(2):294–301. https://doi.org/10.1016/j.molcel.2008.06.016
35. Vanamee ÉS, Santagata S, Aggarwal AK (2001) FokI requires two specific DNA sites for cleavage. J Mol Biol 309(1):69–78. https://doi.org/10.1006/jmbi.2001.4635
36. Miller JC, Holmes MC, Wang J, Guschin DY, Lee Y-L, Rupniewski I, Beausejour CM, Waite AJ, Wang NS, Kim KA et al (2007) An improved zinc-finger nuclease architecture for highly specific genome editing. Nat Biotechnol 25(7):778–785. https://doi.org/10.1038/nbt1319
37. Szczepek M, Brondani V, Büchel J, Serrano L, Segal DJ, Cathomen T (2007) Structure-based redesign of the dimerization interface reduces the toxicity of zinc-finger nucleases. Nat Biotechnol 25(7):786–793. https://doi.org/10.1038/nbt1317

38. Guo J, Gaj T, Barbas CF (2010) Directed evolution of an enhanced and highly efficient FokI cleavage domain for zinc finger nucleases. J Mol Biol 400(1):96–107. https://doi.org/10.1016/j.jmb.2010.04.060

39. Beumer KJ, Trautman JK, Bozas A, Liu J-L, Rutter J, Gall JG, Carroll D (2008) Efficient gene targeting in Drosophila by direct embryo injection with zinc-finger nucleases. Proc Natl Acad Sci 105(50):19821–19826. https://doi.org/10.1073/pnas.0810475105

40. Doyon Y, McCammon JM, Miller JC, Faraji F, Ngo C, Katibah GE, Amora R, Hocking TD, Zhang L, Rebar EJ et al (2008) Heritable targeted gene disruption in zebrafish using designed zinc-finger nucleases. Nat. Biotechnol 26(6):702–708

41. Meng X, Noyes MB, Zhu LJ, Lawson ND, Wolfe SA (2008) Targeted gene inactivation in zebrafish using engineered zinc-finger nucleases. Nat Biotechnol 26(6):695–701. https://doi.org/10.1038/nbt1398

42. Geurts AM, Cost GJ, Freyvert Y, Zeitler B, Miller JC, Choi VM, Jenkins SS, Wood A, Cui X, Meng X et al (2009) Knockout rats via embryo microinjection of zinc-finger nucleases. Science (80-.) 325(5939):433. https://doi.org/10.1126/science.1172447

43. Mashimo T, Takizawa A, Voigt B, Yoshimi K, Hiai H, Kuramoto T, Serikawa T (2010) Generation of knockout rats with X-linked severe combined immunodeficiency (X-SCID) using zinc-finger nucleases. PLoS One 5(1):e8870. https://doi.org/10.1371/journal.pone.0008870

44. Osakabe K, Osakabe Y, Toki S (2010) Site-directed mutagenesis in Arabidopsis using custom-designed zinc finger nucleases. Proc Natl Acad Sci 107(26):12034–12039. https://doi.org/10.1073/pnas.1000234107

45. Cost GJ, Freyvert Y, Vafiadis A, Santiago Y, Miller JC, Rebar E, Collingwood TN, Snowden A, Gregory PD (2010) BAK and BAX deletion using zinc-finger nucleases yields apoptosis-resistant CHO cells. Biotechnol Bioeng 105(2):330–340. https://doi.org/10.1002/bit.22541

46. Liu PQ, Chan EM, Cost GJ, Zhang L, Wang J, Miller JC, Guschin DY, Reik A, Holmes MC, Mott JE et al (2010) Generation of a triple-gene knockout mammalian cell line using engineered zinc-finger nucleases. Biotechnol Bioeng 106(1):97–105. https://doi.org/10.1002/bit.22654

47. Cermak T, Doyle EL, Christian M, Wang L, Zhang Y, Schmidt C, Baller JA, Somia NV, Bogdanove AJ, Voytas DF (2011) Efficient design and assembly of custom TALEN and other TAL effector-based constructs for DNA targeting. Nucleic Acids Res 39(12):e82. https://doi.org/10.1093/nar/gkr218

48. Reyon D, Tsai SQ, Khgayter C, Foden JA, Sander JD, Joung JK (2012) FLASH assembly of TALENs for high-throughput genome editing. Nat Biotechnol 30(5):460–465. https://doi.org/10.1038/nbt.2170

49. Briggs AW, Rios X, Chari R, Yang L, Zhang F, Mali P, Church GM (2012) Iterative capped assembly: rapid and scalable synthesis of repeat-module DNA such as TAL effectors from individual monomers. Nucleic Acids Res 40(15):e117. https://doi.org/10.1093/nar/gks624

50. Schmid-Burgk JL, Schmidt T, Kaiser V, Höning K, Hornung V (2013) A ligation-independent cloning technique for high-throughput assembly of transcription activator-like effector genes. Nat Biotechnol 31(1):76–81. https://doi.org/10.1038/nbt.2460

51. Kim Y, Kweon J, Kim A, Chon JK, Yoo JY, Kim HJ, Kim S, Lee C, Jeong E, Chung E et al (2013) A library of TAL effector nucleases spanning the human genome. Nat Biotechnol 31(3):251–258. https://doi.org/10.1038/nbt.2517

52. Gaj T, Guo J, Kato Y, Sirk SJ, Barbas CF (2012) Targeted gene knockout by direct delivery of zinc-finger nuclease proteins. Nat Methods 9(8):805–807. https://doi.org/10.1038/nmeth.2030

53. Ishino Y, Shinagawa H, Makino K, Amemura M, Nakatura A (1987) Nucleotide sequence of the iap gene, responsible for alkaline phosphatase isoenzyme conversion in Escherichia coli, and identification of the gene product. J Bacteriol 169(12):5429–5433

54. Mojica FJM, Díez-Villaseñor C, Soria E, Juez G (2000) Biological significance of a family of regularly spaced repeats in the genomes of archaea, bacteria and mitochondria. Mol Microbiol 36(1):244–246. https://doi.org/10.1046/j.1365-2958.2000.01838.x

55. Jansen R, Van Embden JDA, Gaastra W, Schouls LM (2002) Identification of genes that are associated with DNA repeats in prokaryotes. Mol Microbiol 43:1565–1575. https://doi.org/10.1046/j.1365-2958.2002.02839.x

56. Barrangou R, van der Oost J (2013) CRISPR-Cas Systems: RNA-mediated adaptive immunity in bacteria and archaea. Springer, Berlin/Heidelberg

57. Barrangou R, Marraffini LA (2014) CRISPR-cas systems: prokaryotes upgrade to adaptive immunity. Mol Cell. https://doi.org/10.1016/j.molcel.2014.03.011

58. Tang T-H, Bachellerie J-P, Rozhdestvensky T, Bortolin M-L, Huber H, Drungowski M, Elge T, Brosius J, Huttenhofer A (2002) Identification of 86 candidates for small non-messenger RNAs from the archaeon Archaeoglobus fulgidus. Proc Natl Acad Sci. https://doi.org/10.1073/pnas.112047299

59. Haft DH, Selengut J, Mongodin EF, Nelson KE (2005) A guild of 45 CRISPR-associated (Cas) protein families and multiple CRISPR/cas subtypes exist in prokaryotic genomes. PLoS Comput Biol. https://doi.org/10.1371/journal.pcbi.0010060

60. Makarova KS, Aravind L, Wolf YI, Koonin EV (2011) Unification of Cas protein families and a simple scenario for the origin and evolution of CRISPR-Cas systems. Biol Direct. https://doi.org/10.1186/1745-6150-6-38

61. Brouns SJJ, Jore MM, Lundgren M, Westra ER, Slijkhuis RJH, Snijders APL, Dickman MJ, Makarova KS, Koonin EV, Van Der Oost J (1993) Small Crispr Rnas guide antiviral defense in prokaryotes. Cancer Epidemiol Biomarkers Prev

62. Hale CR, Zhao P, Olson S, Duff MO, Graveley BR, Wells L, Terns RM, Terns MP (2009) RNA-guided RNA cleavage by a CRISPR RNA-Cas protein complex. Cell. https://doi.org/10.1016/j.cell.2009.07.040

63. Mojica FJM, Díez-Villaseñor C, García-Martínez J, Soria E (2005) Intervening sequences of regularly spaced prokaryotic repeats derive from foreign genetic elements. J Mol Evol. https://doi.org/10.1007/s00239-004-0046-3

64. Pourcel C, Salvignol G, Vergnaud G (2005) CRISPR elements in Yersinia pestis acquire new repeats by preferential uptake of bacteriophage DNA, and provide additional tools for evolutionary studies. Microbiology. https://doi.org/10.1099/mic.0.27437-0

65. Bolotin A, Quinquis B, Sorokin A, Dusko Ehrlich S (2005) Clustered regularly interspaced short palindrome repeats (CRISPRs) have spacers of extrachromosomal origin. Microbiology. https://doi.org/10.1099/mic.0.28048-0

66. Makarova KS, Grishin NV, Shabalina SA, Wolf YI, Koonin EV (2006) A putative RNA-interference-based immune system in prokaryotes: computational analysis of the predicted enzymatic machinery, functional analogies with eukaryotic RNAi, and hypothetical mechanisms of action. Biol Direct. https://doi.org/10.1186/1745-6150-1-7

67. Deltcheva E, Chylinski K, Sharma CM, Gonzales K, Chao Y, Pirzada ZA, Eckert MR, Vogel J, Charpentier E (2011) CRISPR RNA maturation by trans-encoded small RNA and host factor RNase III. Nature. https://doi.org/10.1038/nature09886

68. Garneau JE, Dupuis M-È, Villion M, Romero DA, Barrangou R, Boyaval P, Fremaux C, Horvath P, Magadán AH, Moineau S (2010) The CRISPR/Cas bacterial immune system cleaves bacteriophage and plasmid DNA. Nature. https://doi.org/10.1038/nature09523

69. Jinek M, Chylinski K, Fonfara I, Hauer M, Doudna JA, Charpentier E (2012) A programmable dual-RNA-guided DNA endonuclease in adaptive bacterial immunity. Science (80-.). https://doi.org/10.1126/science.1225829

70. Gasiunas G, Barrangou R, Horvath P, Siksnys V (2012) Cas9-crRNA ribonucleoprotein complex mediates specific DNA cleavage for adaptive immunity in bacteria. Proc Natl Acad Sci. https://doi.org/10.1073/pnas.1208507109

71. Sander JD, Joung JK (2014) CRISPR-Cas systems for editing, regulating and targeting genomes. Nat Biotechnol. https://doi.org/10.1038/nbt.2842

72. Cong L, Ran FA, Cox D, Lin S, Barretto R, Habib N, Hsu PD, Wu X, Jiang W, Marraffini LA et al (2013) Multiplex genome engineering using CRISPR/Cas systems. Science (80-.). https://doi.org/10.1126/science.1231143

73. Mali P, Yang L, Esvelt KM, Aach J, Guell M, DiCarlo JE, Norville JE, Church GM (2013) RNA-guided human genome engineering via Cas9. Science (80-.). https://doi.org/10.1126/science.1232033

74. Cho SW, Kim S, Kim JM, Kim JS (2013) Targeted genome engineering in human cells with the Cas9 RNA-guided endonuclease. Nat Biotechnol. https://doi.org/10.1038/nbt.2507

75. Hsu PD, Scott DA, Weinstein JA, Ran FA, Konermann S, Agarwala V, Li Y, Fine EJ, Wu X, Shalem O et al (2013) DNA targeting specificity of RNA-guided Cas9 nucleases. Nat Biotechnol. https://doi.org/10.1038/nbt.2647

76. Nishimasu H, Ran FA, Hsu PD, Konermann S, Shehata SI, Dohmae N, Ishitani R, Zhang F, Nureki O (2014) Crystal structure of Cas9 in complex with guide RNA and target DNA. Cell. https://doi.org/10.1016/j.cell.2014.02.001

77. Wang H, Yang H, Shivalila CS, Dawlaty MM, Cheng AW, Zhang F, Jaenisch R (2013) One-step generation of mice carrying mutations in multiple genes by CRISPR/cas-mediated genome engineering. Cell. https://doi.org/10.1016/j.cell.2013.04.025

78. Shalem O, Sanjana NE, Hartenian E, Shi X, Scott DA, Mikkelsen TS, Heckl D, Ebert BL, Root DE, Doench JG et al (2014) Genome-scale CRISPR-Cas9 knockout screening in human cells. Science (80-.). https://doi.org/10.1126/science.1247005

79. Barrangou R, Doudna JA (2016) Applications of CRISPR technologies in research and beyond. Nat Biotechnol. https://doi.org/10.1038/nbt.3659

80. Djebali S, Davis CA, Merkel A, Dobin A, Lassmann T, Mortazavi A, Tanzer A, Lagarde J, Lin W, Schlesinger F et al (2012) Landscape of transcription in human cells. Nature. https://doi.org/10.1038/nature11233

81. Zhu S, Li W, Liu J, Chen CH, Liao Q, Xu P, Xu H, Xiao T, Cao Z, Peng J et al (2016) Genome-scale deletion screening of human long non-coding RNAs using a paired-guide RNA CRISPR-Cas9 library. Nat Biotechnol. https://doi.org/10.1038/nbt.3715

82. Qi LS, Larson MH, Gilbert LA, Doudna JA, Weissman JS, Arkin AP, Lim WA (2013) Repurposing CRISPR as an RNA-γuided platform for sequence-specific control of gene expression. Cell. https://doi.org/10.1016/j.cell.2013.02.022

83. Gilbert LA, Horlbeck MA, Adamson B, Villalta JE, Chen Y, Whitehead EH, Guimaraes C, Panning B, Ploegh HL, Bassik MC et al (2014) Genome-scale CRISPR-mediated control of gene repression and activation. Cell. https://doi.org/10.1016/j.cell.2014.09.029

84. Konermann S, Brigham MD, Trevino AE, Joung J, Abudayyeh OO, Barcena C, Hsu PD, Habib N, Gootenberg JS, Nishimasu H et al (2015) Genome-scale transcriptional activation by an engineered CRISPR-Cas9 complex. Nature. https://doi.org/10.1038/nature14136

85. Tanenbaum ME, Gilbert LA, Qi LS, Weissman JS, Vale RD (2014) A protein-tagging system for signal amplification in gene expression and fluorescence imaging. Cell. https://doi.org/10.1016/j.cell.2014.09.039

86. Delorme EJ, Alexander P (1964) Treatment of primary FIBROSARCOMA in the rat with immune lymphocytes. Lancet. https://doi.org/10.1016/S0140-6736(64)90126-6

87. Morgan DA, Ruscetti FW, Gallo R (1976) Selective in vitro growth of T lymphocytes from normal human bone marrows. Science (80-.). https://doi.org/10.1126/science.181845

88. Eberlein TJ, Rosenstein M, Rosenberg SA (1982) Regression of a disseminated syngeneic solid tumor by systemic transfer of lymphoid cells expanded in interleukin 2. J Exp Med

89. Rosenberg SA, Mulé JJ, Spiess PJ, Reichert CM, Schwarz SL (1985) Regression of established pulmonary metastases and subcutaneous tumor mediated by the systemic administration of high-dose recombinant interleukin 2. J Exp Med

90. Donohue JH, Rosenstein M, Chang AE, Lotze MT, Robb RJ, Rosenberg SA (1984) The systemic administration of purified interleukin 2 enhances the ability of sensitized murine lymphocytes to cure a disseminated syngeneic lymphoma. J Immunol

91. Berendt MJ, North RJ (1980) T-cell-mediated suppression of anti-tumor immunity. An explanation for progressive growth of an immunogenic tumor. J Exp Med

92. Rosenberg SA, Lotze MT, Muul LM, Leitman S, Chang AE, Ettinghausen SE, Matory YL, Skibber JM, Shiloni E, Vetto JT et al (1985) Observations on the systemic administration of

autologous lymphokine-activated killer cells and recombinant interleukin-2 to patients with metastatic cancer. N Engl J Med 313:1485–1492

93. Dudley ME, Gross CA, Somerville RPT, Hong Y, Schaub NP, Rosati SF, White DE, Nathan D, Restifo NP, Steinberg SM et al (2013) Randomized selection design trial evaluating CD8 + −enriched versus unselected tumor-infiltrating lymphocytes for adoptive cell therapy for patients with melanoma. J Clin Oncol. https://doi.org/10.1200/JCO.2012.46.6441

94. Khammari A, Knol A-C, Nguyen J-M, Bossard C, Denis M-G, Pandolfino M-C, Quéreux G, Bercegeay S, Dréno B (2014) Adoptive TIL transfer in the adjuvant setting for melanoma: long-term patient survival. J Immunol Res. https://doi.org/10.1155/2014/186212

95. Rosenberg SA, Spiess P, Lafreniere R (1986) A new approach to the adoptive immunotherapy of cancer with tumor-infiltrating lymphocytes. Science (80-.). https://doi.org/10.1126/science.3489291

96. Muul LM, Spiess PJ, Director EP, Rosenberg SA (1987) Identification of specific cytolytic immune responses against autologous tumor in humans bearing malignant melanoma. J Immunol

97. Rosenberg SA, Packard BS, Aebersold PM, Solomon D, Topalian SL, Toy ST, Simon P, Lotze MT, Yang JC, Seipp CA et al (1988) Use of tumor-infiltrating lymphocytes and Interleukin-2 in the immunotherapy of patients with metastatic melanoma. N Engl J Med 319:1676–1680

98. Lawrence MS, Stojanov P, Polak P, Kryukov GV, Cibulskis K, Sivachenko A, Carter SL, Stewart C, Mermel CH, Roberts SA et al (2013) Mutational heterogeneity in cancer and the search for new cancer-associated genes. Nature. https://doi.org/10.1038/nature12213

99. Dudley ME, Wunderlich JR, Robbins PF, Yang JC, Hwu P, Schwartzentruber DJ, Topalian SL, Sherry R, Restifo NP, Hubicki AM et al (2002) Cancer regression and autoimmunity in patients after clonal repopulation with antitumor lymphocytes. Science (80-.). https://doi.org/10.1126/science.1076514

100. Kessels HWHG, Wolkers MC, Van Den Boom MD, Van Den Valk MA, Schumacher TNM (2001) Immunotherapy through TCR gene transfer. Nat Immunol. https://doi.org/10.1038/ni1001-957

101. Morgan RA, Dudley ME, Wunderlich JR, Hughes MS, Yang JC, Sherry RM, Royal RE, Topalían SL, Kammula US, Restifo NP et al (2006) Cancer regression in patients after transfer of genetically engineered lymphocytes. Science (80-.). https://doi.org/10.1126/science.1129003

102. Kochenderfer JN, Wilson WH, Janik JE, Dudley ME, Stetler-Stevenson M, Feldman SA, Maric I, Raffeld M, Nathan DAN, Lanier BJ et al (2010) Eradication of B-lineage cells and regression of lymphoma in a patient treated with autologous T cells genetically engineered to recognize CD19. Blood. https://doi.org/10.1182/blood-2010-04-281931

103. Hodi FS, O'Day SJ, McDermott DF, Weber RW, Sosman JA, Haanen JB, Gonzalez R, Robert C, Schadendorf D, Hassel JC et al (2010) Improved survival with ipilimumab in patients with metastatic melanoma. N Engl J Med. https://doi.org/10.1056/NEJMoa1003466

104. Duong CPM, Yong CSM, Kershaw MH, Slaney CY, Darcy PK (2015) Cancer immunotherapy utilizing gene-modified T cells: from the bench to the clinic. Mol Immunol. https://doi.org/10.1016/j.molimm.2014.12.009

105. Thaxton JE, Li Z (2014) To affinity and beyond: harnessing the T cell receptor for cancer immunotherapy. Hum Vaccin Immunother. https://doi.org/10.4161/21645515.2014.973314

106. Johnson LA, Morgan RA, Dudley ME, Cassard L, Yang JC, Hughes MS, Kammula US, Royal RE, Sherry RM, Wunderlich JR et al (2009) Gene therapy with human and mouse T-cell receptors mediates cancer regression and targets normal tissues expressing cognate antigen. Blood. https://doi.org/10.1182/blood-2009-03-211714

107. Chmielewski M, Hombach A, Heuser C, Adams GP, Abken H (2014) T cell activation by antibody-like immunoreceptors: increase in affinity of the single-chain fragment domain above threshold does not increase T cell activation against antigen-positive target cells but decreases selectivity. J Immunol. https://doi.org/10.4049/jimmunol.173.12.7647

108. Jena B, Dotti G, Cooper LJN (2010) Redirecting T-cell specificity by introducing a tumor-specific chimeric antigen receptor. Blood. https://doi.org/10.1182/blood-2010-01-043737

109. Gross G, Waks T, Eshhar Z (1989) Expression of immunoglobulin-T-cell receptor chimeric molecules as functional receptors with antibody-type specificity. Proc Natl Acad Sci 86:10024–10028

110. Maher J, Brentjens RJ, Gunset G, Rivière I, Sadelain M (2002) Human T-lymphocyte cytotoxicity and proliferation directed by a single chimeric TCRζ/CD28 receptor. Nat Biotechnol. https://doi.org/10.1038/nbt0102-70

111. Imai C, Mihara K, Andreansky M, Nicholson IC, Pui CH, Geiger TL, Campana D (2004) Chimeric receptors with 4-1BB signaling capacity provoke potent cytotoxicity against acute lymphoblastic leukemia. Leukemia. https://doi.org/10.1038/sj.leu.2403302

112. Song DG, Ye Q, Carpenito C, Poussin M, Wang LP, Ji C, Figini M, June CH, Coukos G, Powell DJ (2011) In vivo persistence, tumor localization, and antitumor activity of CAR-engineered T cells is enhanced by costimulatory signaling through CD137 (4-1BB). Cancer Res. https://doi.org/10.1158/0008-5472.CAN-11-0422

113. Davila ML, Riviere I, Wang X, Bartido S, Park J, Curran K, Chung SS, Stefanski J, Borquez-Ojeda O, Olszewska M et al (2014) Efficacy and toxicity management of 19-28z CAR T cell therapy in B cell acute lymphoblastic leukemia. Sci Transl Med. https://doi.org/10.1126/scitranslmed.3008226

114. Brentjens RJ, Rivière I, Park JH, Davila ML, Wang X, Stefanski J, Taylor C, Yeh R, Bartido S, Borquez-Ojeda O et al (2011) Safety and persistence of adoptively transferred autologous CD19-targeted T cells in patients with relapsed or chemotherapy refractory B-cell leukemias. Blood. https://doi.org/10.1182/blood-2011-04-348540

115. Maude SL, Frey N, Shaw PA, Aplenc R, Barrett DM, Bunin NJ, Chew A, Gonzalez VE, Zheng Z, Lacey SF et al (2014) Chimeric antigen receptor T cells for sustained remissions in leukemia. N Engl J Med. https://doi.org/10.1056/NEJMoa1407222

116. Lee DW, Kochenderfer JN, Stetler-Stevenson M, Cui YK, Delbrook C, Feldman SA, Fry TJ, Orentas R, Sabatino M, Shah NN et al (2015) T cells expressing CD19 chimeric antigen receptors for acute lymphoblastic leukaemia in children and young adults: a phase 1 dose-escalation trial. Lancet. https://doi.org/10.1016/S0140-6736(14)61403-3

117. Couzin-Frankel J (2013) Cancer immunotherapy. Science (80-.) 342:1432–1433

118. Morgan RA, Yang JC, Kitano M, Dudley ME, Laurencot CM, Rosenberg SA (2010) Case report of a serious adverse event following the administration of t cells transduced with a chimeric antigen receptor recognizing ERBB2. Mol Ther. https://doi.org/10.1038/mt.2010.24

119. Klebanoff CA, Gattinoni L, Torabi-Parizi P, Kerstann K, Cardones AR, Finkelstein SE, Palmer DC, Antony PA, Hwang ST, Rosenberg SA et al (2005) Central memory self/tumor-reactive CD8+ T cells confer superior antitumor immunity compared with effector memory T cells. Proc Natl Acad Sci. https://doi.org/10.1073/pnas.0503726102

120. Gattinoni L, Powell DJ, Rosenberg SA, Restifo NP (2006) Adoptive immunotherapy for cancer: building on success. Nat Rev Immunol. https://doi.org/10.1038/nri1842

121. Gattinoni L, Klebanoff CA, Palmer DC, Wrzesinski C, Kerstann K, Yu Z, Finkelstein SE, Theoret MR, Rosenberg SA, Restifo NP (2005) Acquisition of full effector function in vitro paradoxically impairs the in vivo antitumor efficacy of adoptively transferred CD8+ T cells. J Clin Invest. https://doi.org/10.1172/JCI24480

122. Muranski P, Borman ZA, Kerkar SP, Klebanoff CA, Ji Y, Sanchez-Perez L, Sukumar M, Reger RN, Yu Z, Kern SJ et al (2011) Th17 cells are long lived and retain a stem cell-like molecular signature. Immunity. https://doi.org/10.1016/j.immuni.2011.09.019

123. Buchholz VR, Flossdorf M, Hensel I, Kretschmer L, Weissbrich B, Gräf P, Verschoor A, Schiemann M, Höfer T, Busch DH (2013) Disparate individual fates compose robust CD8+ T cell immunity. Science (80-.). https://doi.org/10.1126/science.1235454

124. Gerlach C, Rohr JC, Perié L, Van Rooij N, Van Heijst JWJ, Velds A, Urbanus J, Naik SH, Jacobs H, Beltman JB et al (2013) Heterogeneous differentiation patterns of individual CD8+T cells. Science (80-.). https://doi.org/10.1126/science.1235487

125. Rosenberg SA, Yang JC, Sherry RM, Kammula US, Hughes MS, Phan GQ, Citrin DE, Restifo NP, Robbins PF, Wunderlich JR et al (2011) Durable complete responses in heavily pretreated patients with metastatic melanoma using T-cell transfer immunotherapy. Clin Cancer Res. https://doi.org/10.1158/1078-0432.CCR-11-0116

126. Gattinoni L, Lugli E, Ji Y, Pos Z, Paulos CM, Quigley MF, Almeida JR, Gostick E, Yu Z, Carpenito C et al (2011) A human memory T cell subset with stem cell-like properties. Nat Med. https://doi.org/10.1038/nm.2446

127. Tran E, Turcotte S, Gros A, Robbins PF, Lu YC, Dudley ME, Wunderlich JR, Somerville RP, Hogan K, Hinrichs CS et al (2014) Cancer immunotherapy based on mutation-specific CD4+ T cells in a patient with epithelial cancer. Science (80-.). https://doi.org/10.1126/science.1251102

128. Mitsuyasu RT, Anton PA, Deeks SG, Scadden DT, Connick E, Downs MT, Bakker A, Roberts MR, June CH, Jalali S et al (2000) Prolonged survival and tissue trafficking following adoptive transfer of CD4zeta gene-modified autologous CD4(+) and CD8(+) T cells in human immunodeficiency virus-infected subjects. Blood

129. Deeks SG, Wagner B, Anton PA, Mitsuyasu RT, Scadden DT, Huang C, Macken C, Richman DD, Christopherson C, June CH et al (2002) A phase II randomized study of HIV-specific T-cell gene therapy in subjects with undetectable plasma viremia on combination antiretroviral therapy. Mol Ther. https://doi.org/10.1006/mthe.2002.0611

130. Scholler J, Brady TL, Binder-Scholl G, Hwang WT, Plesa G, Hege KM, Vogel AN, Kalos M, Riley JL, Deeks SG et al (2012) Decade-long safety and function of retroviral-modified chimeric antigen receptor T cells. Sci Transl Med. https://doi.org/10.1126/scitranslmed.3003761

131. Hütter G, Nowak D, Mossner M, Ganepola S, Müßig A, Allers K, Schneider T, Hofmann J, Kücherer C, Blau O et al (2009) Long-term control of HIV by CCR5 Delta 32/Delta32 stem-cell transplantation. N Engl J Med 360:692–698

132. Liang M, Kamata M, Chen KN, Pariente N, An DS, Chen ISY (2010) Inhibition of HIV-1 infection by a unique short hairpin RNA to chemokine receptor 5 delivered into macrophages through hematopoietic progenitor cell transduction. J Gene Med. https://doi.org/10.1002/jgm.1440

133. Perez EE, Wang J, Miller JC, Jouvenot Y, Kim KA, Liu O, Wang N, Lee G, Bartsevich VV, Lee YL et al (2008) Establishment of HIV-1 resistance in CD4 + T cells by genome editing using zinc-finger nucleases. Nat Biotechnol. https://doi.org/10.1038/nbt1410

134. Walter EA, Greenberg PD, Gilbert MJ, Finch RJ, Watanabe KS, Thomas ED, Riddell SR (2002) Reconstitution of cellular immunity against cytomegalovirus in recipients of allogeneic bone marrow by transfer of T-cell clones from the donor. N Engl J Med. https://doi.org/10.1056/nejm199510193331603

135. Louis CU, Straathof K, Bollard CM, Ennamuri S, Gerken C, Lopez TT, Huls MH, Sheehan A, Wu MF, Liu H et al (2010) Adoptive transfer of EBV-specific T cells results in sustained clinical responses in patients with locoregional nasopharyngeal carcinoma. J Immunother. https://doi.org/10.1097/CJI.0b013e3181f3cbf4

136. Neuenhahn M, Albrecht J, Odendahl M, Schlott F, Dössinger G, Schiemann M, Lakshmipathi S, Martin K, Bunjes D, Harsdorf S et al (2017) Transfer of minimally manipulated CMV-specific T cells from stem cell or third-party donors to treat CMV infection after Allo-HSCT. Leukemia. https://doi.org/10.1038/leu.2017.16

137. O'Reilly RJ, Prockop S, Hasan AN, Koehne G, Doubrovina E (2016) Virus-specific T-cell banks for 'off the shelf' adoptive therapy of refractory infections. Bone Marrow Transplant. https://doi.org/10.1038/bmt.2016.17

Bioinformatics Tools for Epitope Prediction

4

Mohini Jaiswal, Shafaque Zahra, and Shailesh Kumar

Abstract

Immunological protection is conferred by immune cells, i.e., B and T cells, which can efficiently develop pathogen-specific memory and thus involved in adaptive immunity. More specifically, these immune cells can recognize a specific portion of their respective antigens termed as epitopes which possess their own significant values. There is a noble reason to identify the antigenic region of an antigen as it is having a great empirical cause, which includes exploration of disease etiology, the advancement of diagnosis assays, immune monitoring, and to design epitope-based vaccines. It requires detection and prediction of epitopes which is a considerable concern in the preparation of a peptide-based vaccine that is the centralized issue of immunoinformatics. Experimental screening is involved for large arrays of probable epitope candidates; thereby it is pricey and tedious. There is a requirement of more-advanced immunoinformatics tools as a prodigious amount of information has accumulated because of the onset of next-generation sequencing approaches for collection, analysis, and interpretation of data. Further, development of in silico epitope prediction methods has substantially reduced the difficulties related to epitope mapping by shortening potential epitope candidates list for experimental testing. These software tools have diverse applications in diagnosis of infectious diseases and allergies, understanding immune system function, vaccine designing, and prognosis of cancer. This chapter presents an outlook on how these tools are capable to predict epitopes of various antigens.

Keywords

B-cell and T-cell epitopes · Immunoinformatics · Immunological protection · Vaccine designing

M. Jaiswal · S. Zahra · S. Kumar (✉)
Data Sciences Laboratory, National Institute of Plant Genome Research (NIPGR),
New Delhi, India
e-mail: shailesh@nipgr.ac.in

© Springer Nature Singapore Pte Ltd. 2020
S. Singh (ed.), *Systems and Synthetic Immunology*,
https://doi.org/10.1007/978-981-15-3350-1_4

4.1   Introduction

The adaptive immune system is also termed as acquired immune system as it is acquired during the lifetime rather than the inherited one and is considered as a subsystem of the global immune system whose constituents are highly specialized systemic cells and processes that help out in elimination of pathogens as well as in their growth prevention. Due to the existence of acquired immunity, immunological memory creates an initial response for each specific pathogen which results in a strong anamnestic response at the time of subsequent exposure to that particular pathogen. Vaccination is based on this particular feature of acquired immunity. B and T cells are involved in adaptive immunity which is responsive for humoral- and cell-mediated immunity, respectively. They recognize a specific portion of protein residing on the surface of pathogen rather than pathogens as a whole and that protein is termed as an antigen. Distinct receptors residing on the surface of B and T cells designated as B-cell and T-cell receptors (BCR & TCR) consist of membrane-bound immunoglobulins helping in the recognition of the solvent-exposed antigens. There is a remarkable difference between perceptions by B and T cells [30]. Different functions are triggered from antibodies released by B cells upon binding with their respective antigens. As a result, toxins and pathogens get neutralized and labeled as for destruction [20].

Besides this, cell surface-residing T-cell receptor (TCR) presented by T cells assist recognition of antigen-presenting cells (APCs) displayed antigens bounded with major histocompatibility complex (MHC) molecules. MHC I and II molecules are involved in T-cell epitopes presentation. Co-receptor CD4 expressed by helper T cells assists in the perception of antigen in the context of MHC class II, while antigen displayed by MHC class I molecules is acknowledged by cytotoxic CD8+ T cells as per the immunological dogma. Subsequently, CD8 and CD4 T-cell epitopes exist. Meanwhile, CD4 T cells can act as a helper or regulatory T cells [20]. The immune response is amplified by helper T cells which are divided into three major subclasses that include Th1 involved in cell-mediated immunity against intracellular pathogens, Th2 involved in antibody-mediated immunity, and Th17 showing inflammatory response as well as defense across extracellular bacteria [37].

Along with the advancement in recombinant DNA technology, bioinformatics tools development and information of host immune response that acts as the genetic background of pathogen has led to the advancement of new vaccines which are more efficient, secure, and inexpensive in contrast to conventional vaccines. Conservation of chosen epitopes in a vaccine is a prerequisite event across distinct stages of pathogen and its variants. Intracellular antigen processing is required for cytotoxic T-cell-intervened response for which linear epitopes act as a prevailing target. In this respect, the binding affinity of selected epitopes should be with more than one major histocompatibility complex allele for a particular vaccine.

To identify B-cell and T-cell epitopes for vaccine designing is a decisive step as it requires to construct overlapping peptides based on experimental scanning result of epitope-active regions that span complete sequence of a protein antigen, and it is again a pricey and tedious job. Therefore, to elicit an immune response, in silico techniques are a perfect substitute to identify protein domains out of thousands of plausible candidates [29]. This chapter gives an insight regarding some of the commonly used bioinformatics tools developed for B-cell and T-cell epitope prediction.

4.2   Tools for B-Cell Epitopes Prediction

B-cell epitope anticipation tools aim to contribute to the detection of the specific antigenic peptide (epitope), and thus it has a significant purpose as it acts as a substitute of antigen for antibody production.

However, linear and conformational epitopes are the two groups based on B-cell epitopes classification. Sequential residues in primary sequence constitute a segment of linear epitope, whereas a cluster of antigen residues placed at a distance from each other in their primary sequence is regarded as conformational epitope that is brought to spatial vicinity because of polypeptide folding [1]. Thereby, linear and conformational B-cell epitopes are equally termed as continuous and discontinuous B-cell epitopes, respectively. This means that denatured antigens can be identified by antibodies which are used to identify linear B-cell epitopes, while in case of conformational B-cell epitopes, denaturation leads to recognizance failure. Unlike linear epitopes, conformational epitopes prediction depends on the three-dimensional structure of the protein. Linear B-cell epitopes are possessed by only a few of the native antigens; otherwise, approximately 90% of them are conformational [26].

4.2.1   Linear B-Cell Epitopes Anticipation

In spite of being a trivial one, linear B-cell epitopes can act as a substitute for immunization and antibody production. Thus, their anticipation received major attention. It has been predicted via methods based on a sequence from the primary sequence of antigens. Earlier computational methods were rooted on propensity scales of simplified amino acids featuring physicochemical characteristics for B-cell epitopes. For example, residue hydrophilicity calculations were implemented by Hopp and Wood to predict B-cell epitopes [11, 12] on the basis of the hypothesis that hydrophilic regions preferentially reside on the protein surface and are probably antigenic. For developing diverse prediction tools datasets, algorithms and training features used to differ.

Currently, accessible linear B-cell epitopes envision tools involve BcePred indulged in anticipation of linear B-cell epitopes as per their physicochemical attributes. Another one is Lbtope based on Immune Epitope Database (IEDB)-derived data of experimentally approved non-B-cell epitopes [39]. Analogous positive data of B-cell epitopes is required for training of artificial neural networks (ANNs) algorithm that has been implemented in Lbtope yet vary on negative data of non-B-cell epitopes.

Another one is BepiPred, which involves random forests algorithm-based training of B-cell epitopes derived from the three-dimensional architecture of antigen-antibody complexes. It is involved in the prediction of both varieties of B-cell epitopes [14]. On the whole, B-cell epitope prediction methods implementing machine learning algorithm outperformed other methods rooted on the basis of amino acid propencities.

4.2.2 Conformational B-Cell Epitopes Anticipation

It has been already mentioned that preferentially B-cell epitopes are conformational, even though linear B-cell epitopes anticipation is ahead of them, for that two major empirical approaches exist. Firstly, the requirement of conformational B-cell epitopes prediction is whole information of protein 3D structure which is available only for a few proteins [31]. The second one is the complicated task of discontinuous B-cell epitopes isolation from their corresponding protein frame to formulate a particular antibody. Its necessity is suitable scaffolds for epitope grafting. In spite of these difficulties, various mechanisms exist to envisage conformational B-cell epitopes.

One of them is CBTOPE which relies on Support Vector Machine (SVM) algorithm. Physicochemical characteristics and sequence-derived attributes are utilized for training of conformational B-cell epitopes, and a benchmark dataset of conformational epitopes derived from 3D structures of antibody-protein complexes is used for their assessment along with 86.59% accuracy from cross-validation experiments [1]. This tool is involved in predicting discontinuous B-cell epitope of an antigen based on its primary sequence by overcoming the first difficulty.

Another one is ElliPro that depends on the geometrical properties of protein structure. In addition to CBTOPE, ElliPro also assessed on the same benchmark dataset derivative of 3D structures of antibody-protein complexes [24].

There is a significant role of bioinformatics tools for each of the B-cell epitopes envision in peptide-based vaccine designing and disease identification [9, 22].

Although there are various tools for each of the B-cell epitope prediction, the five most commonly highly utilized tools are described in Table 4.1.

Table 4.1 Some freely accessible B-cell epitope anticipation tools

B-cell types	Tools	Method	Server (URL)	References
Continuous	BcePred	Physicochemical properties	http://www.imtech.res.in/raghava/bcepred/	[28]
	Lbtope	ML (ANN)	http://www.imtech.res.in/raghava/lbtope/	[35]
Discontinuous	ElliPro	Structure-based method (geometrical properties)	http://tools.iedb.org/ellipro/	[24]
	CBTOPE	Sequence based (SVM)	http://www.imtech.res.in/raghava/cbtope/submit.php	[1]
Both	BepiPred-2.0	ML (DT)	http://www.cbs.dtu.dk/services/BepiPred/	[14]

4.2.3   Description of Various Tools and Their Overall Performance Enlisted in Table 4.1

4.2.3.1 BcePred Server

BcePred server assists in envision of linear B-cell epitope rooted on physicochemical characteristics of amino acids. These properties comprised of mobility, turns, flexibility, exposed surface, accessibility, hydrophilicity, polarity, and antigenicity of any particular antigen. To quantify these properties, attributes value is allocated to all of the 20 natural amino acids. The user can opt for any combination of physicochemical attributes for epitopes prediction.

PERL version 5.03 is used for writing a common gateway interface (CGI) script. Sun Server (420E) with a UNIX (Solaris 7) environment is used for their installation.

Submission Form Using the Following Steps for BcePred Server

- Input data is in the form of sequence that should be written in submission form by using one-letter amino acid code: "acdefghiklm-npqrstvwy" or "ACDEFGHIKLMNPQRSTVWY." Other letters get transformed into "X" which were reviewed as obscure amino acids.
- Threshold values lie in the range of −3 to +3. As per the outstanding sensitivity and specificity value gained, default thresholds for various parameters have been opted.
- After pressing "Submit sequence" button, a WWW page will return as a result that delivers summarized information about entered query sequence in graphical (Fig. 4.1a) as well as in tabular and in overlap display format (Fig. 4.1b). The tabular format provides a normalized score of opted attributes with the respective amino acid residue of a protein as well as minimum, maximum, and average values of integrated methods opted.
- Quick picturing of B-cell epitope on protein is achieved when residue properties are plotted along protein backbone. A particular amino acid residue will be reviewed as expected B-cell epitope when their peak is having value above threshold (default value is 2.38 in the combined approach).

Pros and Cons

- By using BcePred server, prediction of B-cell epitopes can be made based on two or more physicochemical properties at a time. So it would be more accurate.
- However, there is no autonomous assessment or benchmarking of prevailing procedures in this server; thereby, the decision of much better residue property or method is a difficult task.

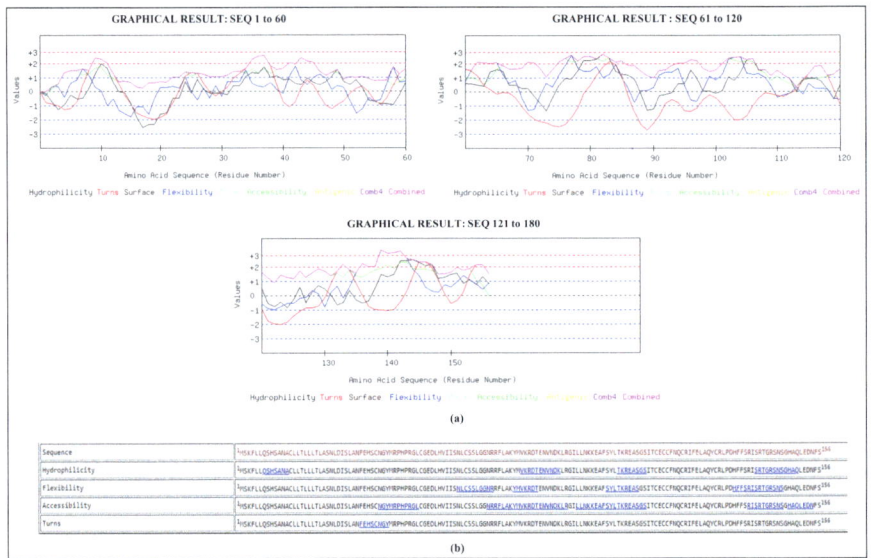

Fig. 4.1 BcePred server showing B-cell epitope regions in insulin precursor sequence (length is 156 aa) of *Aplysia californica*. (**a**) Graphical result. (**b**) Overlap display in which selected programs are hydrophilicity, flexibility, accessibility, and turns having threshold value as 1.9, 2.0, 1.9, and 2.4, respectively. Predicted B-cell epitopes are shown in blue color and are underlined

4.2.3.2 Lbtope

Lbtope is a tool designed to predict linear B-cell epitope. PHP 5.2.9, HTML, and JavaScript have been used to develop its front end. Further, Red Hat Enterprise Linux 6 server environment has been utilized for its installation. Along with experimentally certified B-cell epitopes, non-B-cell epitopes can be also retrieved from Immune Epitope Database (IEDB) which include five datasets termed as Lbtope_Fixed, Lbtope_Fixed_non_redundant, Lbtope_Variable, Lbtope_Variable_non_redundant, and Lbtope_Confirm dataset. Various models have been developed based on these datasets to discriminate B-cell epitopes from non-epitopes.

In Lbtope, SVM[light] package is used for implementing SVM technique in association with Weka implemented Ibk.

Working Steps

I. Input data is the primary amino acid sequences in fasta format (Fig. 4.2a).
II. Overlapping peptides containing 20 amino acids and 5–30 amino acids are developed for Lbtope fixed dataset model and for variable datasets, respectively, for prediction of linear epitopes. Due to the very high specificity, nonredundant model is introduced as well.
III. Antigen sequences profiled with B-cell epitopes having probability scale of 20–80% comes as an output data (Fig. 4.2a).
IV. A higher score is meant for a higher possibility of a peptide to behave as B-cell epitope.

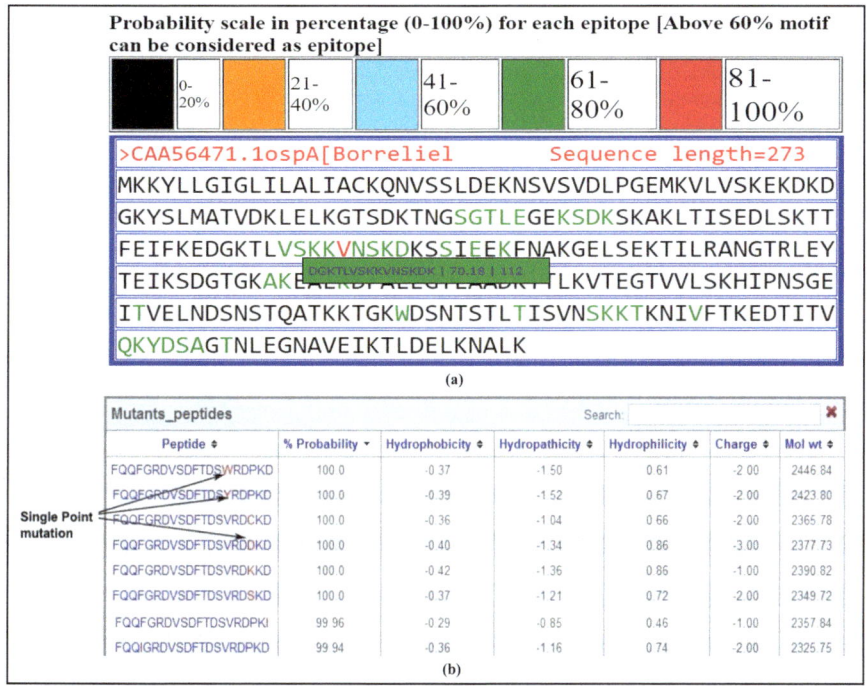

Fig. 4.2 (**a**) Sequence of OspA from *Borrelia burgdorferi* taken as input showing highlighted text as the predicted B-cell epitope along with probability scale. (**b**) Output data from peptide submission and mutant generation

Pros and Cons

- In addition to B-cell epitope prediction, this server exhibits a peptide mutation tool. It helps to create all plausible single-point mutations of a given peptide (Fig. 4.2b) and to predict its other properties. The further probability score is calculated based on a particular algorithm. Thereby, mutation tool is useful in the creation of peptide mutants and examination of its epitopic and other desired probability as well.
- Model based on Lbtope_Confirm dataset executed in an improved way as a comparison to mock-up established on Lbtope_Variable dataset. However, these model's activity decreased on nonredundant datasets.

4.2.3.3 ElliPro

ElliPro is a Web server obtained from Ellipsoid and Protrusion, that executes a modified version of Thornton's method according to which identification of continuous epitopes from protruding regions of protein globular surface becomes possible [38]. In addition to a residue clustering algorithm, the MODELLER program [8] and a Jmol viewer (Fig. 4.3b) are implemented in ElliPro as well. Due to this

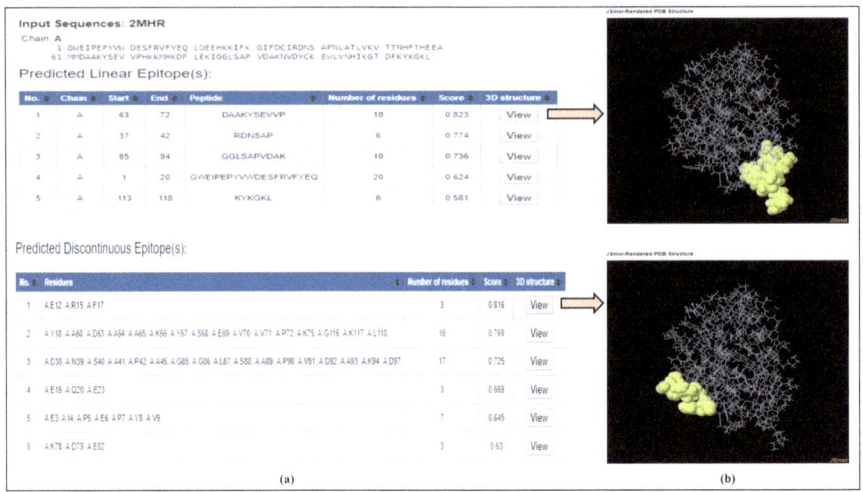

Fig. 4.3 (**a**) ElliPro prediction result for myohemerythin as an input sequence having sequence ID as 2MHR. (**b**) Epitope 3D structures for 2MHR via Jmol viewer program

implementation, envision of antibody epitopes as well as its visualization becomes possible in protein sequences as well as in structures. From 3D structures of antibody-protein complexes, a benchmark dataset of epitopes has been derived which is used to train ElliPro having the Area Under the ROC Curve (AUC) value as 0.732 [23].

Three algorithms are introduced in ElliPro to perform some major objectives that include an understanding of protein shape as an ellipsoid, estimation of residue protrusion index (PI), and grouping of neighboring residues as per their PI values.

Working Steps

I. Input data is either a protein structure or its primary amino acid sequence.
II. The sequence in fasta format or single-letter codes or their SwissProt/UniProt ID can be entered as a query in case the only sequence is available. To design a 3D structure of the submitted sequence, the selection of both a threshold for BLAST e-value and structural templates from PDB are required.
III. In case of structure, either a four-character PDB ID is entered in required space or a PDB file in PDB format can be uploaded (Fig. 4.3a). If submitted framework possesses more than one protein chain, then a specific chain has to be selected by the user on which calculation would be based.
IV. Threshold values are changeable based on parameters utilized by server to predict epitope, like minimum residue score (protrusion index), referred as S, that ranges in between 0.5 and 1.0 and maximum distance, termed as R, that ranges from 4 to 8 Å.

Pros and Cons

- ElliPro proves to be a helpful server for recognition of antibody epitopes from protein antigens and is helpful in identifying protein-protein interactions.
- A procedure that relies on geometrical attributes of protein structure has been introduced in this server which doesn't require training as well, so it is unable to properly differentiate between epitopes and non-epitopes.

4.2.3.4 CBTOPE

CBTOPE is a user-friendly Web server. It is established to anticipate conformational B-cell epitopes from antigen's amino acid sequence rather than based on their tertiary structure. A CGI script is written in Perl and HTML. Sun Server (420E) is used for installation under UNIX (Solaris 7) environment [1]. Development of this server is evident for envisioning of antigen's conformational B-cell epitope in which their primary amino acid sequences play a possible role.

Methodology

(a) For prediction via CBTOPE, main dataset is created by obtaining 526 antigenic sequences in combination with IEDB database as well as benchmark dataset [23] which is comprised of 161 protein chains derived from 144 antigen-antibody complex structures.
(b) Sequence redundancy is excluded by using program CD-HIT [16] at 40% cutoff.
(c) Finally, a nonredundant set of 187 antigens is gained. This set is devoid of sequences with the sequence identity of more than 40%.
(d) A different pattern is created. Standard procedure for assigning patterns is that if there would be any interaction between central residues and antibody, a positive value is assigned otherwise defined as negative (Fig. 4.4).
(e) By using patterns like the binary profile of pattern (BPP) and physiochemical profile of patterns (PPP), several models have been developed by using SVM as a classifier. It gained a maximal value of MCC as 0.22 and 0.17, respectively.
(f) Conventional characteristics of binary and physicochemical profiles are used and further assessed via fivefold cross-validation.

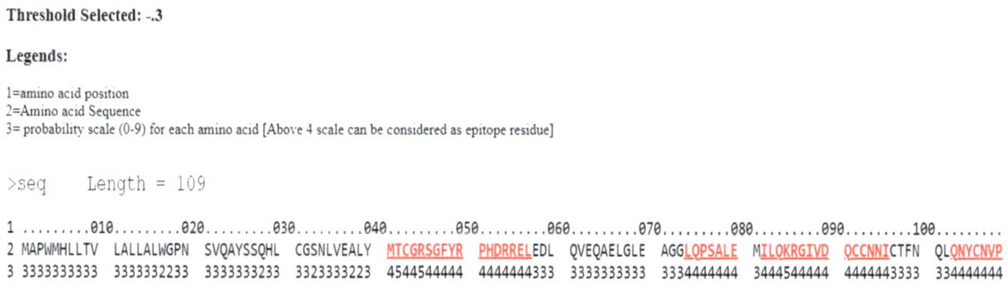

```
Threshold Selected: -.3

Legends:

1=amino acid position
2=Amino acid Sequence
3= probability scale (0-9) for each amino acid [Above 4 scale can be considered as epitope residue]

>seq    Length = 109

1 .........010.........020.........030.........040.........050.........060.........070.........080.........090.........100.........
2 MAPWMHLLTV LALLALWGPN SVQAYSSQHL CGSNLVEALY MTCGRSGFYR PHDRRELEDL QVEQAELGLE AGGLOPSALE MILOKRGIVD OCCNNICTFN QLONYCNVP
3 3333333333 3333332233 3333333233 3323333223 4544544444 4444444333 3333333333 3334444444 3444544444 4444443333 334444444
```

Fig. 4.4 CBTOPE prediction result for insulin sequence of *Octodon degus* as an input. Predicted B-cell epitope is shown in red color

(g) The number of non-redundant protein chains is 187 comprising of 2261 antibody-interacting B-cell epitope residues that are used for training and assessment of all SVM models.

Working Steps

I.   Input data is amino acid sequences in fasta format.
II. Total of 19 window patterns for each of the submitted sequences is created via server. The further amino acid composition is calculated to predict residues interacting with the antibody.
III. Amino acid sequence mapped with probability scale that ranges in between zero and nine comes as an output data for all amino acids where zero signifies the unusual possibility of residue to be a part of B-cell epitope and nine is the most plausible one (Fig. 4.4).
IV. For extraordinary precision (high-confidence) prediction, higher threshold value should be selected as per suggestion along with compromising the sensitivity of prediction. Nonetheless, lower threshold value should opt for maximum prediction of antibody-interacting residues.
V.   The default threshold value is fixed at −0.3 as sensitivity and specificity are found to be equivalent at this value during CBTOPE development.

Pros and Cons

•   Structure determination of a protein via techniques like X-ray crystallography proves to be costly, prolix, and time-consuming. Due to development of CBTOPE, one can predict conformational B-cell epitopes of antigens with ease which is lacking their tertiary structures with better sensitivity and AUC than other structure-based methods on same benchmark dataset as CPP composition-based SVM model is used in this server which outperformed others.
•   Limitation of CBTOPE is its ineptitude for determination of number and distance required to obtain an epitope segment from antigen sequence.

4.2.3.5 BepiPred-2.0

BepiPred-2.0 is a Web server based on random forest algorithm for estimation of B-cell epitope, and annotated epitopes extracted from a dataset are used for its training which is composed of 649 antigen-antibody crystal structures and is derived from Protein Data Bank (PDB). Antibody molecules of each complex are recognized via HMM models.

Methodology

(a) Random Forest Regression (RF) algorithm is assessed on a dataset to determine the plausibility of a given antigen residue so that it can be a part of an epitope with the usage of the fivefold cross-validation strategy.

(b) All of the residues is encrypted with the help of its polarity, hydrophobicity, computed volume along with secondary structure (SS), and relative surface accessibility (RSA) as anticipated by NetSurfP [21].

(c) The overall volume of antigen is gained via the addition of respective volumes of entire antigen's residues for almost 46 variables.

(d) Rolling average of window 9 is implemented on RF output to acquire concluding BepiPred-2.0 predictions.

Working Steps

  I. Input data is protein sequences of interest having size more than 10 amino acids and lesser than 6000 in fasta format that can be entered into textbox either by pasting them or via uploading as a single file.

 II. When predictions get completed, the user is automatically redirected to output page (Fig. 4.5) that has a navigation bar containing distinct tabs like "Summary" showing the result of each of the individual sequence in horizontal as well as in the form of a vertical table. Optionally, an email address can be given by the user so that after the job gets finished, result page link will be emailed.

III. "E" in "Epitopes" line is indicated as predictions higher than the user-defined threshold which is by default 0.5 above itself the protein sequence and is used to select the background color for protein sequences. Epitope classifications are alterable as per desire with the usage of "Epitope Threshold" slider.

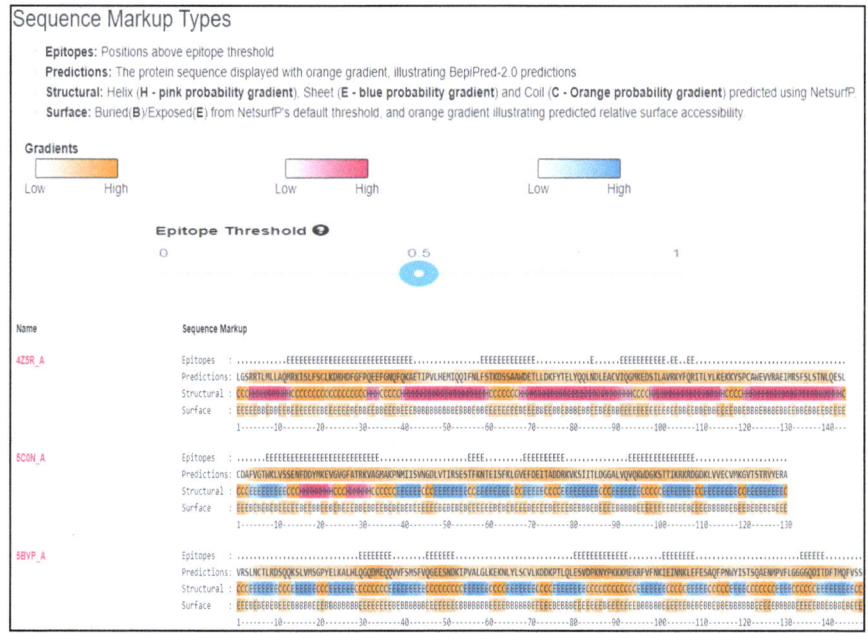

Fig. 4.5 Sequence markup table of epitope predictions for three antigenic sequences to visualize the predictions on sequences in advanced output mode

IV. Predictions result are downloadable as JSON or CSV format via dropdown tab "Downloads." Besides this, by clicking the "All Downloads" tab, a short descriptive file can be found as well.

Pros and Cons

- BepiPred-2.0 attains a considerably better positive predictive value (PPV) and a moderately better true positive rate (TPR) in comparison to other methods. Also, it outperforms other available tools like BepiPred-1.0 and Lbtope for sequence-based epitope prediction relies on dataset retrieved from solved 3D structures or of a large collection of linear epitopes downloadable from IEDB database.
- The result format is informative as well as convenient.
- Limitation of BepiPred-2.0 is that it doesn't respond to nucleic acid sequences.

4.3 Tools for T-Cell Epitopes Prediction

Recognition of shortest peptides within an antigen is the main objective of T-cell epitope prediction which possesses immunogenicity, meaning capable to incite either CD4 or CD8 T cells. Immunogenicity is mainly based on three essential events which include processing of antigen and its binding with MHC molecules and acceptance from its respective TCR.

Amid all steps, MHC-peptide binding is the most discerning to delineate T-cell epitopes [13, 15]. Subsequently, the peptide-MHC binding prediction is the substantive baseline for prediction of T-cell epitopes.

4.3.1 Peptide-MHC Binding Anticipation

For peptide-MHC binding prediction, there should be an overview of already known peptide sequences that adhere with MHC molecules such as the existence of specified epitope databases, for instance, antigen [32], EPIMHC [18], and IEDB [39].

At the level of 3D structures of groove-resided bound peptides, resemblance exists between MHC I and II molecules, even though there is a major distinction between their binding grooves. For MHC I molecules, its peptide binding cleft consists of a single α chain; thereby, it is closed due to which their binding peptide length is reduced to 9 to 11 amino acid residues whose N- and C-terminal ends continue to stick by means of a linkage of hydrogen bonds with preserved residues of MHC I molecules [17, 36]. Tight physicochemical preferences also exist in addition to deep binding pockets in their peptide-binding groove that assist binding predictions. Alternative binding pockets exist for the same MHC I molecule which is often used by peptides of distinct sizes. Hence, there is a requirement of a fixed peptide length for the prophecy of MHC I-binding peptides. As mostly ligands have 9–11 residues, it can be the desired length.

On the contrary, open peptide-binding cleft is found in MHC II molecules, that allows expansion of peptide's N- and C-terminal ends beyond its binding groove [17, 36] which results in diversification of their peptide-binding length (9–22 residues). However, peptide-binding cleft allows to reside merely a core of nine residues, termed as peptide-binding core, into them. Consequently, the target of peptide-MHC II binding anticipation tools is to recognize peptide-binding cores mainly. The reason behind this imprecise forecasting of peptides that bind with MHC II molecule is their shallower and less demanding binding pockets than that of MHC I molecules [30].

Apart from this, peptide antigens derived from endogenous and exogenous pathway are offered by MHC I and MHC II molecules, respectively. Endosomal compartments are used for degradation and loading endocytosed antigens onto MHC II molecule [7], while antigens degraded via cytosolic pathway are transported via TAP to the endoplasmic reticulum and further loaded onto MHC I molecules. Before loading, peptides mostly go for trimming with the aid of ERAAP N-terminal aminopeptidases [10].

Along with MHC I and II-peptide binding anticipation tools, various tools are there to envisage even TAP binding that has been designed by training distinct algorithms on peptides having a significant affinity with TAP [3].

Consistently occurring amino acids are present in peptides at particular positions that bind with MHC molecules, termed as anchor residues thought to be liable for its binding with MHC molecule. However, later, it has been shown that along with anchor residues, peptide binding to a given MHC molecule is facilitated by non-anchor residues as well [27]. Accordingly, development of motif matrices (MM) helps in the assessment of input for each and all peptide positions of MHC molecule binding [19, 25].

Several ML algorithm has been used to solve mainly two distinct problems which are trained on datasets having peptides of known kinship to MHC molecules. First and foremost is the discernment of MHC binders from non-binders, and the second one is to envisage peptides binding affinity with MHC molecules.

MHC polymorphism is the major challenge in T-cell epitopes prediction. Human leukocyte antigen (HLA) is a term for MHC molecules in case of humans, and hundreds of their allelic variants exist which bind to peptide variants that need distinctive models to predict peptide-MHC binding. These variants are expressed at immensely diverse frequencies due to which HLA polymorphism creates hindrance in the advancement of T-cell epitope-based vaccines for distinct ethnic groups. In spite of all obstruction, there are various tools accessible for prediction of peptide-MHC binding. Some of them are described in Table 4.2.

Table 4.2 Some freely accessible T-cell epitope anticipation tools

MHC class	Tools	Method	Server (URL)	References
MHC I	nHLAPred	ANN	http://www.imtech.res.in/raghava/nhlapred/	[4]
	ProPred1	QAM	http://www.imtech.res.in/raghava/propred1/	[34]
	TAPPred	SVM	http://www.imtech.res.in/raghava/tappred	[6]
MHC II	ProPred	QAM	http://www.imtech.res.in/raghava/propred/	[33]
	EpiDOCK	SB	http://epidock.ddg-pharmfac.net	[2]

4.3.2 Description of Various Tools for T-Cell Epitope Prediction Enlisted in Table 4.2

4.3.2.1 nHLAPred

nHLAPred is a hybrid approach-based Web server which includes, firstly, a quantitative matrix (QM)-rooted technique in which involvement of each residue has been taken into consideration rather than just anchor residues and is formulated for 47 MHC class I alleles for which minimal 15 binders are accessible from MHCBN version 1.1 [5]. Secondly, an artificial neural network (ANN)-based method is implemented for 30 alleles out of 47 MHC alleles featuring at least 40 binders approachable from the database. Mutual approach (ANN and QM) has been used for the anticipation of 30 MHC alleles (Fig. 4.6), while the prediction of the remaining 37 alleles relies on QM [4]. The average accuracy of prediction is 92.8% that has ameliorated by 6% compared to each individual means with the development of this amalgam approach.

Sun Server 420R is used for installation under the Solaris environment. There is a partitioning of server in two substantial parts, ComPred and ANNPred, amid which ComPred enables for estimation of binders for 67 MHC class I alleles. Along with that, proteasomal matrices have been utilized by both parts to anticipate proteasomal cleavage site possessing MHC binders at C-terminal.

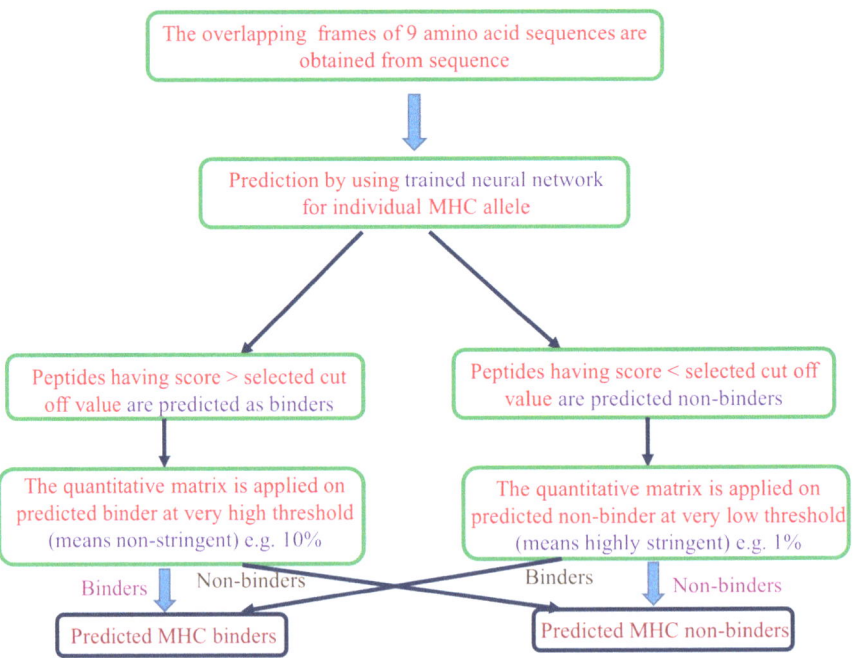

Fig. 4.6 Diagrammatic representation of combining ANNs and QM

Working Steps

  I. ReadSeq developed by Dr. Don Gilbert has been implemented in the server, so input data can be the protein sequence query of any standard format.

  II. For 47 MHC class I alleles, quantitative matrices are developed that are further assessed via jackknife validation test.

  III. For each amino acid from point one to nine, coefficient value has been calculated via allocating the possibility of an amino acid at an exact point in binders as well as in non-binders.

  IV. For prophecy of proteasomal cleavage sites which befall at the midpoint of 12mer peptides mainly six amino acids away from N-terminal, proteasomal and immunoproteasomal matrices are acquired from ProPred I server [34].

Pros and Cons

- The server is user-friendly, and its outcome demonstration format (HTML-II) is helpful in tracing promiscuous MHC-binding regions as of antigenic sequence with fair accuracy.
- However, certain limitations are also there like the incapability to handle non-linearity in data because of significant confinement of quantitative matrix-based method. Also, the ANN-based method requires a large dataset for training.
- Proteasome cleavage site prediction procedures are less authentic due to extensive specificity of the proteasome in comparison of MHC-peptide binding specificity. Proteasome digested data are present in limited amount as well. Moreover, cleavage specificity depends on cleavage site-residing residues as well as on neighboring residues equally.

4.3.2.2 ProPred1

ProPred1 is an online matrix-based Web server in order to predict peptide binding to 47 MHC class I alleles. Matrices implemented have been acquired from BIMAS server as well as from literature. Results are in a user-friendly format that helps out users to identify promiscuous MHC binders in an antigen sequence.

The server enables users to predict MHC binders in an antigenic sequence along with their usual proteasome and immunoproteasome cleavage sites at C terminus simultaneously which results in identifying T-cell epitope with high potency.

PERL is used for writing a common gateway interface (CGI) script and is launched via Apache Web server. Further, Sun Server (420E) with a UNIX (Solaris 7) environment is used for installation.

Working Steps

  I. Input data is the primary amino acid sequence of protein query in any frequently used sequence formats as the server uses ReadSeq to analyze input sequence (Fig. 4.7a).

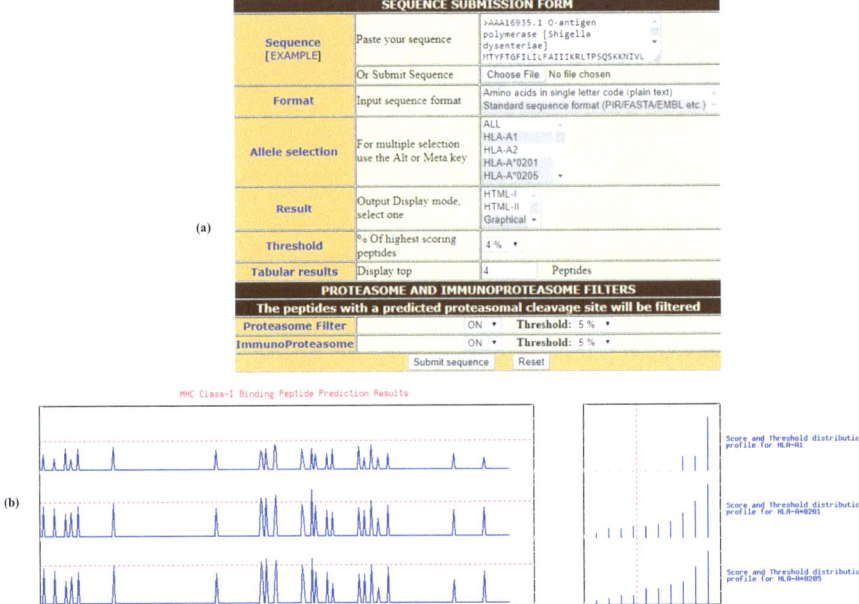

Fig. 4.7 (**a**) Sequence submission form of ProPred1 server showing protein sequence of O-antigen polymerase of *Shigella dysenteriae* as an input. (**b**) Prediction result in graphical format

II. There is an independency to select a threshold value for prediction.

III. Representation of output data in graphical (Fig. 4.7b) or text form provides assistance to the user in appropriate recognition of promiscuous MHC-binding domains in their query sequence.

IV. Firstly, for a given antigen sequence, all probable overlying 9mer peptides are produced followed by a quantitative matrix-based score calculation of selected MHC alleles. A peptide is designated as predicted binder if their score would be superior to a particular threshold value (e.g., at 4%) for selected MHC allele.

V. In an effort to forecast proteasome cleavage sites in an antigenic sequence, overlying 12mer peptides were developed for sequence followed by their score calculation with the usage of weight matrix of the proteasome.

VI. Further peptides having score superior to a certain threshold value (e.g., at 5%) are deemed as peptides featuring proteasome cleavage site at their midpoint positions (6-position left and 6-position right) as per prediction.

VII. Prediction of the immunoproteasome cleavage site of peptides shares analogy with proteasome cleavage site prediction.

VIII. Concurrent anticipation of MHC binders and proteasome cleavage sites results in removal of MHC binders not retaining cleavage site at C terminus.

Pros and Cons

- Purpose of ProPred1 development is to efficaciously attenuate wet lab experiments number indulged in to identify effective T-cell epitopes and thereby develop relevant vaccines.
- However, due to lack of sufficient data for MHC non-binders, calculation of threshold value is little bit crucial.

4.3.2.3 TAPPred

TAPPred is a user-friendly, support vector machine (SVM)-based Web server designed to predict TAP-binding affinity as well as translocation efficiency of the peptide. The server is initiated via public domain software package Apache on Sun server 420R in Solaris background. HTML is used for writing all the Web pages, while PERL and JavaScript are used for inscription of CGI scripts. By utilizing freely downloadable software, SVM^{light}, SVM has been implemented.

Working Steps

I. Input data is protein sequence as a single-letter amino acid code whose minimum length should be nine that is uploaded as a local sequence file or is pasted in required space, in any of the standard formats because of integration of ReadSeq.

II. Before running prediction sequence, uploaded format must be chosen by the user that it is in either plain or formatted form as server acknowledges both formatted and unformatted raw antigenic sequences which results in erroneous prediction if the selected format is false.

III. Prediction of binding affinity of the peptide has given permission by the server on the basis of two variants of SVM. Simple SVM involves prediction relied on sequential knowledge of peptides and is quicker than cascade SVM which includes characteristics of amino acids along with its sequential knowledge.

IV. Two tiers exist for prediction. Initially via joining characteristics of amino acids with sequential information, preliminary results are gained. Later on, the results of the first tier are further filtered. Despite having a slower rate of prediction, cascade SVM is more trustworthy as compared to simple SVM. Only a single approach can be selected for prediction at a time.

V. Results are depicted in two user-friendly formats. In the first format, the result is presented by coloring the residues. N-terminal is demarcated by the green color background of residues. Rest of the residues are represented with the violet-blue background (Fig. 4.8a).

VI. Type of peptides can be chosen to be displayed in the result.

VII. Tabular format display (Fig. 4.8b) has four alternatives. Only one output display can be selected at a time.

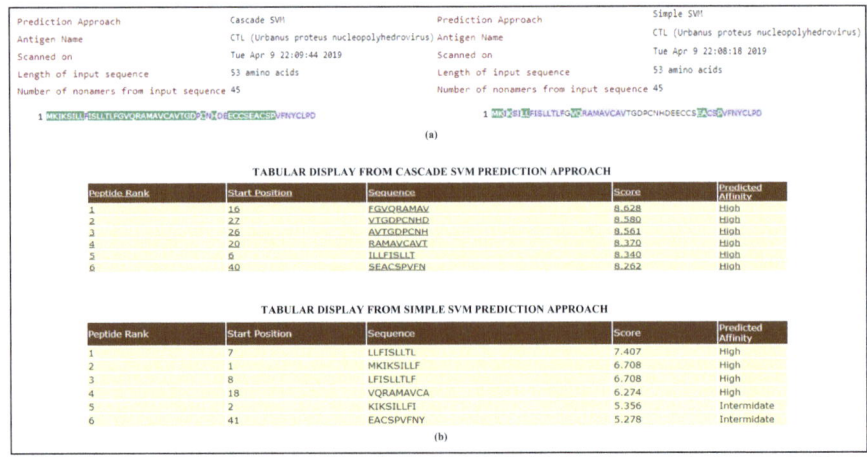

Fig. 4.8 Prediction results from TAPPred server for CTL as an input sequence. (**a**) Displaying result in the form of colors. (**b**) Tabular display format

VIII. Only one output display can be selected by the user at a time that includes primarily a header and has data about the length of the peptide sequence, about nonamers obtained, as well as the date of prediction.

Pros and Cons

- The user can select parameters of their choice in this server.
- However, due to insufficient data for TAP-binding peptides, limited algorithms are there. Also, the minimum length of the query sequence should be nine; otherwise, it won't be accepted for prediction.

4.3.2.4 ProPred

ProPred is a graphics-based Web server in which matrix-based prediction algorithm has been deployed along with the implementation of amino acid or position coefficient table inferred from literature in order to foretell binding domain for MHC class II in antigenic sequences. Either as peaks in graphical interface or as colored residues in HTML interface, predicted binders can be envisioned. It has been developed mainly for 51 HLA-DR alleles whose matrices have been extracted from a pocket profile database defined by Sturniolo et al. in 1999 [33].

Working Steps

I. Input data is protein sequences in fasta or PIR format which are generally used as standard sequence formats and can be uploaded as a file.
II. In order to attain desirable results, selection of alleles, threshold, and other parameters are customizable.

III. An output as text or graphics is generated from the analysis of sequence data in which two choices have been provided by text display: the first choice in which binding regions of antigenic sequences are displayed by different colors thus providing easier detection. An option of representing binding score in a commonly used tabular format is also there that has been calculated from the matrix.

IV. The second choice involves the representation of coinciding regions independently on discrete lines; thus, delineation of specific regions from display becomes easier.

V. GDPlot library established by Lincoln D. Stein is used for graphics formulation in GIF format. HLA-DR-binding tendency laterally with the primary structure of a protein is represented as an output along with their binding strength. Consequently, it has an advantage over text presentation.

VI. Besides this, an alternative method is there for plotting threshold versus binding peptides, i.e., threshold profile, which renders assistance in the selection of a reasonable threshold value for finding promiscuous binders.

Pros and Cons

• All HLA-DR alleles are evaluated by server independently, and output is posted on a single screen that helps out the user in rapid visualization of promiscuous binders. Henceforth, it can be considered as a useful tool.

• Binding strength for all peptide frames in an opted subsequence can be computed by this server.

However, it is less expressive in representing overlapping binding regions.

4.3.2.5 EpiDOCK

EpiDOCK is the first structure-based server for prediction of peptide binding to 23 utmost common human MHC class II proteins which include 5 HLA-DP, 6 HLA-DQ, and 12 HLA-DR proteins. These alleles are the composition of more than 95% of the human population. The server is implicated to identify 90% of true binders as well as 76% of true non-binders, with a global precision of 83%.

Working Steps

I. Input data is protein sequence in fasta format. Multi-fasta protein format is likely reinforced.

II. Selection of HLA class II protein of concern is the next step that can be a single protein or all proteins.

III. Peptide-binding core is composed of nine adjacent residues due to which a collection of overlapping nonamers is formed as a result of input sequence conversion. A docking score-based quantitative matrix (DS-QM) is used for assessment of all nonamers retrieved for certain HLA class II protein and allotted a specific score.

IV. For any DS-QM, thresholds are defined with utmost certainty. Peptides having higher scores than the threshold or equal to them are expected to be binders, else considered as non-binders.

V. After that, if prophesied nonamer binder is a portion of recognized binder sequence, only then it will be categorized as an accurately foretold binder, else referred to as a false binder. Data is reported either in xls or csv formats.

VI. To validate anticipations, a test set of 7050 identified binders to HLA-DR, HLA-DQ, and HLA-DP proteins is implicated that originates from 1195 proteins, which is collected from Immune Epitope Database.

VII. Assigned values for specificity, sensitivity, accuracy, and AUC are 0.759, 0.903, 0.831, and 0.892, respectively.

Pros and Cons

- Structure-based approaches require information about peptide-MHC protein complex centered on their X-ray structure only rather than extensive preexisting experimental data.
- It is authentic and credible.
- Because of high resource implications of experimental testing at the time of scanning large proteome, a number of false positives can be more in contrast to a large number of false negatives which is a major problem to be dealt with.
- Amino acids having negative coefficients decrease the affinity of peptides for HLA-DRB1.

References

1. Ansari HR, Raghava GP (2010) Identification of conformational B-cell epitopes in an antigen from its primary sequence. Immunome Res 6:6. https://doi.org/10.1186/1745-7580-6-6
2. Atanasova M, Patronov A, Dimitrov I et al (2013) EpiDOCK: a molecular docking-based tool for MHC class II binding prediction. Protein Eng Des Sel 26:631–634. https://doi.org/10.1093/protein/gzt018
3. Bhasin M, Raghava GPS (2004) Analysis and prediction of affinity of TAP binding peptides using cascade SVM. Protein Sci 13:596–607. https://doi.org/10.1110/ps.03373104
4. Bhasin M, Raghava GPS (2007) A hybrid approach for predicting promiscuous MHC class I restricted T cell epitopes. J Biosci 32:31–42. https://doi.org/10.1007/s12038-007-0004-5
5. Bhasin M, Singh H, Raghava GPS (2003) MHCBN: a comprehensive database of MHC binding and non-binding peptides. Bioinformatics (Oxford, UK) 19:665–666
6. Bhasin M, Lata S, Raghava GPS (2007) TAPPred prediction of TAP-binding peptides in antigens. Methods in molecular biology (Clifton, NJ):381–386
7. Blum JS, Wearsch PA, Cresswell P (2013) Pathways of antigen processing. Annu Rev Immunol 31:443–473. https://doi.org/10.1146/annurev-immunol-032712-095910
8. Eswar N, Webb B, Marti-Renom MA et al (2006) Comparative protein structure modeling using MODELLER. Curr Protoc Bioinformatics 15:5.6.1–5.6.30. https://doi.org/10.1002/0471250953.bi0506s15

9. Gershoni JM, Roitburd-Berman A, Siman-Tov DD et al (2007) Epitope mapping. BioDrugs 21:145–156. https://doi.org/10.2165/00063030-200721030-00002
10. Hammer GE, Gonzalez F, Champsaur M et al (2006) The aminopeptidase ERAAP shapes the peptide repertoire displayed by major histocompatibility complex class I molecules. Nat Immunol 7:103–112. https://doi.org/10.1038/ni1286
11. Hopp TP, Woods KR (1981) Prediction of protein antigenic determinants from amino acid sequences. Proc Natl Acad Sci U S A 78:3824–3828
12. Hopp TP, Woods KR (1983) A computer program for predicting protein antigenic determinants. Mol Immunol 20:483–489. https://doi.org/10.1016/0161-5890(83)90029-9
13. Jensen PE (2007) Recent advances in antigen processing and presentation. Nat Immunol 8:1041–1048. https://doi.org/10.1038/ni1516
14. Jespersen MC, Peters B, Nielsen M, Marcatili P (2017) BepiPred-2.0: improving sequence-based B-cell epitope prediction using conformational epitopes. Nucleic Acids Res 45:W24–W29. https://doi.org/10.1093/nar/gkx346
15. Lafuente E, Reche P (2009) Prediction of MHC-peptide binding: a systematic and comprehensive overview. Curr Pharm Des 15:3209–3220. https://doi.org/10.2174/138161209789105162
16. Li W, Godzik A (2006) Cd-hit: a fast program for clustering and comparing large sets of protein or nucleotide sequences. Bioinformatics 22:1658–1659. https://doi.org/10.1093/bioinformatics/btl158
17. Madden DR (1995) The three-dimensional structure of peptide-MHC complexes. Annu Rev Immunol 13:587–622. https://doi.org/10.1146/annurev.iy.13.040195.003103
18. Molero-Abraham M, Lafuente EM, Reche P (2014) Customized predictions of peptide–MHC binding and T-cell epitopes using EPIMHC. Methods in molecular biology (Clifton, NJ):319–332
19. Nielsen M, Lundegaard C, Worning P et al (2004) Improved prediction of MHC class I and class II epitopes using a novel Gibbs sampling approach. Bioinformatics 20:1388–1397. https://doi.org/10.1093/bioinformatics/bth100
20. Paul WE (2013) Fundamental immunology. Wolters Kluwer Health/Lippincott Williams & Wilkins, Philadelphia
21. Petersen B, Petersen T, Andersen P et al (2009) A generic method for assignment of reliability scores applied to solvent accessibility predictions. BMC Struct Biol 9:51. https://doi.org/10.1186/1472-6807-9-51
22. Pomés A (2010) Relevant B cell epitopes in allergic disease. Int Arch Allergy Immunol 152:1–11. https://doi.org/10.1159/000260078
23. Ponomarenko JV, Bourne PE (2007) Antibody-protein interactions: benchmark datasets and prediction tools evaluation. BMC Struct Biol 7:64. https://doi.org/10.1186/1472-6807-7-64
24. Ponomarenko J, Bui H-H, Li W et al (2008) ElliPro: a new structure-based tool for the prediction of antibody epitopes. BMC Bioinformatics 9:514. https://doi.org/10.1186/1471-2105-9-514
25. Reche PA, Reinherz EL (2007) Definition of MHC supertypes through clustering of MHC peptide-binding repertoires. Methods in molecular biology (Clifton, NJ):163–173
26. Regenmortel MHV (2009) What is a B-cell epitope? Methods in molecular biology (Clifton, NJ):3–20
27. Ruppert J, Sidney J, Celis E et al (1993) Prominent role of secondary anchor residues in peptide binding to HLA-A2.1 molecules. Cell 74:929–937. https://doi.org/10.1016/0092-8674(93)90472-3
28. Saha S, Raghava GPS (2004) BcePred: prediction of continuous B-cell epitopes in antigenic sequences using physico-chemical properties. Springer, Berlin/Heidelberg, pp 197–204
29. Saha S, Raghava GPS (2007) Prediction methods for B-cell epitopes. Humana Press, pp 387–394
30. Sanchez-Trincado JL, Gomez-Perosanz M, Reche PA (2017) Fundamentals and methods for T- and B-cell epitope prediction. J Immunol Res 2017. https://doi.org/10.1155/2017/2680160
31. Scaiewicz A, Levitt M (2015) The language of the protein universe. Curr Opin Genet Dev 35:50–56. https://doi.org/10.1016/j.gde.2015.08.010

32. Singh SP, Mishra BN (2016) Major histocompatibility complex linked databases and prediction tools for designing vaccines. Hum Immunol 77:295–306. https://doi.org/10.1016/j.humimm.2015.11.012
33. Singh H, Raghava GPS (2001) ProPred: prediction of HLA-DR binding sites. Bioinformatics 17:1236–1237. https://doi.org/10.1093/bioinformatics/17.12.1236
34. Singh H, Raghava GPS (2003) ProPred1: prediction of promiscuous MHC Class-I binding sites. Bioinformatics 19:1009–1014. https://doi.org/10.1093/bioinformatics/btg108
35. Singh H, Ansari HR, Raghava GPS (2013) Improved method for linear B-cell epitope prediction using antigen's primary sequence. PLoS One 8:e62216. https://doi.org/10.1371/journal.pone.0062216
36. Stern LJ, Wiley DC (1994) Antigenic peptide binding by class I and class II histocompatibility proteins. Structure (London, UK: 1993) 2:245–251. https://doi.org/10.1016/S0969-2126(00)00026-5
37. Sun B, Zhang Y (2014) Overview of orchestration of CD4+ T cell subsets in immune responses. Adv Exp Med Biol 841:1–13
38. Thornton JM, Edwards MS, Taylor WR, Barlow DJ (1986) Location of "continuous" antigenic determinants in the protruding regions of proteins. EMBO J 5:409–413
39. Vita R, Overton JA, Greenbaum JA et al (2015) The immune epitope database (IEDB) 3.0. Nucleic Acids Res 43:D405–D412. https://doi.org/10.1093/nar/gku938

A Chronological Journey of Breg Subsets: Implications in Health and Disease

5

Hamid Y. Dar, Lekha Rani, Leena Sapra, Zaffar Azam, Niti Shokeen, Asha Bhardwaj, Gyan C. Mishra, and Rupesh K. Srivastava

Abstract

B cells play a multidimensional role in host immunity. Regulatory B (Breg) cells are a class of B lymphocytes with immunomodulatory properties that play an important role in maintaining immunological tolerance along with dampening harmful immune responses. Bregs suppress various immune pathologies through the production of interleukin (IL)-10, IL-35, and transforming growth factor-β (TGF-β). They act by inhibition of T helper 1 (Th1) and Th17 cells proliferation, suppression of dendritic cell (DC), differentiation and simultaneous enhancement of the expression and differentiation of fork head transcription factor P3-positive regulatory T cells (FoxP3$^+$ Tregs). In this chapter, we discuss the induction, function, and phenotypes of the various Breg cell subsets defined in both mice and humans along with their proposed mechanism of action in various immune responses.

Keywords

Regulatory B cells (Bregs) · Plasma Bregs · BR2 Bregs · B10 Bregs · T2-MZP Bregs · TIM1$^+$ Bregs · B1 B cells · Br1 Bregs · Plasmablast · iBregs · IgA$^+$ Bregs · GrB$^+$Bregs

Hamid Y. Dar, Lekha Rani and Leena Sapra have equally contributed to this chapter.

H. Y. Dar · L. Sapra · N. Shokeen · A. Bhardwaj · R. K. Srivastava (✉)
Department of Biotechnology, All India Institute of Medical Sciences (AIIMS), New Delhi, India

L. Rani · G. C. Mishra
National Centre for Cell Science (NCCS), Savitribai Phule University Campus, Pune, Maharashtra, India
e-mail: gcmishra@nccs.rcs.in

Z. Azam
Dr. Harisingh Gour Central University, Sagar, Madhya Pradesh, India

© Springer Nature Singapore Pte Ltd. 2020
S. Singh (ed.), *Systems and Synthetic Immunology*,
https://doi.org/10.1007/978-981-15-3350-1_5

Abbreviations

Bregs	B regulatory cells
BCR	B cell receptor
TLR	Toll-like receptor
PAMPs	pathogen-associated molecular patterns
EAE	experimental autoimmune encephalomyelitis
IL	interleukin
LPS	lipopolysaccharide
TGF-β	transforming growth factor Beta
Tregs	T regulatory cells
MHC	major histocompatibility complex
AIA	antigen-induced arthritis
Th	T helper cells
STAT	Signal Transducer and Activator of Transcription
IFN-γ	interferon gamma
TNF-α	tumor necrosis factor alpha
mAbs	monoclonal antibodies
T2-MZP	transitional 2 marginal-zone precursor
TIM-1	T-cell Ig mucin domain-1
CTLA-4	cytotoxic T lymphocyte-associated protein 4
iBreg	induced B regulatory cells
IDO	indoleamine 2,3-dioxygenase
MS	multiple sclerosis
SLE	systemic lupus erythematosus
RA	rheumatoid arthritis
NOD	non-obese diabetic
RANKL	receptor activator of nuclear factor-κB ligand
OPG	osteoprotegerin
T1D	Type 1 diabetes
Tr1	T regulatory type 1

5.1   Discovery of Breg Cells

The concept of B cells regulating immune responses dates back to 1974, when the suppressive nature of B cells in modulating delayed type hypersensitivity in guinea pigs was described [1]. Wolf et al. suggested a regulatory subset of B cells (Bregs) exhibiting immunomodulatory properties in an experimental autoimmune encephalomyelitis (EAE) model of mice in 1996 [2]. From 2002 to 2003, Fillatreau et al., Mizoguchi et al., and Mauri et al. through independent studies demonstrated that B cells produce IL-10 and suppress inflammatory conditions such as EAE, inflammatory bowel disease and collagen-induced arthritis respectively [3–5]. Further, Parekh et al. were the first to show a IL-10-independent mechanism of action in 2003, demonstrating TGF-β-dependent B cell–mediated regulation of CD8$^+$ T cell responses, though they did not name these as Bregs at the time [6]. It was only after 3 years that Mizoguchi and Bhan proposed the concept of Bregs while studying their role in colitis, demonstrating that B cell–deficient mice experienced higher severity of colitis than normal [7]. Moreover, Mizoguchi et al. also established that a specific B cell subset induced in gut-associated lymphoid tissue was secreting higher levels of IL-10 and had increased CD1d expression during intestinal inflammatory condition [4]. Till date, numerous studies have been carried out to illustrate the role of various Breg subsets via IL-10-dependent or IL-10-independent manner in modulating host immunity. In 2008, Yanaba et al. also showed the role of CD1d^{hi}CD5$^+$ cells in negatively regulating T-cell responses through IL-10 in contact hypersensitivity model [8]. Dittel et al. observed that mice with B cell deficiency have reduced numbers of both Foxp3$^+$ regulatory T cells (Tregs) and IL-10 levels in EAE and demonstrated a novel IL-10, B7, and MHC class II-independent regulatory role for B cells in suppressing autoimmunity by the maintenance of Tregs via glucocorticoid-induced TNFR family–related gene ligands [9, 10]. In 2010, Amu et al. reported that helminths-induced Bregs were responsible for Treg induction that could suppress allergic airway inflammation (AAI) in the murine model [11]. Carter et al. demonstrated the unique ability of Bregs in inhibiting Th1/Th17 cells during arthritic conditions in mice [12]. Strikingly, the regulatory function of B cells is mediated by the production of various regulatory cytokines such as IL-10, IL-35, and TGF-β1, which are responsible for suppressing autoreactive B cells and pathogenic T cells in a cytokine or cell-cell contact-dependent manner [7, 13]. Another mechanism of immune regulation by B cells involve expression of FAS ligand on CD5$^+$ B cells, known as killer B cells that regulate effector immune responses by inducing cell death [14]. Kaku et al. showed a population of B cells that express both CD73 and CD39, ectoenzymes responsible for the production of adenosine, which inhibited the severity of colitis [15]. Khan et al. described additional phenotype of Bregs, PD-L1^{hi} B cells, which regulate humoral immunity through their interaction with CD4$^+$CXCR5$^+$PD-1$^+$ follicular helper T cells and ameliorate EAE [16]. Recently, Oleinika et al. reported a novel role of CD1d$^+$ T2-MZP Bregs in the induction of immunosuppressive iNKT cells that downregulate excessive Th1/Th17 responses partially via secreting IFN-γ and limit inflammation in experimental arthritis [17]. Together, these studies indicate that Bregs suppress inflammation by inhibiting the differentiation of pro-inflammatory cells and inducing a population of immunosuppressive cells. In addition, studies on exacerbation of colitis and development of psoriasis in patients treated with anti-CD20 mAb (rituximab) suggest the regulatory function of B cells in human subjects [18, 19]. Bregs constitute fewer than 10% of immature B cells in healthy individuals and play an important role in functioning of the immune system by maintaining tolerance and immune homeostasis [20]. Over the last decade, numerous studies in both mice and human have extensively shown the importance of Bregs in regulating various diseases, including inflammatory disorders, autoimmunity, and cancer [21, 22] Bregs with their wide range of immunomodulatory functions can thus be exploited for therapy in various B cell–mediated diseases. Thus, it is important to exhaustively consider the known Breg cell phenotypes, their induction, and function in a chronological manner (Fig. 5.1 and Table 5.1).

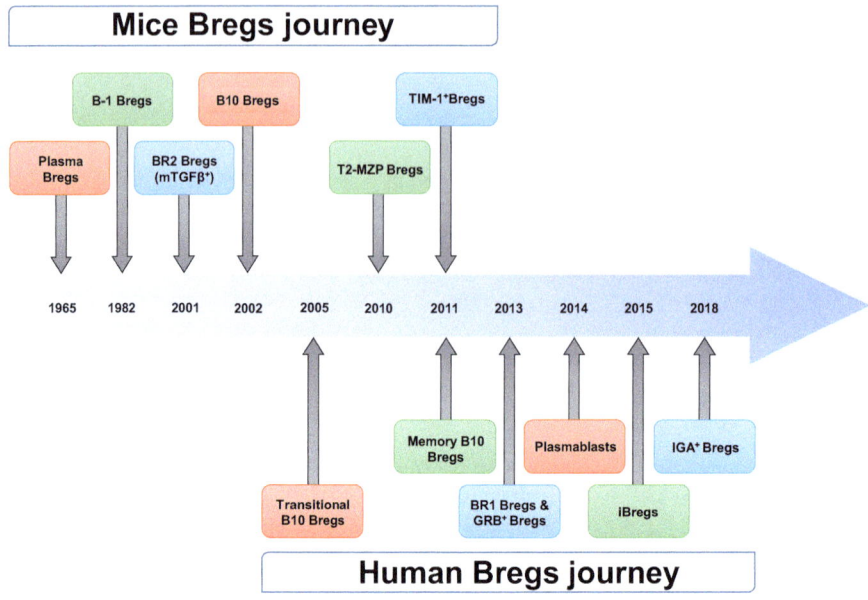

Fig. 5.1 **Chronological journey of Bregs**. This timeline represents the important events in the journey of Bregs discovery, establishing them as a functionally and developmentally distinct cell lineage

5.2 Identification and Phenotypes of Breg Cells

B cell subsets with strong immunomodulatory functions have been reported both in vitro and in vivo (Figs. 5.2 and 5.3) (Table 5.1). Phenotypic identification of Breg cells using the immunomodulatory cytokine IL-10 continues to be a matter of debate due to difficulties in assessing the functionality of Bregs, because IL-10 detection requires intracellular staining. Therefore, other surrogate markers have been employed to identify various Breg subsets. Different overlapping markers are presently being used to describe these cells. Here we discuss both murine and human Breg subsets under separate heads for clarity and distinction among these subsets.

5.2.1 Mouse Breg Subsets

In mice, Plasma B cells, B-1 cells, $CD5^+CD1d^{hi}$ B10 B cells, $CD21^{hi}CD23^{hi}CD24^{hi}$ transitional type 2 marginal zone precursors (T2-MZP) Breg cells, and $TIM-1^+$ B cells have been proposed with regulatory functions in a variety of infections, in autoimmune and transplantation settings [21, 23]. $IL-10^+$ Bregs have also been observed to inhibit IFN-γ production in hepatitis B virus (HBV) infection by modulating $CD8^+$ T cell responses [24, 25]. Furthermore, $IL-10^+$ Bregs inhibit TNF-α production by activated monocytes following stimulation with LPS and bacterial

Table 5.1 Breg cell subsets in mice and humans

S. no.	Types of Breg subset	Phenotype	Year of discovery	Mechanism of action	Refs.
Mice Breg subsets					
1.	Plasma Bregs	CD138+MHC-11lo B220+	1965	Found in bone marrow, spleen, mucosa-associated lymphoid tissues (MALT), or lymph nodes. They secrete IL-10 and IL-35, and play a key role in host defense against infection	[28, 29]
2.	B-1 Bregs	CD5+	1982	Found in bone marrow, lymph node, spleen, and blood, leading to innate adaptive regulation via IL-10 production	[28–30, 33]
3.	BR2 (mTGFβ+Bregs)	CD40+TGFβ1	2001	Found in spleen, lymph node, and blood, express membrane TGF-β1, and cause anergy and hyporesponsiveness in CD8+ T cells	[6]
4.	B10 Bregs	CD19hiCD1dhiCD5+	2002	Found in spleen and blood, produce IL-10, and inhibit expression of effector CD4+ T cells, DCs, and monocytes	[4]
5.		TIM-1+CD19+	2011	Found in the spleen and suppress the expression of effector CD4+ T cells through IL-10 production	[5, 10]
5.	T2-MZP Bregs	CD19+CD21hiCD23hiCD24hi	2010	Found in the spleen, produce IL-10, enhance Treg cells, and inhibit the expression of effector CD4+ and CD8+ T cells	[65]
Human Breg subsets					
1.	CD19+CD24hiCD38hi Bregs	Transitional B10 cells	2005	Found in blood and support the development of Tregs through IL-10 and TGFβ production	[144]
2.	CD19+CD24hiCD27+ Bregs	Memory B10 cells	2011	Inhibit the proliferation of TNF-α and IFNγ producing CD4+ T cells, DCs, and monocytes via IL-10-dependent and -independent pathways	[22]
3.	Br1 Bregs	CD19+CD25+CD71+CD73-	2013	Found in blood and produce IL-10	[72]
4.	GrB+ Bregs	CD19+CD38+CD1d+IgM+CD147+	2013	Inhibit the proliferation of T cells through the expression of Granzyme B	[88]
5.	Plasmablasts	CD19+CD27intCD38+	2014	Found in draining lymph nodes in mice and in blood from humans and inhibit DCs and effector T cells through IL-10 expression	[73]
6.	iBregs	IDO, TGFβ	2015	Through IDO and TGFβ production, they induce natural Tregs as well as TGFβ and IL-10–producing Tregs	[92]
7.	IgA+ Bregs	IgA+	2018	Induce the differentiation of T cells more toward a regulatory phenotype through the expression of IL-10 and PD-L1	[93]

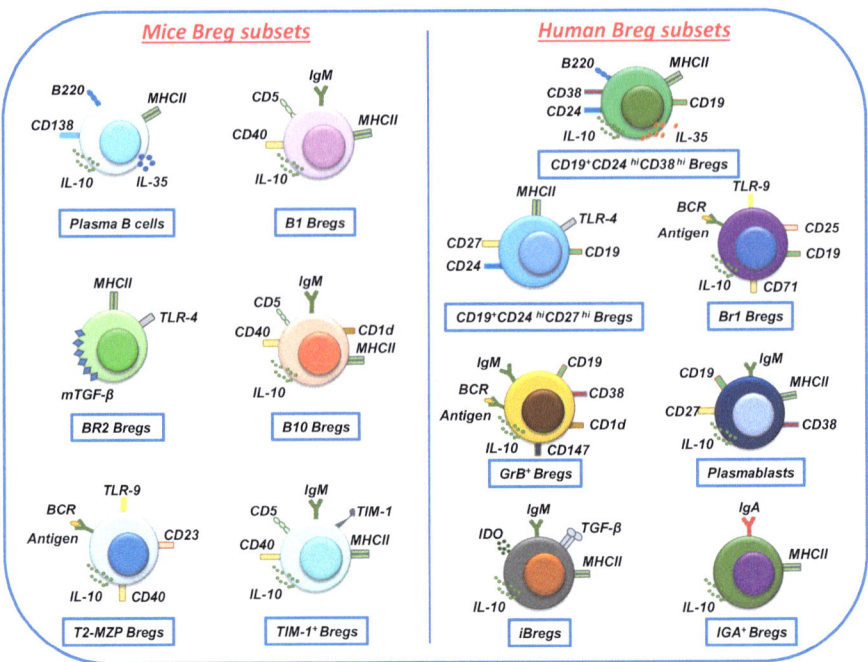

Fig. 5.2 Breg subsets in mice and humans. Mice have a total of five defined Breg subsets: Plasma B cells (CD138+MHC-11lo B220+), B1 Bregs (CD5+), BR2 Bregs (CD40+TGFβ1), B10 Bregs (CD19hiCD1dhiCD5+), and T2-MZP Bregs (CD19+CD21hiCD23hiCD24hi). Humans, on the contrary, have seven defined human Breg subsets: Br1 Bregs (CD19+CD25+CD71+ CD73−), CD19+CD24hiCD38hi Bregs, CD19+CD24hiCD27hi Bregs, Plasmablasts (CD19+ CD27intCD38+), iBregs (IDO, TGFβ), GrB+Bregs (CD19 +CD38+CD1d+IgM+CD147+), and IGA+Bregs (IgA+)

CpG DNA [9, 22]. Bacterial components such as LPS and CpG are known to induce the expansion, differentiation, and activation of murine Bregs through TLR signaling in vitro [26, 27]. Furthermore, mice harboring TLR2- or TLR4-deficient B cells fail to recover from EAE. Alltogether these studies clearly indicate that inflammation acts as stimuli for the activation and differentiation of Bregs.

5.2.1.1 Plasma Bregs

Plasma B cells are representative antibody-secreting cells (ASCs) [28] present in all lymphoid organs. Plasma cells have also been found to occur in significant numbers in the bone marrow compared to their lower numbers in the spleen. Indeed, the bone marrow is primarily responsible for the long-term maintenance of plasma cells arising from immunization [29]. Recently, Lino et al. described a subset of resident Plasma B cells specialized for producing IL-10 upon TLR stimulation and are found to occur naturally, i.e., prior to antigenic challenge [30]. Genome-wide approaches have shown that this Breg lineage is triple-positive for the following markers: IL-10+LAG-3+CD138hi. The lymphocyte activation gene 3 (LAG-3+) helps in

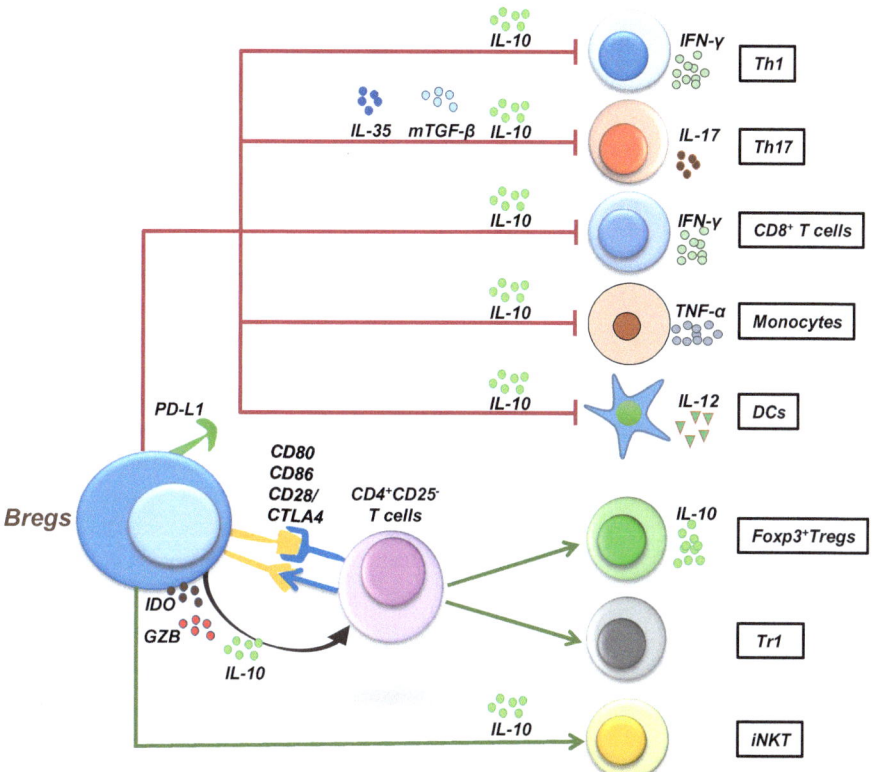

Fig. 5.3 Regulatory mechanisms of Bregs in various immune responses. Bregs lead to the suppression and inhibition of pro-inflammatory lymphocytes such as Th1, Th17, cytotoxic CD8+ T cells, monocytes, and IL-12-producing dendritic cells through the production of various factors like IL-10, IL-35, TGF-β, IDO, GZB, and so on. IL-10 production by Bregs is primarily responsible for restoring the Th1/Th2 balance, where it is shifted toward Th2. One more mechanism of inhibiting inflammatory cascades is via tweaking the Treg/Th17 balance, leading to suppression of Th17 cells. The Breg population is reportedly responsible for enhancing the differentiation of Foxp3+Treg cells and helps in the maintenance of iNKT cells

regulating humoral immunity and in maintaining immunological tolerance toward endogenous T-independent type 2 antigens, which are normally not detected by CD4+Foxp3+ T regulatory cells. Unlike conventional plasma cell differentiation, which requires several days for proliferation, the detection of IL-10+LAG-3+CD138hi plasma cells at day one post-infection with *Salmonella typhimurium* in the spleen of mice, confirmed that this subset is derived from already existing cells LAG-3+CD138hi cells. These LAG-3+CD138hi cells are likely induced by self-antigen and remain in a quiescent state. Further, genome-wide methylome, transcriptome, and gene-set enrichment analysis of LAG-3+CD138hi cells in naïve mice and at day one post-*Salmonella* infection showed that after antigenic challenge, LAG-3+CD138hi cells express IL-10 and become IL-10+LAG-3+CD138hi plasma Bregs [30].

Thus, these results indicate that plasma Bregs provide a first layer of immune regulation in response to stimuli. In contrast, Matsumoto et al. showed that mice lacking genes such as Prdm1 and IRF4, which are required for plasma cell differentiation, develop a severe form of EAE compared to control mice. This study suggested that Bregs are inducible in nature. Thus, these studies clearly establish both the innate and inducible nature of Bregs. During EAE, plasma B cells are known to be the main source of IL-35 and facilitate recovery from EAE. IL-35 secreted by plasma Bregs exhibits anti-inflammatory properties by expanding the immunosuppressive $CD4^+CD25^+$ Tregs population which inhibits $CD4^+CD25^-$ T effector cell proliferation when cultured in vitro [31]. IL-35 also inhibits the differentiation of inflammatory Th17 cells. Recent studies have indicated the role of BATF/IRF-4/IRF-8 axis in regulating IL-35 and IL-10 expression in activated B cells [32]. IL-35 cytokine can act as a potential target in the treatment of both autoimmune and inflammatory conditions. Interestingly, declined populations of LAG-3$^+$CD138^{hi} cells have been reported in mice deficient in CD19 or Bruton's tyrosine kinase [33], further establishing that differentiation of LAG-3$^+$CD138^{hi} cells to plasma cells is under the control of BCR. Taken together, these studies establish that B cell differentiation into LAG-3$^+$CD138^{hi} cells is a steady-state process driven primarily by BCR signaling rather than TLR-mediated signaling or T cells.

5.2.1.2 B1 Bregs

B-1 cells represent a class of innate immune cells that are responsible for higher antibody production, especially IgMs for mounting rapid immune responses against pathogens [34]. This subset of CD5$^+$ B cells was initially identified in the early 90s in mice, as a set of distinctive fetal B cells to differentiate them from B-2 cells that usually develop in the adult bone marrow [35, 36]. B-1 cells represent a population of B cells found predominantly in the pleural and peritoneal cavities (35–70%). A smaller number of B-1 cells are also found in the spleen [37], bone marrow, mucosal sites, lymph nodes, and blood [38]. Despite their very low frequency in lymphoid tissues, B-1 cells are important regulators of immune defense and tissue homeostasis. B-1 B cells are chiefly produced in the absence of any antigen exposure [39, 40] and are a major source (>80%) of naturally occurring antibodies [41]. Higher levels of natural IgMs are produced by B-1 cells residing in the spleen and bone marrow [38]. These polyreactive [42, 43] antibodies help in recognizing self as well as foreign antigens [44, 45], act as the first line of defense, and are analogously linked to innate immune responses. B-1 cells are categorized into different functioning subsubsets based on the relative CD5 expression. B-1a represents a class of CD5$^+$(Ly-1) B-1 cells that chiefly express IL-10 upon innate activation [46] whereas B-1b represents a class of CD5$^-$ B-1 cells [34, 45]. B-1a cells are major producers of B-cell-derived IL-10 [46], and their activation and expansion are regulated by cross-regulatory cytokines such as IL-12 and IFN-γ [47]. Using Schistosomal infection model, Vellupillai P et al. demonstrated that the outgrowth of IL-10 producing B-1 after infection is genetically restricted and regulated by polylactosamine sugars. Interestingly, it has also been shown that B-cell defect in BALB.Xid mice impart susceptibility to develop filariosis and is associated with lack of antibody production and IL-10 production in response to dominant surface molecule of invading pathogen [48]. B-1a cells were shown to inhibit TLR-mediated excessive inflammation in neonatal mice in an IL-10-dependent manner [49]. Another subset of B-1a, FAS ligand expressing B-1a cells also known as killer B cells, has been shown to mediate T cell apoptosis during schistosomal infection and prevent granulomatous inflammation [14]. Interestingly, the regulatory role of IgM-producing B-1a cells has also been associated with the suppression of colitis in mice that were kept in conventional facility as compared to mice kept under specific pathogen free facility [50]. Thus, B-1a cells play an important role in immune regulation and tissue homeostasis.

5.2.1.3 BR2 (mTGFβ⁺) Bregs

Here, we propose a novel subset of Bregs called "BR2" Bregs. These Bregs were first reported and studied by Parekh et al. in 2003. They found that B cells activated via T-independent mechanisms such as LPS showed membrane expression of TGFβ1, leading to CD8⁺ T cell anergy. These Bregs thus have the unique phenotype of mTGFβ⁺ Bregs. This manner of B cell activation is a major factor influencing CD8⁺ T cell responses as T-dependent activated B cells provide higher stimulatory properties to CD8⁺ T cells [6]. Membrane expression of TGFβ1 was found to be solely responsible for conferring these B cells with regulatory properties, thus influencing CD8⁺ T cell responses. Thus, we now name these Bregs as BR2 (mTGFβ⁺Bregs), with regulatory properties governed by membrane TGFβ expression. These findings provide insights into the immune evasion strategies adopted by retroviruses and gram-negative bacteria that target toll-like receptor-4 (TLR-4) signaling in B cells. Recent reports have also shown that Bregs producing TGF-β induce Tregs for promoting transplantation tolerance [51]. These results illuminate the importance of novel modes of B-cell activation in the development of therapeutic strategies to modulate the balance between active immunity and tolerance [6].

5.2.1.4 B10 Bregs

B10 cells are defined by their ability to express IL-10 following ex vivo stimulation with PMA and ionomycin and are enriched within CD1d^{hi}CD5⁺ B cell subset [8]. Mouse B10 cells represent around 1–3% of cells in the spleen. Other tissues like the lymph nodes, central nervous system, Peyer's patches, and intestinal tissues comprise a very small number of B10 cells. Their presence in peritoneal cavity is also prominent [29, 52, 53]. Mouse B10 cells have a typical phenotype as IgD^{lo}IgM^{hi} cells, although a very small number of B10 cells are also reported to co-express IgA or IgG [54]. B10 cells secrete polyreactive or Ag-specific IgMs and IgGs upon differentiation [53, 54]. T-cell Ig mucin domain-1 (TIM-1) is a type of transmembrane glycoprotein responsible for immunomodulatory responses [55], and its expression was found to be important for the induction and maintenance of IL-10-producing B cells, whereas a defect in TIM-1 expression leads to increased production of proinflammatory cytokines such as IL-1 and IL-6 [56]. During allotransplantation, TIM-1 is particularly responsible for Breg stimulation to prolong allograft survival. TIM-1⁺ B cells usually express IL-4 and IL-10 and promote Th2 responses with subsequent allograft tolerance [57]. Numerous studies have shown the potential of B10 cells in inhibiting disease initiation and subsequent pathology after their adoptive transfer in models of contact hypersensitivity [8], EAE [3, 52, 58], lupus [59], IBD [53, 60], and graft-versus-host disease [61]. Mauri et al. were the first to elucidate the therapeutic potential of B cells using agonistic CD40 mAbs for treating mice with collagen-induced arthritis [5, 62]. Depletion of B10 cells can have either therapeutic or detrimental effects in the course of various human pathological mouse models. Depletion of IL-10-producing B cells is known to enhance the innate, humoral, and cellular immune responses in mice [62, 63]. This intensifies the severity of disease-related symptoms in various autoimmune diseases in mice such as EAE, skin transplant rejection, and contact hypersensitivity [27, 58, 64].

5.2.1.5 T2-MZP Bregs

The T2-MZP Breg cell subset was discovered by Evans et al. in 2007 [65]. T2-MZP Bregs are immature transitional B cells found in the spleen with a CD19$^+$CD21^{hi}C D23^{hi}CD24^{hi}IgM^{hi}IgD^{hi}CD1d^{hi} phenotype. Among the different B-cell subsets residing in the spleen of mice with arthritis, this specific Breg cell type is responsible for IL-10 production after collagen stimulation. T2-MZP Bregs were discovered to have decisive suppressing properties both in vitro and in vivo, and the mechanism of suppression includes inhibition of pathogenic Th1 responses via producing IL-10 [65]. IL-10-producing T2-MZP B cells are shown to exert immunomodulatory properties in various immune-mediated pathologies, including autoimmune diseases, cancer, and allergy [21, 65, 66]. Recently, Oleinika et al. reported a novel role of CD1d$^+$ T2-MZP Bregs in the induction of immunosuppressive invariant Natural Killer T (iNKT)-cells that downregulate excessive Th1/Th17 responses partially via secreting IFN-γ and limit inflammation in experimental arthritis [17]. Recently, T2-MZP Breg cells have been linked as the precursors of B10 Bregs, but the interrelation between these two Breg subsets needs to be further established [21].

5.2.2 Human Breg Subsets

Similar to mouse Bregs, human Breg cells also play an important role in the maintenance of tissue homeostasis. Mauri et al. in an extensive study demonstrated that CD19$^+$CD24^{hi}CD38^{hi} B cells with a phenotype very similar to immature B cells produce the highest fraction of IL-10 in healthy human peripheral blood upon CD40 stimulation [20]. Separately, Tedder et al. also categorized human Breg cells as CD24^{hi}CD27$^+$, a phenotype related to memory B cells [22]. Furthermore, Bosma et al. reported that due to altered CD1d recycling in B cells, defect in B-cell-mediated iNKT expansion was observed in SLE patients [67]. Human Bregs exert immunomodulatory properties through their actions on various immune cell types such as inhibiting cytokine production in monocytes [22]; inducing immunosuppressive NKT cells [67], restraining IFN-α production from pDCs [68]; and regulating CD4$^+$ T cell proliferation [69], inhibition of Th1 and Th17 differentiation, and conversion of CD4[+] T-cells into CD4[+]CD25[+] cells along with enhancing FOXP3 and PD-1 expression on Tregs [20, 70, 71]. In humans, research on Bregs is mainly restricted due to lack of access to the human spleen, the primary site of the Bregs population. Thus, the majority of identified human Bregs are from peripheral blood where Bregs ranging from immature B cells to differentiated plasmablasts are found. Other phenotypes of human Bregs comprise CD19[+]CD25[+]CD71[+]CD73[−] B regulatory 1 (Br1) cells [72], CD19[+]CD27[int]CD38[+] plasmablasts [73]. Furthermore, human Bregs (i.e., equivalent to B10 of mice) with the CD19[+]CD24[hi]CD27[+] phenotype along with Tim1[+] Bregs are preferentially found in the transitional B cells [22, 74]. Thus, it is important to describe different defined subsets of human Bregs.

5.2.2.1 CD19[+]CD24[hi]CD38[hi] Bregs

Human B cells with regulatory function have been described in CD19[+]CD24[hi]CD38[hi] immature subset of peripheral blood B cells. After CD40 stimulation, this subpopulation isolated from peripheral blood of healthy individuals is known to inhibit the differentiation of Th1 cells via IL-10 production and CD80 and CD86 engagement [20]. However, CD24[hi]CD38[hi] cells isolated from SLE patients lacked regulatory capacity [20]. Recently, in patients with SLE, an expanded population of CD19[+]CD24[hi]CD38[hi] Bregs was observed with deficient IL-10R expression, which is correlated with compromised Breg function despite showing enhanced IL-10 expression [75]. Thus, targeting the 'Bregs/IL-10/IL-10R' axis may prove to be a novel therapeutic approach in the treatment of SLE. In addition to inhibiting Th1 and Th17 differentiation, these cells also convert CD4[+]CD25[−] into Tregs [70]. Both numerical and functional impairment has been observed in a number of autoimmune diseases such as SLE [20, 75] and RA [70]. Recent studies showing reduced capacity of CD19[+]CD24[hi]CD38[hi] Bregs to secrete IL-10 in GVHD patients as compared to transplant tolerant and healthy controls indicated their important role in preventing graft rejection by promoting tolerance. Moreover, Cherukuri et al. in 2014 found low IL-10/TNF-α ratio by CD19[+]CD24[hi]CD38[hi] transitional B cells in renal patients with graft rejection when compared with healthy controls, further highlighting their role in establishing transplant tolerance [76] TIM-1 is also a marker for IL-10[+] Bregs and around 50% of IL-10[+] B cells were TIM-1[+]. On evaluating TIM-1 expression on human B cell subsets, this transitional subset was enriched in TIM-1[+] subset [74]. In the same study, authors found a decreased number as well as impaired function of TIM-1[+] in patients with systemic sclerosis [74]. In 2015, Kristensen et al. stated that in humans, 40% of IL-10[+] B10 cells expressed TIM-1 [77]. Supporting this study, Liu et al. found that compared to HIV-infected patients, healthy controls have more than 75% of peripheral B10 cells expressing TIM-1. These studies highlight the role of TIM-1 as a marker of Bregs and will open new avenues for the isolation of Bregs that could be utilized for achieving immune homeostasis.

5.2.2.2 CD19[+]CD24[hi]CD27[hi] Bregs

The IL-10-producing B cells, named B10 in humans, are predominantly CD19[+]CD24[hi]CD27[+] memory subset of B cells, known to be a major source of IL-10 after stimulation with LPS or CpG along with CD40 ligation B cells. B10 cells also express CD48, and CD148 [22]. IL-21 has the potential to further induce IL-10 production from CpG- or LPS-treated CD19⁺CD27⁺ memory B10 cells [78]. Among other subsets, B10 cells are also present in the tonsils, spleen, and newborn cord blood [76]. Interestingly, an increase in the number of B10 cells was observed in a number of autoimmune diseases [22, 79, 80]. In patients with RA, B10 cells are highly capable of expressing receptor activator of nuclear factor-κB ligand (RANKL) compared to those in the healthy controls, suggesting a possible mechanism by which B10 cells are involved in RA pathogenesis [81]. At the molecular level, Zheng et al. in 2017 reported that microRNA-155 (miR-155) positively regulates IL-10 expression in B10 cells, which is impaired in patients with Crohn's disease (CD), leading to miR-155-induced expression of TNF-α by monocytes. These findings further suggest a novel miRNA-mediated approach in developing Breg-based strategies to control the progression of autoimmune diseases.

5.2.2.3 Br1 Bregs

This subset of human Bregs with the CD19⁺CD25⁺CD71⁺CD73⁻ phenotype was identified by Van de Veen et al. in 2013. These IL-10-producing Br1 Bregs share homology with the Tr1 subtype of T cells. Due to the low CD73 expression on their surface, the immunosuppressive function of Br1 cells was considered to be independent of adenosine and could thus be IL-10 dependent. In support of this, further studies substantiated the role of IL-10 in imparting immunosuppressive functions to Br1 cells. This IL-10⁺ subset of Bregs is reported to induce tolerance toward allergens by repressing the proliferation of allergen-specific CD4⁺ T cells as well as by producing allergen-specific anti-inflammatory IgG4 antibodies [72], thus contributing to peripheral tolerance. This subset of Bregs can induce tolerance against bee venom allergen and PLA2 (phospholipase A2) in an IL-10-dependent manner and also showed tolerance toward various food allergens like casein (cow milk protein). Van de Veen et al. used flow cytometry and whole-genome sequencing to further show that human Br1 cells express the inhibitory ligand PD-L1 (programmed death ligand-1), which binds PD-1 on T cells to inhibit T cell activation and promote the maintenance of Tregs cells.

5.2.2.4 Plasmablasts

This subset of Bregs is known to be derived from both naïve and immature B cells in humans with the CD19⁺CD27ⁱⁿᵗCD38⁺phenotype, which secretes IL-10 [73]. In the presence of IL-2, IL-6, CpG, and IFN-α, immature B cells undergo differentiation, leading to expansion of plasmablasts with increased expression of IRF4, Blimp1, and XBP1 [73]. In normal tissues, CD30 expression is limited to a few T and B cells, whereas in B cell lymphoma, CD30 expression is upregulated on B cells. Recently, in a mouse model of B cell lymphoma, higher CD30 expression on B cells was found to promote the differentiation of plasma B cells to plasmablasts via NF-κB activation and enhanced phosphorylation of STAT3, STAT6, and nuclear factor IRF4 [82]. Interestingly, exacerbation of inflammatory symptoms in MS patients upon treatment with Atacicept, which deplete antibody-secreting cells, further suggests the regulatory function of plasmablasts [83]. Patients with immunoglobulin G4 (IgG4)–related disorder (IgG4-RD), primary Sjögren's syndrome [84, 85], and SLE [86] have increased plasmablast number, indicating their expansion could be the result of inflammatory conditions. In 2019, Arbore et al. further reported that microRNA-155 (miR-155) plays an important role in the survival and proliferation of plasmablast B cells [87].

5.2.2.5 Granzyme B (GrB⁺) Bregs

Granzyme B–expressing Bregs are known to display the characteristic phenotype of $CD19^+CD38^+CD1d^+IgM^+CD147^+$ [88]. Expression of Granzyme B on Bregs (GrB^+ Bregs) mediates their inhibitory effect on T cells by suppressing their proliferation and inducing apoptosis. In various inflammatory conditions such as SLE [89] and in acute viral infections [90], the percentage of GrB^+ Bregs is relatively high. Peripheral B cells stimulated in the presence of IL-21 are reported to produce and secrete GrB. These cells mediate their suppressive function by repressing T cell proliferation, partly via downregulation of the TCR zeta chain, thereby promoting T cell apoptosis [88]. In the case of RA, the proportion of GrB^+ Bregs is significantly reduced due to the lowered expression of IL-21R, which in turn impairs the negative regulation of Th1/Th17 by GrB^+ Bregs [91], suggesting that impaired GrB^+ Bregs are associated with RA pathogenesis.

5.2.2.6 iBregs (Induced Bregs)

B cells like other immunosuppressive cells differentiate into induced Breg (iBreg) cells when subjected to certain stimuli and express indoleamine 2,3-dioxygenase (IDO) and TGFβ. T cells expressing cytotoxic T lymphocyte–associated protein 4 (CTLA-4) enhance the induction of iBregs, which then convert T cells into TGF-β- and IL-10-producing Tregs, thereby modulating various immune responses [92].

5.2.2.7 IgA⁺ Bregs

This subset of Bregs has been identified recently by Fehres et al. in 2019. They described that overexpression of APRIL (A Proliferation-Inducing Ligand) instead of BAFF induces activation of $IL-10^+$ human Bregs that further repress inflammatory immune reactions. These APRIL-induced IgA^+ Bregs suppress the effector function of T cells and macrophages and induce Tregs via IL-10 and PD-L1 expression [93]. These findings collectively suggest the importance of the novel APRIL-induced Breg subset with IgA^+ phenotype, both in the immunopathology and homeostasis of immunological reactions. In colorectal cancer patients, a higher proportion of IgA^+ Bregs was observed at the tumor site due to lowered expression of microRNA15A (miRNA15A) and microRNA16–1 (miRNA16–1). These microRNAs exhibit the ability to regulate proliferation, drug resistance, and apoptosis. These studies thus concluded that microRNAs and IgA^+ Bregs are negatively correlated and that a lower level of microRNAs along with higher proportion of IgA^+ Bregs reduces the survival rates in cancer patients [94].

5.3   Bregs in Health and Diseases

The discovery of various defined subsets of Bregs has now compelled researchers to revisit the understanding of B cell biology in the context of various immune-mediated diseases. Vaccines have been ideally responsible for eradicating several diseases via the specific activation of B cells. Similarly, cancer immunotherapies demonstrate their course of action via production of different B cells. Moreover, B cell deficiencies lead to various devastating impacts on health and immunity. It is now well established that B lymphocytes produce antibodies and are associated with various immunomodulatory properties. Bregs are now extensively studied for their novel immune-regulatory roles, as mice deprived of B cells are reported to demonstrate higher incidences of immune-related disorders. Bregs are known to produce various cytokines and immunomodulatory factors responsible for proper functioning of the host immune system [95]. A cohort study indicated that targeted depletion of B cell populations serves as a treatment in autoantibody-mediated autoimmune disorders such as SLE [96]. Thus, Bregs undoubtedly play an important role in host pathology, thereby opening Pandora's Box in harnessing the potential of Bregs in mediating health. In the following sections, we focus on the role of Bregs in selected diseases/pathologies.

5.3.1   Multiple Sclerosis

Multiple sclerosis (MS) is an autoimmune disorder occurring due to T and B cell hyperactivation, leading to demyelination and axonal damage in the central nervous system (CNS). Apart from the role of B cells as pathogenic cells, they also modulate immune responses in MS. IL-10-producing Bregs were first observed in MS patients infected with helminthes; these Bregs were found to suppress the proliferation and IFN-γ production in T cells in vitro [97]. The role of Bregs in MS was further substantiated by diminished levels of IL-10 production in MS patients. In relapsing-remitting MS patients, a significantly reduced number of IL-10-producing naïve Bregs were observed compared to that in the controls [98]. Further, treatment of MS patients with IFN-β, fingolimod, or alemtuzumab is reported to increase the number and function of Bregs [99, 100]. In EAE, one of the most widely studied animal model of MS, the importance of Bregs in alleviating EAE progression was recently illustrated [52, 58, 101, 102]. The suppressive functioning of Bregs involves binding to the BCR co-receptor CD19, which plays an inhibitory role in the development of EAE by modulating the Th1/Th2 cytokine balance [103]. Fillatreau et al. found that B-cell-depleted mice have a persistent type I immune response in EAE and that their recovery was dependent on myelin oligodendrocyte glycoprotein (MOG)-specific IL-10-producing B cells [3]. Further studies indicate that Bregs with the CD1d^{hi}CD5$^+$ phenotype are effective in inhibiting EAE progression. CD1d^{hi}CD5$^+$ Bregs possess highly decisive immunomodulatory properties in controlling the pathogenesis of the initial and late phase of EAE [52, 58]. Further, depletion of CD20$^+$ B cell enhances the pathogenesis of EAE. This was evident from a simultaneous increase in the expression of various inflammatory cytokines in the CNS and an increased number of autoreactive CD4+ T cells due to absence of the IL-10-producing CD1^{hi}CD5+ Bregs subset [52, 58].

5.3.2   Inflammatory Bowel Disease

Inflammatory bowel disease (IBD) is a highly deteriorating inflammatory condition of the intestine, usually represented by Crohn's disease (CD) and ulcerative colitis (UC) [104, 105]. Recently, an alarming rise in the prevalence and incidence of IBD has been observed globally [105]. Numerous studies have reported the functions of Bregs in regulating intestinal inflammation. Mizoguchi et al. [106] credited B cells and autoantibody production as important factors in protecting T cell receptor (TCR) α chain-deficient (TCRα$^{-/-}$) mice, which are highly susceptible to develop chronic colitis. They showed that CD1+ B cells producing higher levels of IL-10 upon induction in the gut-associated lymphoid tissues in TCRα$^{-/-}$ mice reduced the intestinal inflammation and disease incidence [4]. IL-10-producing Bregs have now been linked with downregulating the inflammatory cascade associated with IL-1 and signal transducer and activator of transcription 3 (STAT3) without tweaking T cell responses. Wei et al. demonstrated that adoptive transfer of B cells from mesenteric lymph nodes could repress IBD by enhancing the Tregs population [107, 108]. A numerical (number/percentage of Bregs) defect in IL-10-producing Bregs has also been described in patients with both CD and UC [109].

5.3.3   Systemic Lupus Erythematosus

Systemic lupus erythematosus (SLE) is designated as a systemic multigene autoimmune disorder characterized by higher production of autoantibodies with simultaneous deposition of immune complexes, resulting in tissue inflammation and damage to the skin, kidneys, and joints. This phenomenon results in proteinuria and large-scale renal tubule inflammation (glomerulonephritis), which eventually affects the immune system [110, 111]. Both B- and T-cell abnormalities have been found to be responsible for the occurrence of SLE in mammals [112]. SLE-affected individuals usually show a reduced number as well as decreased functional activity of circulating Bregs. This defect usually arises as immature B cells (CD19+CD24^{hi}CD38^{hi}) fail to differentiate into Bregs [20, 68, 113]. Various mouse models have been identified to study the role of regulatory B cells in spontaneous lupus. Recently two well-defined models, New Zealand Black (NZB) × New Zealand White (NZW) F1 hybrid (NZB/W) mice and MRL/lpr mice, have been used to investigate the inhibitory role of Bregs in regulating the severity of SLE [59, 112]. Depletion of Bregs in infant mice resulted in higher severity of SLE, whereas deletion of Bregs from adult mice did not affect SLE progression. Thus, Bregs have been found as predominantly effective during the initiation phase of SLE rather than during disease progression [59, 112]. Additionally, the higher therapeutic interventions of Bregs have come into play due to their role in enhancing the number of Tregs after the transfer of splenic CD1d^{hi} CD5$^+$ B cells from wild-type NZB/W F1 mice to CD19 $^{-/-}$ NZB/W F1 [95]. Blair et al. further observed that anti-CD40-induced T2 Breg cells significantly improved the survival rate in MPL/lpr mice via higher expression of IL-10. Collectively, these findings indicate that T2-MZP B cells as well as B10 cells effectively help in protecting mice from severe SLE [21].

5.3.4   Rheumatoid Arthritis

Rheumatoid arthritis (RA) is a systemic autoimmune disease with a worldwide prevalence of 0.3–1%. It is responsible for increased societal dependency with simultaneous reduction of mobility and working ability [114]. RA is characterized by autoimmune inflammatory responses at synovial membranes and joint capsules, resulting in significant morbidity and mortality due to synovial proliferation, cartilaginous injury, and bone erosion [115]. B cells produce various factors including autoantibodies like anti-citrullinated protein antibodies (ACPAs) and rheumatoid factor (RF) that are responsible for severe disease activity in RA [116]. Moreover, reduced numbers of Bregs such as IL-10-producing Bregs, CD19$^+$TIM-1$^+$IL-10$^+$ Bregs, CD19$^+$CD5$^+$CD1d^{hi} B cells, and CD19$^+$CD5$^+$CD1d$^+$IL-10$^+$ Bregs were observed in RA patients compared to those in healthy controls upon stimulation with CpG or LPS along with phorbol myristate acetate and ionomycin [117, 118]. Further, the function of Bregs was found to be impaired in RA. One study demonstrated that CD24^{hi}CD38^{hi} Breg cells from healthy individuals inhibited Th1 and Th17 differentiation and favored the conversion of CD4$^+$CD25$^-$ T cells to Tregs via IL-10 expression. In contrast, CD19$^+$CD24^{hi}CD38^{hi} cells from RA patients were unable to reduce Th17 development and induce Tregs differentiation [70]. In 2017, Banko et al. showed that CD19$^+$CD27$^+$IL-10$^+$ Bregs are significantly reduced in RA patients compared to those in the controls and that the existing Bregs showed a reduced ability to suppress IFN-γ production by T helper cells. Breg-deficient mice demonstrate higher incidences of autoimmune arthritic conditions due to enhanced induction of Th1 and Th17 cells along with simultaneous suppression of Treg cells [113]. Bregs have thus been found instrumental in suppressing inflammation via restoring or modulating the Th1/Th2 balance in various T-cell-mediated autoimmune diseases such as EAE and RA [113].

5.3.5   Type 1 Diabetes

Type 1 diabetes (T1D) is an autoimmune disease caused by the obliteration of insulin-producing pancreatic β cells mediated by CD4$^+$ and CD8$^+$ T cells [119]. Onset of T1D usually occurs around 13–15 weeks of age in non-obese diabetic (NOD) mice, a model of human T1D. The prevalence of T1D in NOD mice is higher in females with about 80% females and 20% males affected by this disease by 30 weeks [120]. B cells are particularly found to be responsible for the development of pathogenesis of T1D. B cell penetration into the pancreatic islets of NOD mice results in selective propagation of T cells within lymphoid structures, leading to an increased number of autoreactive B cells [121]. Treatment of 5-week-old NOD female mice with anti-CD20 mAbs was found to deplete 95% of B cells, thereby arresting insulitis; however, at 15 weeks, the same treatment was inefficient to hinder the progression of T1D [8, 122]. Grey et al. found that the increased population of CD4$^+$CD25$^+$Foxp3$^+$ Treg cells due to B cell depletion reduced the occurrence of diabetes [123]. Smith and Tedder further postulated that B-cell-depleted NOD mice remained free from diabetes even after reconstitution with B cells [124]. Among various types of B cells, IL-10-expressing B cells have been primarily found to be responsible for decreasing the pathogenicity of insulitis and reducing T1D incidence. Simultaneously, various Th1 immune-related responses were curbed, leading to the diversion of CD4$^+$ T cells toward the Th2 phenotype upon introduction of activated B cells in pre-diabetic NOD mice [125]. Tian and colleagues further established that LPS-activated B cells mediate apoptosis of diabetogenic Th1 cells in NOD mice via expression of FasL and secretion of TGF-β [24]. These findings provide new insights into treating human T1DM via targeting the T cell-B cell interaction. Reduced numbers of IL-10-producing Bregs have been reported in patients with T1D [126]. There is substantial evidence that Bregs are either insufficient in number and/or functionally compromised in autoimmune diseases. Thus, further studies are needed to understand their mechanisms of action in these diseases.

5.3.6 Infectious Diseases

The role of B cells in infectious diseases has been studied extensively. In contrast, the role of Bregs in intracellular infections is unclear. Studies on Bregs in infections will uncover the valuable targets/potent markers in developing therapeutic interventions to treat various infectious diseases. Recent studies have shown that successful treatment of *Mycobacterium tuberculosis* infection induces Bregs with the ability to express FasL and IL-5RA in TB patients. Thus, these molecules could be potentially utilized as indicators of monitoring treatment responses during infections [127, 128]. Various studies have demonstrated the suppressive role of Bregs in chronic hepatitis B virus infection. Das et al. [129] first demonstrated that Bregs are responsible for regulating antigen-specific CD8$^+$ T cells in hepatitis B virus infection. They also found that inhibition of IL-10 may reestablish HBV-specific CD8$^+$ T cells in vitro. Various studies have reported that in HIV infection, Bregs impaired T cells via expression of IL-10 and programmed death (PD)-L1, contributing to immune dysfunction [130]. In 2014, Jiao et al. found that the frequency of Bregs in HIV patients was negatively correlated with the CD4$^+$ T cell count but was positively correlated with the viral load. Supporting this, it is also observed that following anti-retroviral treatment, the frequency of Bregs was decreased along with a concomitant step-wise increase in the CD4$^+$ T cell count.

5.3.7 Allergy and Asthma

Bregs also exert protection against allergic airway inflammation [131]. Through antigen- specific/non-specific immunomodulatory mechanisms, it is apparent that Bregs demonstrate allergen tolerance and contribute to suppress allergic diseases. Allergic inflammation is reported to be suppressed by IL-10-producing Bregs and involves a delicate balance between IL-10 induced parasite responses and detrimental IL-4-mediated allergic responses [132]. Br1 and Br3 cells increase in response to casein in milk-tolerant individuals [133] but not in milk-allergic individuals. Thus, both Br1 and Br3 cell types are critical for immune tolerance in non-IgE-mediated food allergies related to atopic dermatitis. Patients with allergic asthma and allergic rhinitis have a decreased number of IL-10-producing $CD24^{hi}CD27^+$ Bregs [134]. In a similar manner, beekeepers also develop tolerance against bee venom allergen, i.e., Phospholipase Az (PLAz)–specific to BR1 cells producing IgG4 antibodies by suppressing T cell responses in an IL-10-dependent manner [71, 135]. In allergic asthma, treatment with oral corticosteroids (OCS) significantly affects the frequency of Bregs as well as their ability to express IL-10 in a Breg subset–specific manner [136].

5.3.8 Osteoporosis

Osteoporosis represents one of the most common bone loss conditions, leading to higher fragility and bone fractures often related to advanced age and post-menopausal conditions [137, 138]. Osteoporosis is often a neglected disease with more than 200 million affected individuals worldwide, thus also referred as a "silent killer" [139, 140]. In the bone marrow, B cells are a major source of the osteoclastogenesis inhibitor osteoprotegerin (OPG), in the presence of activated T cells signaled by CD40L-CD40 interaction on B cells. Moreover, a CD40L-CD40-deficient mice showed reduced bone mass compared to the control mice. B cells also express RANKL along with OPG, which in the long run affects bone physiology. Furthermore, mice with B cell deficiency show suppressed OPG production and high prevalence of osteoporosis [141]. Bregs suppress various proinflammatory cytokines such as IL-1 and TNF-α, which are osteoclastogenic in nature, therefore leading to enhanced bone loss. The ratio of Th1/Th2 is an important parameter defining bone strength [142], including the rate of bone resorption and the resulting bone loss. Moreover, several subtypes of Bregs have now been reported with the suppression of Th1-, Th2-, or Th17-mediated autoimmune responses with a subsequent increase in Foxp3+ Treg cells along with conversion of effector T cells into Tr1 cells (CD4+ Foxp3+IL-10+ Treg 1 cells). Bregs have also been observed to suppress the expression of Th17 cells [59, 109], which are responsible for enhanced osteoclastogenesis and bone loss [142]. Recent observations (unpublished) from our lab clearly demonstrate the role of $CD19^{hi}CD1d^{hi}CD5^{hi}IL-10^{hi}$ Bregs in modulating bone health. Thus, further research is needed to establish the precise role of Bregs in regulating bone health.

5.4   Therapeutic Potential of Bregs: From Bench to Bedside

The present global scenario arising from various studies using experimental models and human disorders validate the vital role of Bregs in several diseases. Together, these studies indicate that Bregs have the potential to modulate a number of immune pathologies. Tedder et al. demonstrated that Bregs are involved in autoimmune responses and also provide protection to host tissues during the immunopathogenesis of infectious diseases [143]. More importantly, understanding the basic principle underlying the induction of Bregs will help in tweaking cellular tolerance and amend the influence of disease. As a small number of Bregs are inefficient in inhibiting inflammation, mechanisms that can enhance both the number and effector functions of Bregs can result in enhanced immune-suppressive functions. In the context of immunological conditions such as autoimmunity and transplantation, long-term usage of immunosuppressive drugs increases the likelihood of life-threatening infections. In certain conditions such as during graft transplantation, autoimmune diseases, and so on, expansion of the immunosuppressive Bregs population is needed. Thus, strategies that can be exploited by therapeutically targeting Bregs can open new avenues in treating various immune-mediated diseases such as the following: (a) ex vivo expansion of Bregs: stimulation of B cells in patient-derived PBMCs, leading to expansion of Bregs, followed by adoptive transfer of Bregs sorted by FACS may suppress the inflammation and re-induce tolerance. (b) in vivo modulation of Bregs for expansion: stimuli that can shift the differentiation of B cells toward immunosuppressive regulatory B cells. Some evidence suggest that pro-inflammatory cytokines such as B cell–activating factor (BAFF), IL-1β, IL-6, IL-21, IFN-α, and IFN- γ [23, 68] are the key cytokines that expand the Bregs population upon exposure. Interestingly, in arthritic mice, the gut microbiota has the potential to induce the expression of IL-1β and IL-6, which further promote Bregs differentiation and production of IL-10 cytokine [23]. (c) Depletion of Bregs: B cell depletion therapies (viz. rituximab), usage of targeted B cell therapies, that can target a specific subtype of B cells is more advantageous than total B cell depletion. Thus, further in-depth studies are required to develop Breg-dependent immunotherapies and to enhance their applications in treating various immune disorders and pathologies.

Acknowledgment This work was financially supported by projects: DST-SERB (EMR/2016/007158), Govt. of India and intramural project from All India Institute of Medical Sciences (AIIMS), New Delhi-India sanctioned to RKS; National Academy of Sciences (NASI), Allahabad-India sanctioned to GCM. HYD, LS, ZA, AB and RKS acknowledge the Department of Biotechnology, AIIMS, New Delhi-India for providing infrastructural facilities. ZA thanks Dr. Harisingh Gour Central University, Sagar (MP). LR and GCM acknowledge National Centre for Cell Science (NCCS), Pune-India for providing infrastructural facilities. HYD thanks ICMR for research fellowship. LR thanks NASI for research fellowship. LS and ZA thank the UGC and CSIR for their respective research fellowships.

Author Contributions RKS and GCM suggested the focus and outline of the review and wrote the review. HYD, LR, LS and ZA participated in writing of the review. RKS suggested and HYD and ZA created the illustrations.

Conflicts of Interest The authors declare no conflicts of interest.

References

1. Katz S, Parker D, Turk J (1974) B-cell suppression of delayed hypersensitivity reactions. Nature 251:550. https://doi.org/10.1038/1550a0
2. Wolf SD, Dittel BN, Hardardottir F, Janeway CA (1996) Experimental autoimmune encephalomyelitis induction in genetically B cell–deficient mice. J Exp Med 184:2271–2278. https://doi.org/10.1084/jem.184.6.2271
3. Fillatreau S, Sweenie CH, Mcgeachy MJ, Gray D, Anderton SM (2002) B cells regulate autoimmunity by provision of IL-10. Nat Immunol 3:944. https://doi.org/10.1038/ni833
4. Mizoguchi A, Mizoguchi E, Takedatsu H, Blumberg RS, Bhan AK (2002) Chronic intestinal inflammatory condition generates IL-10-producing regulatory B cell subset characterized by CD1d upregulation. Immunity 16:219–230. https://doi.org/10.1016/S1074-7613(02)00274-1
5. Mauri C, Gray D, Mushtaq N, Londei M (2003) Prevention of arthritis by interleukin 10–producing B cells. J Exp Med 197:489–501. https://doi.org/10.1084/jem.20021293
6. Parekh VV, Prasad DV, Banerjee PP, Joshi BN, Kumar A, Mishra GC (2003) B cells activated by lipopolysaccharide, but not by anti-Ig and anti-CD40 antibody, induce anergy in CD8+ T cells: role of TGF-β1. J Immunol 170:5897–5911. https://doi.org/10.4049/jimmunol.170.12.5897
7. Mizoguchi A, Bhan AK (2006) A case for regulatory B cells. J Immunol 176:705–710. https://doi.org/10.4049/jimmunol.176.2.705
8. Yanaba K, Bouaziz J-D, Haas KM, Poe JC, Fujimoto M, Tedder TF (2008) A regulatory B cell subset with a unique CD1dhiCD5+ phenotype controls T cell-dependent inflammatory responses. Immunity 28:639–650
9. Mann MK, Maresz K, Shriver LP, Tan Y, Dittel BN (2007) B cell regulation of CD4+ CD25+ T regulatory cells and IL-10 via B7 is essential for recovery from experimental autoimmune encephalomyelitis. J Immunol 178:3447–3456. https://doi.org/10.4049/jimmunol.178.6.3447
10. Ray A, Basu S, Williams CB, Salzman NH, Dittel BN (2012) A novel IL-10–independent regulatory role for B cells in suppressing autoimmunity by maintenance of regulatory T cells via GITR ligand. J Immunol 1103354. https://doi.org/10.4049/jimmunol.1103354
11. Amu S, Saunders SP, Kronenberg M, Mangan NE, Atzberger A, Fallon PG (2010) Regulatory B cells prevent and reverse allergic airway inflammation via FoxP3-positive T regulatory cells in a murine model. J Allergy Clin Immunol 125:1114–1124.e8. https://doi.org/10.1016/j.jaci.2010.01.018
12. Carter NA, Vasconcellos R, Rosser EC, Tulone C, Muñoz-Suano A, Kamanaka M et al (2011) Mice lacking endogenous IL-10–producing regulatory B cells develop exacerbated disease and present with an increased frequency of Th1/Th17 but a decrease in regulatory T cells. J Immunol 1100284. https://doi.org/10.4049/jimmunol.1100284
13. Lundy SK (2009) Killer B lymphocytes: the evidence and the potential. Inflamm Res 58:345. https://doi.org/10.1007/s00011-009-0014-x
14. Lundy SK, Boros DL (2008) Fas ligand-expressing B-1a lymphocytes mediate CD4(+)-T-cell apoptosis during schistosomal infection: induction by interleukin 4 (IL-4) and IL-10. Infect Immun 70:812
15. Kaku H, Cheng KF, Al-Abed Y, Rothstein TL (2014) A novel mechanism of B cell-mediated immune suppression through CD73 expression and adenosine production. J Immunol 193:5904

16. Khan AR, Hams E, Floudas A, Sparwasser T, Weaver CT, Fallon PG (2015) PD-L1hi B cells are critical regulators of humoral immunity. Nat Commun 6:5997
17. Oleinika K, Rosser EC, Matei DE, Nistala K, Bosma A, Drozdov I, Mauri C (2018) CD1d-dependent immune suppression mediated by regulatory B cells through modulations of iNKT cells. Nat Commun 9(1):684
18. Goetz M, Atreya R, Ghalibafian M, Galle PR, Neurath MF (2007) Exacerbation of ulcerative colitis after rituximab salvage therapy. Inflamm Bowel Dis 13:1365–1368
19. Dass S, Vital EM, Emery P (2007) Development of psoriasis after B cell depletion wit rituximab. Arthritis Rheum 56:2715–2718
20. Blair PA et al (2010) CD19(+)CD24(hi)CD38(hi) B cells exhibit regulatory capacity in healthy individuals but are functionally impaired in systemic lupus Erythematosus patients. Immunity 32(1):129–140
21. Blair PA, Chavez-Rueda KA, Evans JG, Shlomchik MJ, Eddaoudi A, Isenberg DA et al (2009) Selective targeting of B cells with agonistic anti-CD40 is an efficacious strategy for the generation of induced regulatory T2-like B cells and for the suppression of lupus in MRL/lpr mice. J Immunol 182:3492–3502. https://doi.org/10.4049/jimmunol.0803052
22. Iwata Y, Matsushita T, Horikawa M, Dilillo DJ, Yanaba K, Venturi GM et al (2011) Characterization of a rare IL-10–competent B-cell subset in humans that parallels mouse regulatory B10 cells. Blood 117:530–541. https://doi.org/10.1182/blood-2010-07-294249
23. Rosser EC, Oleinika K, Tonon S, Doyle R, Bosma A, Carter NA et al (2014) Regulatory B cells are induced by gut microbiota–driven interleukin-1β and interleukin-6 production. Nat Med 20:1334. https://doi.org/10.1038/nm.3680
24. Tian J, Zekzer D, Hanssen L, Lu Y, Olcott A, Kaufman DL (2001) Lipopolysaccharide-activated B cells down-regulate Th1 immunity and prevent autoimmune diabetes in nonobese diabetic mice. J Immunol 167:1081–1089. https://doi.org/10.4049/jimmunol.167.2.1081
25. Yoshizaki A, Miyagaki T, Dilillo DJ, Matsushita T, Horikawa M, Kountikov EI et al (2012) Regulatory B cells control T-cell autoimmunity through IL-21-dependent cognate interactions. Nature 491:264. https://doi.org/10.1038/nature11501
26. Bankoti R, Gupta K, Levchenko A, Stäger S (2012) Marginal zone B cells regulate antigen-specific T cell responses during infection. J Immunol 188(8):3961–3971. https://doi.org/10.4049/jimmunol.1102880
27. Yanaba K, Bouaziz J-D, Matsushita T, Tsubata T, Tedder TF (2009) The development and function of regulatory B cells expressing IL-10 (B10 cells) requires antigen receptor diversity and TLR signals. J Immunol 182:7459–7472. https://doi.org/10.4049/jimmunol.0900270
28. Fagraeus A (1948) The plasma cellular reaction and its relation to the formation of antibodies in vitro. J Immunol 58:1–13
29. Kallies A, Hasbold J, Tarlinton DM, Dietrich W, Corcoran LM, Hodgkin PD et al (2004) Plasma cell ontogeny defined by quantitative changes in blimp-1 expression. J Exp Med 200:967–977
30. Lino AC, Lampropoulou V, Welle A, Joedicke J, Pohar J, Simon Q, Thalmensi J, Baures A, Fluhler V, Sakwa I, Stervbo U (2018) LAG-3 inhibitory receptor expression identifies immunosuppressive natural regulatory plasma cells. Immunity 49(1):120–133
31. Niedbala W, Wei XQ, Cai B, Hueber AJ, Leung BP, McInnes IB, Liew FY (2007) IL-35 is a novel cytokine with therapeutic effects against collagen-induced arthritis through the expansion of regulatory T cells and suppression of Th17 cells. Eur J Immunol 37(11):3021–3029. https://doi.org/10.1002/eji.200737810
32. Yu CR, Choi JK, Uche AN, Egwuagu CE (2018) Production of IL-35 by Bregs is mediated through binding of BATF-IRF-4-IRF-8 complex to il12a and ebi3 promoter elements. J Leukoc Biol 104(6):1147–1157
33. Corneth OB, Verstappen GM, Paulissen SM, de Bruijn MJ, Rip J, Lukkes M, Hendriks RW (2017) Enhanced Bruton's tyrosine kinase activity in peripheral blood B lymphocytes from patients with autoimmune disease. Arthritis Rheum 69(6):1313–1324
34. Hardy RR (2006) B-1 B cell development. J Immunol 177:2749–2754. https://doi.org/10.4049/jimmunol.177.5.2749

35. Kantor A (1991) A new nomenclature for B cells. Immunol Today 12:388. https://doi.org/10.1016/0167-5699(91)90135-G
36. Stall AM, Adams S, Herzenberg LA, Kantor AB (1992) Characteristics and development of the murine B-lb (Ly-1 B sister) cell population. Ann N Y Acad Sci 651:33–43. https://doi.org/10.1111/j.1749-6632.1992.tb24591.x
37. Kawahara T, Ohdan H, Zhao G, Yang Y-G, Sykes M (2003) Peritoneal cavity B cells are precursors of splenic IgM natural antibody-producing cells. J Immunol 171:5406–5414. https://doi.org/10.4049/jimmunol.171.10.5406
38. Choi YS, Dieter JA, Rothaeusler K, Luo Z, Baumgarth N (2012) B-1 cells in the bone marrow are a significant source of natural IgM. Eur J Immunol 42:120–129. https://doi.org/10.1002/eji.201141890
39. Bos NA, Kimura H, Meeuwsen CG, Visser HD, Hazenberg MP, Wostmann BS et al (1989) Serum immunoglobulin levels and naturally occurring antibodies against carbohydrate antigens in germ-free BALB/c mice fed chemically defined ultrafiltered diet. Eur J Immunol 19:2335–2339. https://doi.org/10.1002/eji.1830191223
40. Haury M, Sundblad A, Grandien A, Barreau C, Coutinho A, Nobrega A (1997) The repertoire of serum IgM in normal mice is largely independent of external antigenic contact. Eur J Immunol 27:1557–1563. https://doi.org/10.1002/eji.1830270635
41. Baumgarth N, Herman OC, Jager GC, Brown L, Herzenberg LA, Herzenberg LA (1999) Innate and acquired humoral immunities to influenza virus are mediated by distinct arms of the immune system. Proc Natl Acad Sci 96:2250–2255. https://doi.org/10.1073/pnas.96.5.2250
42. Kantor AB, Merrill CE, Herzenberg LA, Hillson JL (1997) An unbiased analysis of V (H)-DJ (H) sequences from B-1a, B-1b, and conventional B cells. J Immunol 158:1175–1186
43. Notkins AL (2004) Polyreactivity of antibody molecules. Trends Immunol 25:174–179. https://doi.org/10.1016/j.it.2004.02.004
44. Stewart J (1992) Immunoglobulins did not arise in evolution to fight infection. Immunol Today 13:396–399. https://doi.org/10.1016/0167-5699(92)90088-O
45. Baumgarth N, Tung JW, Herzenberg LA (2005) Inherent specificities in natural antibodies: a key to immune defense against pathogen invasion. Springer Semin Immunopathol 26:347–362
46. O'Garra A, Chang R, Go N, Hastings R, Haughton G, Howard M (1992) Ly-1 B (B-1) cells are the main source of B cell-derived interleukin 10. Eur J Immunol 22:711–717
47. Velupillai P, Sypek J, Harn DA (1996) Interleukin-12 and -10 and gamma interferon regulate polyclonal and ligand-specific expansion of murine B-1 cells. Infect Immun 64:4557–4560
48. Al-Qaoud KM, Fleischer B, Hoerauf A (1998) The Xid defect imparts susceptibility to experimental murine filariosis – association with a lack of antibody and IL-10 production by B cells in response to phosphorylcholine. Int Immunol 10(1):17–25
49. Zhang X, Deriaud E, Jiao X, Braun D, Leclerc C, Lo-Man R (2007) Type I interferons protect neonates from acute inflammation through interleukin 10-producing B cells. J Exp Med 204:1107
50. Shimomura Y, Mizoguchi E, Sugimoto K et al (2008) Regulatory role of B-1 B cells in chronic colitis. Int Immunol 20:729
51. Lee KM, Stott RT, Zhao G, Soohoo J, Xiong W, Lian MM et al (2014) TGF-β-producing regulatory B cells induce regulatory T cells and promote transplantation tolerance. Eur J Immunol 44:1728–1736. https://doi.org/10.1002/eji.201344062
52. Matsushita T, Horikawa M, Iwata Y, Tedder TF (2010) Regulatory B cells (B10 cells) and regulatory T cells have independent roles in controlling experimental autoimmune encephalomyelitis initiation and late-phase immunopathogenesis. J Immunol 1001307. https://doi.org/10.4049/jimmunol.1001307
53. Maseda D, Candando KM, Smith SH, Kalampokis I, Weaver CT, Plevy SE et al (2013) Peritoneal cavity regulatory B cells (B10 cells) modulate IFN-γ+ CD4+ T cell numbers during colitis development in mice. J Immunol 1300649. https://doi.org/10.4049/jimmunol.1300649
54. Maseda D, Smith SH, Dilillo DJ, Bryant JM, Candando KM, Weaver CT et al (2012) Regulatory B10 cells differentiate into antibody-secreting cells after transient IL-10 production in vivo. J Immunol 188:1036–1048. https://doi.org/10.4049/jimmunol.1102500

55. Kuchroo VK, Dardalhon V, Xiao S, Anderson AC (2008) New roles for TIM family members in immune regulation. Nat Rev Immunol 8:577

56. Xiao S, Brooks CR, Sobel RA, Kuchroo VK (2015) Tim-1 is essential for induction and maintenance of IL-10 in regulatory B cells and their regulation of tissue inflammation. J Immunol 194:1602

57. Ding Q, Yeung M, Camirand G, Zeng Q, Akiba H, Yagita H et al (2011) Regulatory B cells are identified by expression of TIM-1 and can be induced through TIM-1 ligation to promote tolerance in mice. J Clin Invest 121. https://doi.org/10.1172/JCI46274

58. Matsushita T, Yanaba K, Bouaziz J-D, Fujimoto M, Tedder TF (2008) Regulatory B cells inhibit EAE initiation in mice while other B cells promote disease progression. J Clin Invest 118:3420–3430. https://doi.org/10.1172/JCI36030

59. Haas KM, Watanabe R, Matsushita T, Nakashima H, Ishiura N, Okochi H et al (2010) Protective and pathogenic roles for B cells during systemic autoimmunity in NZB/W F1 mice. J Immunol 184(9):4789–4800. https://doi.org/10.4049/jimmunol.0902391

60. Yanaba K, Yoshizaki A, Asano Y, Kadono T, Tedder TF, Sato S (2011) IL-10-producing regulatory B10 cells inhibit intestinal injury in a mouse model. Am J Pathol 178:735–743

61. Le Huu D, Matsushita T, Jin G, Hamaguchi Y, Hasegawa M, Takehara K, Tedder TF, Fujimoto M (2013) Donor-derived regulatory B cells are important for suppression of murine sclerodermatous chronic graft-versus-host disease. Blood 121:3274–3283

62. Mauri C, Mars LT, Londei M (2000) Therapeutic activity of agonistic monoclonal antibodies against CD40 in a chronic autoimmune inflammatory process. Nat Med 6:673. https://doi.org/10.1038/76251

63. Horikawa M, Weimer ET, Dilillo DJ, Venturi GM, Spolski R, Leonard WJ et al (2013) Regulatory B cell (B10 cell) expansion during Listeria infection governs innate and cellular immune responses in mice. J Immunol 190:1158–1168. https://doi.org/10.4049/jimmunol.1201427

64. Dilillo DJ, Griffiths R, Seshan SV, Magro CM, Ruiz P, Coffman TM et al (2011) B lymphocytes differentially influence acute and chronic allograft rejection in mice. J Immunol 1002983. https://doi.org/10.4049/jimmunol.1002983

65. Evans JG, Chavez-Rueda KA, Eddaoudi A, Meyer-Bahlburg A, Rawlings DJ, Ehrenstein MR et al (2007) Novel suppressive function of transitional 2 B cells in experimental arthritis. J Immunol 178:7868–7878. https://doi.org/10.4049/jimmunol.178.12.7868

66. Schioppa T, Moore R, Thompson RG, Rosser EC, Kulbe H, Nedospasov S et al (2011) B regulatory cells and the tumor-promoting actions of TNF-α during squamous carcinogenesis. Proc Natl Acad Sci 201100994. https://doi.org/10.1073/pnas.1100994108

67. Bosma A, Abdel-Gadir A, Isenberg DA, Jury EC, Mauri C (2012) Lipid-antigen presentation by CD1d+ B cells is essential for the maintenance of invariant natural killer T cells. Immunity 36:477–490. https://doi.org/10.1016/j.immuni.2012.02.008

68. Menon M, Blair PA, Isenberg DA, Mauri C (2016) A regulatory feedback between plasmacytoid dendritic cells and regulatory B cells is aberrant in systemic lupus erythematosus. Immunity 44:683–697. https://doi.org/10.1016/j.immuni.2016.02.012

69. Bouaziz JD, Calbo S, Maho-Vaillant M, Saussine A, Bagot M, Bensussan A et al (2010) IL-10 produced by activated human B cells regulates CD4+ T-cell activation in vitro. Eur J Immunol 40:2686–2691. https://doi.org/10.1016/j.immuni.2012.02.008

70. Flores-Borja F, Bosma A, Ng D, Reddy V, Ehrenstein MR, Isenberg DA et al (2013) CD19+ CD24hiCD38hi B cells maintain regulatory T cells while limiting TH1 and TH17 differentiation. Sci Transl Med 5:173ra23–173ra23. https://doi.org/10.1126/scitranslmed.3005407

71. Tarique M, Naz H, Kurra SV, Saini C, Naqvi RA, Rai R et al (2018) Interleukin-10 producing regulatory B cells transformed CD4+ CD25− into Tregs and enhanced regulatory T cells function in human leprosy. Front Immunol 9. https://doi.org/10.3389/fimmu.2018.01636

72. Van de Veen W, Stanic B, Yaman G, Wawrzyniak M, Sollner S, Akdis DG et al (2013) IgG4 production is confined to human IL-10–producing regulatory B cells that suppress antigen-specific immune responses. J Allergy Clin Immunol 131(4):1204–1212. https://doi.org/10.1016/j.jaci.2013.01.014

73. Matsumoto M, Baba A, Yokota T, Nishikawa H, Ohkawa Y, Kayama H, Kallies A, Nutt SL, Sakaguchi S, Takeda K, Kurosaki T (2014) Interleukin-10-producing plasmablasts exert regulatory function in autoimmune inflammation. Immunity 41(6):1040–1051. https://doi.org/10.1016/j.immuni.2014.10.016

74. Aravena O, Ferrier A, Menon M, Mauri C, Aguillón JC, Soto L, Catalán D (2017) TIM-1 defines a human regulatory B cell population that is altered in frequency and function in systemic sclerosis patients. Arthritis Res Ther 19(1):8

75. Wang T, Li Z, Li X, Chen L, Zhao H, Jiang C, Song L (2017) Expression of CD19+ CD24highCD38high B cells, IL-10 and IL-10R in peripheral blood from patients with systemic lupus erythematosus. Mol Med Rep 16(5):6326–6333. https://doi.org/10.3892/mmr.2017.7381

76. Cherukuri A, Rothstein DM, Clark B, Carter CR, Davison A, Hernandez-Fuentes M, Baker RJ (2014) Immunologic human renal allograft injury associates with an altered IL-10/TNF-α expression ratio in regulatory B cells. J Am Soc Nephrol 25(7):1575–1585. https://doi.org/10.1681/ASN.2013080837

77. Kristensen B, Hegedus L, Lundy SK, Brimnes MK, Smith TJ, Nielsen CH (2015) Characterization of regulatory B cells in Graves' disease and Hashimoto's thyroiditis. PLoS One 10(5):e0127949

78. Bankó Z, Pozsgay J, Szili D, Tóth M, Gáti T, Nagy G et al (2017) Induction and differentiation of IL-10–producing regulatory B cells from healthy blood donors and rheumatoid arthritis patients. J Immunol 198(4):1512–1520. https://doi.org/10.4049/jimmunol.1600218

79. Amel Kashipaz MR, Huggins ML, Lanyon P, Robins A, Powell RJ, Todd I (2003) Assessment of Be1 and Be2 cells in systemic lupus erythematosus indicates elevated interleukin-10 producing CD5+ B cells. Lupus 12:356–363

80. Llorente L, Richaud-Patin Y, Fior R, Alcocer-Varela J, Wijdenes J, Fourrier BM, Galanaud P, Emilie D (1994) In vivo production of interleukin-10 by non-T cells in rheumatoid arthritis, Sjogren's syndrome, and systemic lupus erythematosus. A potential mechanism of B lymphocyte hyperactivity and autoimmunity. Arthritis Rheum 37:1647–1655

81. Liu HJ, Guo XF, Hu FL, Yan CP, Cui XJ, Yan XL et al (2018) Increased receptor activator of nuclear factor kappa B ligand expressed on B10 cells in rheumatoid arthritis. Beijing da xuexue bao. Yi xue ban. J Peking Univ Health Sci 50(6):968–974. PMID: 30562766

82. Sperling S, Fiedler P, Lechner M, Pollithy A, Ehrenberg S, Schiefer AI et al (2019) Chronic CD30 signaling in B cells results in lymphomagenesis by driving the expansion of plasmablasts and B1 cells. *Blood* 133(24):2597–2609. https://doi.org/10.1182/blood.2018880138

83. Hartung HP, Kieseier BC (2010) Atacicept: targeting B cells in multiple sclerosis. Ther Adv Neurol Disord 3:205–216

84. Lin W, Zhang P, Chen H, Chen Y, Yang H, Zheng W et al (2017) Circulating plasmablasts/plasma cells: a potential biomarker for IgG4-related disease. Arthritis Res Ther 19(1):25. https://doi.org/10.1186/s13075-017-1231-2

85. Kubo S, Nakayamada S, Zhao J, Yoshikawa M, Miyazaki Y, Nawata A et al (2017) Correlation of T follicular helper cells and plasmablasts with the development of organ involvement in patients with IgG4-related disease. Rheumatology 57(3):514–524. https://doi.org/10.1093/rheumatology/kex455

86. Arce E, Jackson DG, Gill MA, Bennett LB, Banchereau J, Pascual V (2001) Increased frequency of pre-germinal center B cells and plasma cell precursors in the blood of children with systemic lupus erythematosus. J Immunol 167:2361–2369

87. Arbore G, Henley T, Biggins L, Andrews S, Vigorito E, Turner M, Leyland R (2019) MicroRNA-155 is essential for the optimal proliferation and survival of plasmablast B cells. Life Sci Alliance 2(3):e201800244

88. Lindner S, Dahlke K, Sontheimer K, Hagn M, Kaltenmeier C, Barth T et al (2013) Interleukin-21-induced granzyme B-expressing B lymphocytes regulate T cells and infiltrate tumors (P1088). Cancer Res. https://doi.org/10.1158/0008-5472.CAN-12-3450

89. Hagn M, Ebel V, Sontheimer K, Schwesinger E, Lunov O, Beyer T, Fabricius D, Barth TF, Viardot A, Stilgenbauer S, Hepp J, Scharffetter-Kochanek K, Simmet T, Jahrsdörfer B (2010)

CD5+ B cells from individuals with systemic lupus erythematosus express granzyme B. Eur J Immunol 40(7):2060–2069

90. Hagn M, Schwesinger E, Ebel V, Sontheimer K, Maier J, Beyer T, Syrovets T, Laumonnier Y, Fabricius D, Simmet T, Jahrsdörfer B (2009) Human B cells secrete Granzyme B when recognizing viral antigens in the context of the acute phase cytokine IL-21. J Immunol 183(3):1838–1845

91. Xu L, Liu X, Liu H, Zhu L, Zhu H, Zhang J et al (2017) Impairment of granzyme B-producing regulatory B cells correlates with exacerbated rheumatoid arthritis. Front Immunol 8(768). https://doi.org/10.3389/fimmu.2017.00768

92. Nouel A, Pochard P, Simon Q, Segalen I, Le Meur Y, Pers J et al (2015) B-cells induce regulatory T cells through TGF-β/IDO production in a CTLA-4 dependent manner. J Autoimmun 59:53–60. https://doi.org/10.1016/j.jaut.2015.02.004

93. Fehres CM, van Uden NO, Yeremenko NG, Fernandez L, Franco Salinas G, Van Duivenvoorde LM et al (2019) APRIL induces a novel subset of IgA+ regulatory B cells that suppress inflammation via expression of IL-10 and PD-L1. Front Immunol 10:1368

94. Liu R, Lu Z, Gu J, Liu J, Huang E, Liu X et al (2018) MicroRNAs 15A and 16–1 activate signaling pathways that mediate chemotaxis of immune regulatory B cells to colorectal tumors. Gastroenterology 154(3):637–651. https://doi.org/10.1053/j.gastro.2017.09.045

95. Porakishvili N, Mageed R, Jamin C, Pers JO, Kulikova N, Renaudineau Y et al (2001) Recent progress in the understanding of B-cell functions in autoimmunity. *Scand J Immunol* 54:30–38

96. Isenberg DA (2006) B cell targeted therapies in autoimmune diseases. J Rheumatol Suppl 33:24–28. PMID: 16652442

97. Correale J, Farez M, Razzitte G (2008) Helminth infections associated with multiple sclerosis induce regulatory B cells. Ann Neurol 64:187–199. https://doi.org/10.1002/ana.21438

98. Knippenberg S, Peelen E, Smolders J, Thewissen M, Menheere P, Cohen Tervaert JW et al (2011) Reduction in IL-10 producing B cells (Breg) in multiple sclerosis is accompanied by a reduced naive/memory Breg ratio during a relapse but not in remission. J Neuroimmunol 239:80–86. https://doi.org/10.1016/j.jneuroim.2011.08.019

99. Grutzke B, Hucke S, Gross CC, Herold MV, Posevitz-Fejfar A, Wildemann BT et al (2015) Fingolimod treatment promotes regulatory phenotype and function of B cells. Ann Clin Transl Neurol 2:119–130. https://doi.org/10.1002/acn3.155

100. Thompson SA, Jones JL, Cox AL, Compston DA, Coles AJ (2010) B-cell reconstitution and BAFF after alemtuzumab (Campath-1H) treatment of multiple sclerosis. J Clin Immunol 30:99–105. https://doi.org/10.1007/s10875-009-9327-3

101. Bouaziz J-D, Yanaba K, Venturi GM, Wang Y, Tisch RM, Poe JC et al (2007) Therapeutic B cell depletion impairs adaptive and autoreactive CD4+ T cell activation in mice. Proc Natl Acad Sci 104:20878–20883. https://doi.org/10.1073/pnas.0709205105

102. Lund FE (2008) Cytokine-producing B lymphocytes key regulators of immunity. Curr Opin Immunol 20:332–338. https://doi.org/10.1016/j.coi.2008.03.003

103. Matsushita T, Fujimoto M, Hasegawa M, Komura K, Takehara K, Tedder TF et al (2006) Inhibitory role of CD19 in the progression of experimental autoimmune encephalomyelitis by regulating cytokine response. Am J Pathol 168:812–821. https://doi.org/10.2353/ajpath.2006.050923

104. Talley NJ, Abreu MT, Achkar J-P, Bernstein CN, Dubinsky MC, Hanauer SB et al (2011) An evidence-based systematic review on medical therapies for inflammatory bowel disease. Am J Gastroenterol 106:S2. https://doi.org/10.1038/ajg.2011.58

105. Molodecky NA, Soon S, Rabi DM, Ghali WA, Ferris M, Chernoff G et al (2012) Increasing incidence and prevalence of the inflammatory bowel diseases with time, based on systematic review. Gastroenterology 142:46–54.e42. https://doi.org/10.1053/j.gastro.2011.10.001

106. Mizoguchi A, Mizoguchi E, Smith RN, Preffer FI, Bhan AK (1997) Suppressive role of B cells in chronic colitis of T cell receptor α mutant mice. J Exp Med 186:1749–1756. https://doi.org/10.1084/jem.186.10.1749

107. Wei B, Velazquez P, Turovskaya O, Spricher K, Aranda R, Kronenberg M et al (2015) Mesenteric B cells centrally inhibit CD4+ T cell colitis through interaction with regulatory T cell subsets. Proc Natl Acad Sci 102:2010–2015. https://doi.org/10.1073/pnas.0409449102

108. Kalampokis I, Yoshizaki A, Tedder TF (2013) IL-10-producing regulatory B cells (B10 cells) in autoimmune disease. Arthritis Res Ther 15:S1. https://doi.org/10.1186/ar3907

109. Oka A, Ishihara S, Mishima Y, Tada Y, Kusunoki R, Fukuba N et al (2014) Role of regulatory B cells in chronic intestinal inflammation: association with pathogenesis of Crohn's disease. Inflamm Bowel Dis 20:315–328. https://doi.org/10.1097/01.MIB.0000437983.14544.d5

110. Blanco P, Palucka AK, Gill M, Pascual V, Banchereau J (2001) Induction of dendritic cell differentiation by IFN-α in systemic lupus erythematosus. Science 294:1540–1543. https://doi.org/10.1126/science.1064890

111. Carroll M (2001) Innate immunity in the etiopathology of autoimmunity. Nat Immunol 2:1089. https://doi.org/10.1038/ni1201-1089

112. Miyagaki T, Fujimoto M, Sato S (2015) Regulatory B cells in human inflammatory and autoimmune diseases: from mouse models to clinical research. Int Immunol 27:495–504. https://doi.org/10.1093/intimm/dxv026

113. Mauri C, Bosma A (2012) Immune regulatory function of B cells. Annu Rev Immunol 30:221–241. https://doi.org/10.1146/annurev-immunol-020711-074934

114. Lawrence RC, Felson DT, Helmick CG, Arnold LM, Choi H, Deyo RA et al (2008) Estimates of the prevalence of arthritis and other rheumatic conditions in the United States: Part II. Arthritis Rheum 58:26–35. https://doi.org/10.1002/art.23176

115. Feldmann M, Brennan FM, Maini RN (1996) Rheumatoid arthritis. Cell 85:1277–1289. https://doi.org/10.1016/s0092-8674(00)81109-5

116. Kuhn KA, Kulik L, Tomooka B, Braschler KJ, Arend WP, Robinson WH et al (2006) Antibodies against citrullinated proteins enhance tissue injury in experimental autoimmune arthritis. J Clin Invest 116:961–973. https://doi.org/10.1172/JCI25422

117. Daien CI, Gailhac S, Mura T, Audo R, Combe B, Hahne M et al (2014) Regulatory B10 cells are decreased in patients with rheumatoid arthritis and are inversely correlated with disease activity. Arthritis Rheum 66:2037–2046. https://doi.org/10.1002/art.38666

118. Cui D, Zhang L, Chen J, Zhu M, Hou L, Chen B et al (2015) Changes in regulatory B cells and their relationship with rheumatoid arthritis disease activity. Clin Exp Med 15:285–292. https://doi.org/10.1007/s10238-014-0310-9

119. Anderson MS, Bluestone JA (2005) The NOD mouse: a model of immune dysregulation. Annu Rev Immunol 23:447–485. https://doi.org/10.1146/annurev.immunol.23.021704.115643

120. Silveira PA, Grey ST (2006) B cells in the spotlight: innocent bystanders or major players in the pathogenesis of type 1 diabetes. Trends Endocrinol Metab 17:128–135. https://doi.org/10.1016/j.tem.2006.03.006

121. Fox CJ, Danska JS (1998) Independent genetic regulation of T-cell and antigen-presenting cell participation in autoimmune islet inflammation. Diabetes 47:331–338. https://doi.org/10.2337/diabetes.47.3.331

122. Xiu Y, Wong CP, Bouaziz J-D, Hamaguchi Y, Wang Y, Pop SM et al (2008) B lymphocyte depletion by CD20 monoclonal antibody prevents diabetes in nonobese diabetic mice despite isotype-specific differences in FcγR effector functions. J Immunol 180:2863–2875. https://doi.org/10.4049/jimmunol.180.5.2863

123. Mariño E, Villanueva J, Walters S, Liuwantara D, Mackay F, Grey ST (2009) CD4+ CD25+ T cells control autoimmunity in the absence of B cells. Diabetes. https://doi.org/10.2337/db08-1504

124. Smith SH, Tedder TF (2009) Targeting B-cells mitigates autoimmune diabetes in NOD mice: what is plan b? Diabetes 58:1479–1481. https://doi.org/10.2337/db09-0497

125. Hussain S, Delovitch TL (2007) Intravenous transfusion of BCR-activated B cells protects NOD mice from type 1 diabetes in an IL-10-dependent manner. J Immunol 179:7225–7232. https://doi.org/10.4049/jimmunol.1201427

126. Kleffel S, Vergani A, Tezza S, Ben Nasr M, Niewczas MA, Wong S et al (2015) Interleukin-10+ regulatory B cells arise within antigen-experienced CD40+ B cells to maintain tolerance to islet autoantigens. Diabetes 64:158–171. https://doi.org/10.2337/db13-1639

127. Van Rensburg IC, Kleynhans L, Keyser A, Walzl G, Loxton AG (2017a) B-cells with a FasL expressing regulatory phenotype are induced following successful anti-tuberculosis treatment. Immun Inflamm Dis 5(1):57–67. https://doi.org/10.1002/iid3.140

128. Van Rensburg IC, Wagman C, Stanley K, Beltran C, Ronacher K, Walzl G, Loxton AG (2017b) Successful TB treatment induces B-cells expressing FASL and IL5RA mRNA. Oncotarget 8(2):2037. https://doi.org/10.18632/oncotarget.12184

129. Das A, Ellis G, Pallant C, Lopes AR, Khanna P, Peppa D, Kennedy PT (2012) IL-10–producing regulatory B cells in the pathogenesis of chronic hepatitis B virus infection. J Immunol 189(8):3925–3935. https://doi.org/10.4049/jimmunol.1103139

130. Siewe B, Stapleton JT, Martinson J, Keshavarzian A, Kazmi N, Demarais PM, Landay A (2013) Regulatory B cell frequency correlates with markers of HIV disease progression and attenuates anti-HIV CD8+ T cell function in vitro. J Leukoc Biol 93(5):811–818. https://doi.org/10.1189/jlb.0912436

131. Lundy SK, Berlin AA, Martens TF, Lukacs NW (2005) Deficiency of regulatory B cells increases allergic airway inflammation. Inflamm Res 54(12):514–521. https://doi.org/10.1007/s00011-005-1387-0

132. Mangan NE, Fallon RE, Smith P, van Rooijen N, McKenzie AN, Fallon PG (2004) Helminth infection protects mice from anaphylaxis via IL-10-producing B cells. J Immunol 173(10):6346–6356. https://doi.org/10.4049/jimmunol.173.10.6346

133. Lee JH, Noh J, Noh G, Choi WS, Cho S, Lee SS (2011) Allergen-specific transforming growth factor-β-producing CD19 (+) CD5 (+) regulatory B-cell (Br3) responses in human late eczematous allergic reactions to cow's milk. J Interf Cytokine Res 31(5):441–449. https://doi.org/10.1089/jir.2010.0020

134. Kamekura R, Shigehara K, Miyajima S, Jitsukawa S, Kawata K, Yamashita K et al (2015) Alteration of circulating type 2 follicular helper T cells and regulatory B cells underlies the comorbid association of allergic rhinitis with bronchial asthma. Clin Immunol 158(2):204–211. https://doi.org/10.1016/j.clim.2015.02.016

135. Mauri C, Menon M (2017) Human regulatory B cells in health and disease: therapeutic potential. J Clin Invest 127(3):772–779. https://doi.org/10.1172/JCI85113

136. Wiest M, Upchurch K, Hasan MM, Cardenas J, Lanier B, Millard M, Joo H (2019) Phenotypic and functional alterations of regulatory B cell subsets in adult allergic asthma patients. Clin Exp Allergy. https://doi.org/10.1111/cea.13439

137. Dar HY, Pal S, Shukla P, Mishra PK, Tomar GB, Chattopadhyay N et al (2018) *Bacillus clausii* inhibits bone loss by skewing Treg-Th17 cell equilibrium in postmenopausal osteoporotic mice model. Nutrition 54:118–128. https://doi.org/10.1016/j.nut.2018.02.013

138. Dar HY, Shukla P, Mishra PK, Anupam R, Mondal RK, Tomar GB et al (2018) *Lactobacillus acidophilus* inhibits bone loss and increases bone heterogeneity in osteoporotic mice via modulating Treg-Th17 cell balance. Bone Rep 8:46–56. https://doi.org/10.1016/j.bonr.2018.02.001

139. Dar HY, Azam Z, Anupam R, Mondal RK, Srivastava RK (2018) Osteoimmunology: the Nexus between bone and immune system. Front Biosci 23:464–492. https://doi.org/10.2741/4600

140. Dar HY, Singh A, Shukla P, Anupam R, Mondal RK, Mishra PK et al (2018) High dietary salt intake correlates with modulated Th17-Treg cell balance resulting in enhanced bone loss and impaired bone-microarchitecture in male mice. Sci Rep 8:2503. https://doi.org/10.1038/s41598-018-20896-y

141. Li Y, Toraldo G, Li A, Yang X, Zhang H, Qian WP, Weitzmann MN (2007) B cells and T cells are critical for the preservation of bone homeostasis and attainment of peak bone mass in vivo. Blood 109(9):3839–3848. https://doi.org/10.1182/blood-2006-07-037994

142. Srivastava RK, Dar HY, Mishra PK (2018) Immunoporosis: immunology of osteoporosis-role of T cells. Front Immunol 9:657. https://doi.org/10.3389/fimmu.2018.00657

143. Dilillo DJ, Matsushita T, Tedder TF (2010) B10 cells and regulatory B cells balance immune responses during inflammation, autoimmunity, and cancer. Ann N Y Acad Sci 1183:38–57. https://doi.org/10.1111/j.1749-6632.2009.05137.x
144. Sims GP, Ettinger R, Shirota Y, Yarboro CH, Illei GG, Lipsky PE (2005) Identification and characterization of circulating human transitional B cells. Blood 105(11):4390–4398. https://doi.org/10.1182/blood-2004-11-4284

T-Cell Activation and Differentiation: Role of Signaling and Metabolic Cross-Talk

Rupa Bhowmick, Piyali Ganguli, and Ram Rup Sarkar

Abstract

Different types of T effector cells function centrally in the immune-regulatory network, which acts as a line of defense for the body and elicits immune response during any diseased condition. At the molecular level, this functioning is maintained by an intricately designed network of signaling and metabolic pathways that function via multiple cross-talks to regulate complex immune responses during different antigenic challenges. These pathways regulate phenomena such as quiescence exit of naïve T cells, their activation, and differentiation into different effector T cells. Signaling properties of these T cells and their response to different cytokine signals have been well studied. Immune-metabolism is comparatively a new area of research that has been identified as driver for immune response. However, to gain a holistic understanding of the activation and differentiation of naïve T cells into the subtypes, the integration of signaling and metabolic pathway information is a prerequisite. The bidirectional mode of regulation between these cross-talking signaling and metabolic pathways governs the differentiation patterns. In this chapter, we review the activation and differentiation pattern of naïve T cells from both signaling and metabolic perspectives and also look into their cross-talk to understand their mutual regulation during differentiation into effector T cells.

R. Bhowmick · P. Ganguli · R. R. Sarkar (✉)
CSIR-National Chemical Laboratory, Pune, India

Academy of Scientific and Innovative Research (AcSIR), Ghaziabad, India
e-mail: rr.sarkar@ncl.res.in

© Springer Nature Singapore Pte Ltd. 2020
S. Singh (ed.), *Systems and Synthetic Immunology*,
https://doi.org/10.1007/978-981-15-3350-1_6

6.1   Introduction

The immune system forms the sentinel of the body that protects it from infectious disease and cancer. The adaptive immune system, composed mainly of the T and B lymphocytes, is responsible for maintaining this defense mechanism of the body as it helps to generate immune responses specific to the type of antigenic challenge that the body encounters [1]. The helper T cells (T_H) form the central orchestrators of the entire immune-regulatory network. They have been known to have an essential role in the recognition of the antigen when presented on the surface of the antigen-presenting cells and secrete cytokines that aid in the proliferation of the cytotoxic T cells and B cells, thereby playing an active role in stimulating both the humoral and the cell-mediated immunity [2]. The effector functions of these immune systems are mediated mainly by the cytokines and other microbicidal molecules secreted by them as a result of the activation of complex biochemical signaling pathways inside the immune cells. The T_H cells themselves produce a high amount of interferon and tumor necrosis factor via TCR and co-receptor mediated pathways that mediates apoptosis of infected and cancerous cells [3, 4].

The differentiation of the helper T cells is primarily influenced by the changes in the micro-environmental conditions that favor the proliferation of a certain subset of T cells that leads to disruption of the balance and ratio of the normal proportions of T-cell subsets present in a healthy individual [5, 6].

Naive T cells circulate in the body surveying for antigens. The metabolic activity of these cells is maintained low by allowing low uptake of glucose enough to fuel the TCA cycle and OXPHOS to produce ATP [7]. These cells are kept in a quiescent state that promotes their survival and persistence. On antigen stimulation, the metabolism of T cells is triggered via increased uptake of glucose, which allows quiescence exit and initiates clonal expansion and effector differentiation primarily by mTOR-mediated signaling responses [8]. Initially, the focus of studies remained on the immune receptors and transcriptional regulators involved in T-cell quiescence and activation, but recent findings highlight cell metabolism as a crucial regulator of these processes [9–12]. Receptor-induced signaling and metabolic networks in naïve T cells are mutually regulated by each other depending on the micro-environmental cues obtained by the cell that also influence quiescence exit. Here we will discuss the bidirectional communication of signaling and metabolic pathways that promotes proliferation, quiescence exit, and activation of naïve T cells and functioning of T cells upon activation. We will take into account the different signaling and metabolic events and their cross-talks that lead to differentiation of naïve T cells into T_{H1}, T_{H2}, T_{H17}, Treg, or Tfh effector cells. Understanding the cross-talks between T-cell signaling and metabolism under different environmental cues will be vital for understanding the differentiation patterns of naïve T cells during different pathogenic conditions. This will provide better prospects of developing novel approaches to modulate protective and pathological T-cell responses in human diseases.

6.2   Signaling and Metabolic Pathways Involved in Activation of Naïve T Cell

The activation of T_H cell is mediated by a complex chain of signaling events that involve the activation of distinct co-stimulators and co-inhibitors present on the surface of the lymphocyte. The interaction between the antigen-bound major histo-compatibility complex (MHC) on the antigen-presenting cells (APCs) and the T-cell receptor (TCR) on T_H cells triggers the TCR-mediated signaling pathway. The phosphorylation of the LAT signalosome by LCK sends signal to three major cell-signaling pathways, viz. NFκB, MAPK, and the calcium-mediated NFAT pathways [13]. Along with the TCR, the T cell also expresses several other co-receptor molecules that can be classified into two major functional groups. The first group consists of co-signaling receptors that have an immunoglobulin (Ig)-like fold in their ectodomains, such as CTLA-4, CD28, PD1, and BTLA [14]. The other co-signaling group belongs to the tumor necrosis factor receptor (TNFR) superfamily and includes DR3, OX40, 41BB, CD27, CD30, and HVEM [14]. Together with the TCR activation, a second signal from the co-stimulatory signal emanating from B7-CD28 interaction is also necessary for the T-cell activation. This is called the "two signal hypothesis" [13]. The B7 molecule present on the APC also binds with the CTLA-4 receptor of the T cell after the clearance of the antigen. This induces T-cell anergy after the antigen is cleared from the system and the T-cell activation is no longer required. The other co-receptor signaling pathway influences the type of cytokine expressed and regulates the T-cell differentiation pattern. Experimental studies have shown CD40-L, expressed on the surface of activated T cells, induces the APC to produce IL-12, thereby stimulating the T_H cells to differentiate into the T_{H1} cells [15, 16]. On the other hand, the TRAF2-mediated OX40 signaling pathway contributes to long-term survival of T_H cells [17]. OX40 has been implicated in the development of memory T cells, clonal expansion, and differentiation. It also mediates suppression of the Treg cells [17, 18]. The negative regulators of T-cell activation are required to maintain homeostasis and deactivate the T cells after the antigen is cleared out. This is mediated by the PD1-PDL axis that provides co-inhibitory signal to the T-cell activation. The T cell also expresses CD45, a phosphatase, that de-phophorylates the carboxyl-terminal tyrosine of p56lck and p59fyn that aids T-cell activation [19]. Apart from these, the T cells express several other co-receptors that serve to regulate the cytokine expression and differentiation of the cell [20].

The calcium pathway also plays a major role in the proliferation of the T_H cell activation [21]. The influx of Ca^{2+} ions from the CRAC channels leads to the activation of the NFAT (Nuclear Factor of Activated T cell) transcription factor that acts as the master regulator of T-cell activation and T-cell anergy [22]. The activation of the calcium pathway in the T cell is initiated by the binding of the TCR with an antigenic peptide presented on MHC complexes of the APC that induces activation of PLC-γ that cleaves PIP2 into IP3 and DAG. This IP3 now activates the IP3-receptors located on the endoplasmic reticulum membranes inside the T cell, which causes the release of intracellular stores of calcium, leading to a transient elevation in cytoplasmic calcium level. This activates the CRAC channels on the T-cell membrane that allows an inward flux of calcium from the extracellular environment. This triggers the calcium-mediated calmodulin-calcineurin pathway, which leads to the de-phosphorylation and nuclear translocation of NFAT proteins where it can cooperate with AP-1 complexes induced by co-stimulatory pathways. The NFAT/AP-1 complexes bind to the sites in the promoters of many cytokine genes to activate their transcription to mediate sustained T-cell activation and survival. In the absence of co-stimulation or in the presence of anergizing stimuli, sustained increases in intracellular calcium concentration activate NFAT proteins. However, in the absence of concomitant AP-1 activation, due to lack of co-stimulatory signals, NFAT proteins dimerize and translocate into the nucleus, inducing the expression of anergy-inducing genes that include E3-ubiquitin ligases, such as Itch, Grail, and Cbl-b that is known to ubiquitinate and inactivate the TCR signalosome and the co-stimulatory CD40-ligand, thereby destabilizing the immunological synapse in the anergic T cell. On the other hand, the calcium/NFAT-dependent activation of the Ikaros transcription factor in anergic T cells leads to the epigenetic changes in the IL-2 promoter by the recruitment of HDACs and other chromatin-modifying complexes, which results in stable silencing of the IL-2 gene expression [22].

Metabolic regulation of T cell is another aspect that determines activation and differentiation of naïve T cells and their functioning upon activation. Naïve T cells utilize glucose and glutamine metabolism for activation, and activation signals increase glucose and glutamine uptake by T cells through GLUT1 and ASCT2, respectively [23, 24]. Thus, both signaling and metabolism cooperate in a bidirectional manner to influence T-cell activation and differentiation. On encountering pathogenic antigens, a cascade of TCR signals and co-stimulatory signals are initiated, which leads to quiescence exit in naïve T cells. The first signal that initiates quiescence exit is the transduction of TCR signaling via PI3K/AKT/mTOR pathway, which induces glycolysis in the naïve T cells [25]. This initiation is marked by a trigger in the metabolism of T cells that suffices the increasing lipid, nucleotide, and amino acid requirement of differentiating cells. During quiescence exit, T cells produce lactate to sustain glycolysis. Lactate is also imported into cells through the monocarboxylate transporters and converted into pyruvate by lactate dehydrogenase A (LDHA). This reaction limits glycolytic programming and proliferation in T cells, potentially owing to the attenuated generation of glycolytic intermediates such as PEP that sustain glycolysis and biosynthesis reactions [26].

Glutamine metabolism regulates T-cell activation in different ways. It has an important role in determining differentiation to T_{H1} and T_{H17} cells. T_{H17} cells utilize both glucose and glutamine to fuel the TCA cycle and OXPHOS, which otherwise is optional for other T effector cells [27]. It regulates leucine uptake via regulation of LAT1-CD98 and together with leucine activates mTORC1 signaling [28]. Other amino acid metabolisms like tryptophan and arginine metabolism and their intermediate metabolites such as kynurenine and ornithine differentially regulate T-cell survival, apoptosis, and proliferation [29–31].

Glucose and glutamine metabolism also induce lipid metabolism via mTORC1-dependent regulation of AMPK [32]. These pathways are metabolically connected to the TCA cycle and OXPHOS, which also affect the redox and oxygen-sensing signals in T cells. The conversion of pyruvate to lactate via NAD^+-NADH-dependent LDH reaction regulates redox signals, and impaired oxygen-sensing machinery of OXPHOS results in the formation of ROS, which induces ROS-dependent signaling that promotes IL-2 productions and induces T-cell proliferation by activating NFAT transcription factor [33].

6.3 T_H-Cell Differentiation and Diversity

The T_H cells display high plasticity that helps them to differentiate into specialized T_H cells according to the type of the antigenic challenge and the micro-environmental conditions (Fig. 6.1). The early events of the T-cell activation play a major role in the determination of the pattern of differentiation of the naïve T cell. The micro-environmental cues, in the form of cytokines, activate the signaling pathways of the T_H cells that eventually lead to the changes at the gene-regulatory levels [34]. The selective activation of specific transcription factors mediates the differentiation of the naïve cells into specialized CD4+ T_H effector cells, viz. T_{H1}, T_{H2}, T_{H17}, etc. (Table 6.1) [35]. Additionally, another type of CD4+ T_H cell called the regulatory T cells (iTreg) has a role in maintaining the T_H cell homeostasis.

The mechanism of T-cell differentiation is governed initially by the strength of the stimulus that the TCR receives from the APC. The strength of stimulus results in differential regulation of phosphatidylinositols that triggers different signaling

Fig. 6.1 Schematic diagram of signature signaling factors, cytokines, metabolites, and metabolic paths, which dictate T_H cell differentiation, proliferation, and effector function

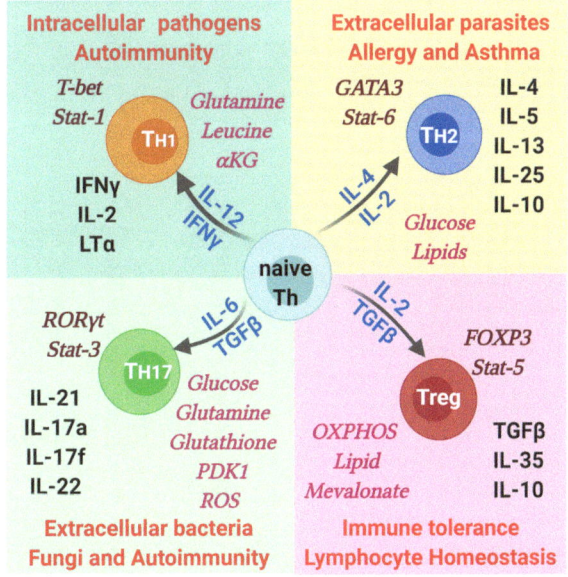

Table 6.1 Summary of T_H cell diversity, factors regulating T-cell plasticity and effector functions of each subtype

CD4+ subset	Polarizing cytokines	Transcription factors	Inhibitory transcription factors	Metabolic signature	Effector functions
T_{H1}	IL12, IFNγ	T bet, STAT1, STAT4, Runx 3, Eomes, Hlx	GATA3	Upregulated glycolysis, glutamine metabolism, leucine uptake, αKG production	Cell-mediated immunity against intracellular pathogens and phagocyte-dependent protective responses
T_{H2}	IL4, IL2	GATA3, STAT6, STAT5, STAT3, Gfi-1, c-Maf, IRF4	T-bet, Runx3	Downregulated glycolysis, upregulated lipid metabolism	Immune response against extracellular parasites, bacteria, allergens, and toxins. Help in activation and maintenance of humoral immune response and tissue repair
T_{H17}	IL6, IL 21, IL 23, TGF-β	RORγt, STAT3, RORα, Runx1, Batf, IRF4, AHR	T-bet+ Runx1, Smad3Runx1+FOXP3	Uptake of both glucose and glutamine, glutathione production, upregulated PDK1 and ROS	Immune response against bacterial and fungal infection
Tfh	IL6, IL21	Bcl6, STAT3	Not known	Not known	Help B cells produce antibody
iTreg	TGF-β, IL2	FOXP3, Smad2, Smad3, STAT5, NFAT	Not known	Inhibited glycolysis, upregulated OXPHOS, lipid and mevalonate metabolism	Suppression of immune response
T_{H9}	TGF-β, IL4	IRF4	Not known	Not known	Promotes mast cell and T-cell growth, stimulates mucous secretion to enhance innate immunity. Plays a role in allergic responses
Tr1	IL27, IL10	c-Maf, AhR	Not known	Not known	Suppression of T effector cells

pathways downstream. It has been observed that while a weak TCR signal generates a high level of PIP2 and lower levels of PIP3, which is required for the activation of the focal adhesion kinase and phosphorylation of AKT_{Thr308}, stronger signal favors the activation of mTORC2, and as a result, elevated level of PIP3 and reduced PIP2 are generated [36]. In vitro experiments have revealed that a stimulus of a lower strength induces the expression of the GATA-3 transcription factor, the master regulator of T_{H2} cells. Simultaneously, the expression of the IL-2 cytokine activates STAT5 that synergizes with GATA-3 to transcribe the IL-4 gene that eventually leads to the differentiation of the naïve cell into the T_{H2} subtype [37]. Recent advances in the field also divulged that during viral infection low TCR signals may also favor the formation of Tfh and memory T cells. On the other hand, a stronger stimulus favors the activation of the T-bet transcription factor that helps in the differentiation into the T_{H1} subtype and triggers the production of IFN-γ and IL12 cytokines. The differentiation of naive CD4$^+$ T cells into T_{H17} cells is induced by TGF-β/IL-6 in combination with TCR stimulation. This triggers the production of IL-23R, which induces the transcription factor RORγt, IL-17, and IL-21. The STAT-3 protein plays an important role in the production of the T_{H17} effector molecules and requires the activation of the ICOS co-stimulatory pathway. However, under the T_{H17}-inducing conditions, the presence of IL2/STAT5 induces the expression of the Foxp3 transcription factor that leads to the differentiation of the naïve cells into iTreg cells. The strength of TCR stimulus also plays a role in the T_{H17}/iTreg determination process, where it has been observed that a weak stimulus favors the differentiation into iTreg cells that is known to have a role in immune-suppression [37].

The effect of signaling in T_H cell differentiation is further augmented by the action of metabolism within these cells. On activation by the upstream TCR and co-stimulatory signals, metabolic pathways trigger the process of T-cell activation with the initiation of glycolysis in most of the cases [38]. The utilization of glucose is maintained nominal in naïve T cells, just to suffice ATP requirement enough to maintain survival during quiescence [39]. However, with the transduction of TCR signals via mTORC1/2 signaling, the rate of glucose utilization increases, leading to quiescence exit and activation of T_H cells [8, 38]. Upon activation, differentiation patterns are regulated by differential expression of metabolic pathways. For example, glutamine metabolism along with leucine induces proliferation and differentiation of T_{H1} and T_{H17} cells [27, 28]. In addition, αKG promotes initial programming in T_{H1} cells [40]. Further, glutaminolysis results in the formation of glutathione, which is required for T_{H17} differentiation [41]. An increase in glucose metabolism induces lipid metabolism to promote T_{H2} differentiation [42]. Inhibition of glycolysis and promotion of OXPHOS along with upregulated lipid and mevalonate metabolism induce Treg proliferation and differentiation [43, 44]. Intermediate metabolites of metabolic pathways, in return, regulate signaling processes as well. For example, tryptophan intermediate, kynurenine, and arginine intermediate ornithine regulate signaling processes in T cells, which have been discussed in the next section.

Each of the T_H sub-type has a specific effector function to perform [34, 35, 37, 45]. A balance between all the T_H cell subtypes is necessary for the proper functioning of the immune system. The effector molecules, in the form of interleukins, interferons, tumor necrosis factor, etc., produced by these diverse groups of immune cells, maintain the integrity of the immune-regulatory network (Table 6.1). However, during any disease condition, this defense mechanism gets subdued. Changes in the micro-environmental conditions lead to alterations in the biochemical reaction network that disrupts the balance between the effector cell populations that favors the progression of the disease. This immune-suppression is observed very frequently in the cases of chronic infections (e.g., chronic *Leishmania* infection) and cancer.

6.4 Signaling and Metabolic Cross-Talk Mediated by mTOR Regulate Differentiation

Activation of naïve T cells is initiated with the tonic signals generated by T-cell receptor (TCR) on their interactions with self-peptides on MHC molecules. There is an intricate design of the signaling and metabolic interactions of these cells, which allow them to proliferate and produce effector molecules (Fig. 6.2). Sensitivity toward TCR signaling in the naïve T cells is partially mediated by the mechanistic target of Rapamycin complex (mTORC1 and mTORC2) [46]. Peripheral naive T cells circulate in the blood and survey antigens. They maintain a low metabolic rate and import a small amount of glucose to fuel the TCA cycle and OXPHOS for ATP production [39]. Naive T-cell homeostasis is disrupted by the activation of mTOR signaling [47]. The activation of mTORC1 signaling enhances glycolytic metabolism in these cells, inducing entry to cell cycle and cell growth. The naive T cells, which otherwise remain in a quiescence state, are activated by the enhanced glycolytic pathway. Different regulators of mTORC affect the process of naive T-cell activation [46].

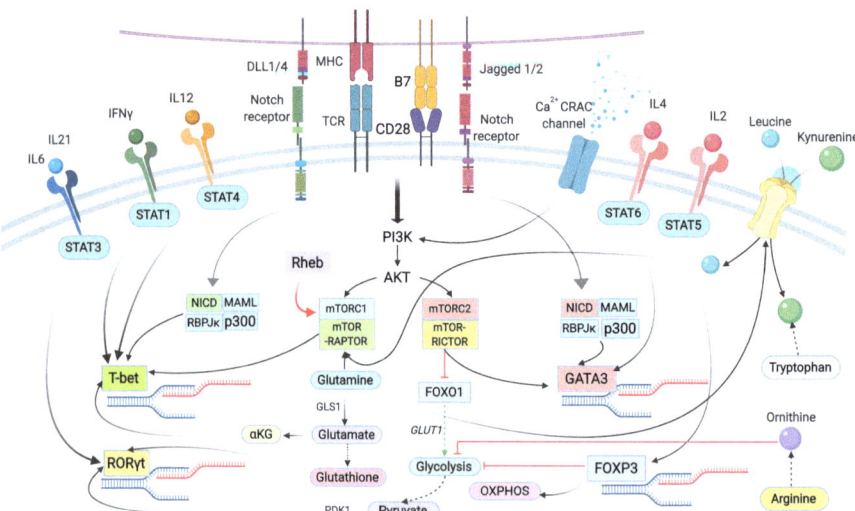

Fig. 6.2 Cross-talks of signaling and metabolic pathways regulating the activation of the T-bet, GATA3, RORγt, and FOXP3 transcription factors that mediate T-cell differentiation

mTOR signaling is regulated by a set of upstream signaling, which determines the formation of mTORC1 and mTORC2 and subsequent signaling. The signal induces upon activation of TCR and subsequently the PI3K/Akt pathway [48]. Raptor and rictor are the main components of mTORC1 and mTORC2 complexes, respectively. mTORC1 signaling is required for differentiation into T_{H1} and T_{H17} effector cells, and an inhibition of mTORC1 has been observed to induce T_{H2} differentiation and prevent T_{H1} and T_{H17} differentiation [25]. However, these observations differ according to the upstream signal received by the complex. Loss of tuberous sclerosis complex 1 (TSC1) results in mTORC1 activation [47]. The metabolic activity of naive T cells can also be enhanced by the exposure to IL-2 released by activated CD4$^+$ effector cells [49]. Inhibition of mTORC1 by the TSC (Tuberous Sclerosis Complex) via Rheb inhibition leads to failure in differentiation into T_{H1} and T_{H17} effector cells [47].

mTORC1 is a master kinase that helps naive T cells to exit quiescence. TCR signaling along with costimulatory and IL-2 signals promote the activation of mTORC1 during quiescence exit. The magnitude and duration of mTORC1 activity likely determine quiescence exit. TCR signals must meet a certain threshold of activation to induce T-cell proliferation. This threshold is determined by the level of mTORC1 activation and expression of IRF4 and c-Myc [50, 51] that regulate anabolic and mitochondrial metabolism. mTORC1 also regulates sterol regulatory element-binding proteins (SREBPs) that has a role in metabolic reprogramming in naive T cells. Metabolism in turn regulates the activity of mTORC1. Leucine and glutamine coordinate with TCR and CD28 signaling to activate mTORC1 and sustain metabolic flux during quiescence exit [27, 28]. T-cell activation demands for the biosynthesis of lipids, cholesterol, nucleotides and amino acids in order to maintain the increase in metabolic rates of the activated cells. These increased demands are facilitated by the upregulation of hexokinase 2 (HK2), which is the rate-limiting enzyme for glycolysis [52, 53]. This induces increased utilization of glucose, which can also activate mTORC1 and inhibit the activation of AMP-activated protein kinase (AMPK) [32, 54]. AMPK induces lipid and cholesterol biosynthesis through the mTORC1-dependent upregulation of SREBP1 and SREBP2 [55]. mTORC1 forms a bridge between signaling and metabolic responses in T cells that senses metabolic cues and mediates signaling regulation over metabolic pathways and vice-versa. Thus, mTORC1-dependent responses are crucial in determining proliferation, activation, and functioning of T cells.

TCR signaling targets the transcription factor, c-Myc, in an mTORC1-dependent manner. It regulates the transcription of metabolic genes critical for T-cell activation. c-Myc induces the transcription factor AP4, which maintains the glycolytic transcriptional program initiated by c-Myc to support T-cell population expansion [50]. However, c-Myc expression is not continually sustained after T-cell activation [56].

Metabolites also influence T cells in an mTORC1-independent manner. For example, post-translational protein modifications by glycolytic, lipid, or mevalonate by-products allow receptors, enzymes, and scaffolding proteins to properly posit at their sites of activity [57, 58]. In T cells, extracellular ATP, glucose, and glutamine modulate AMPK activity to promote T-cell responses against bacteria and viruses

[54]. The glucose metabolite PEP regulates the activation of Ca^{2+}–calcineurin–NFAT signaling [59]. TCR signaling can be altered by cholesterol esters and cholesterol sulfate, which alter TCR clustering or affinity for antigens [60]. Also, N-glycans derived from the hexosamine pathway suppress TCR signaling [61].

mTORC2 also contributes to quiescence exit by enhancing glycolytic pathway. AKT/mTORC2 represses forkhead box protein O1 (FOXO1) function [62], which induces glucose transporter 1 (GLUT1) expression and enhances glycolytic flux [63]. Expression of glucose transporters contribute in determining naïve T-cell survival. IL-7–IL-7R signaling prevents degeneration of quiescent T cells by increasing glucose and amino acid catabolism [64]. Rate or quantity of glucose uptake via the GLUT1 receptor may have a role in determining quiescence versus quiescence exit as its expression is lower on naive T cells than on activated T cells. During quiescence exit, cell growth and clonal proliferation are favored by glucose metabolism upon survival [51].

Duration and strength of TCR signaling mediate both quiescence and activation of T cells. However, based on the type of initiation of these signaling cascades, i.e., tonic or antigen-driven, TCR signals differ in both duration and strength. In antigen-activated T cells, CD28-mediated co-stimulation of TCR signaling induces GLUT1 expression to increase glucose uptake [65]. Expression of the glutamine transporter ASCT2 and of sodium-coupled neutral amino acid transporters (SNATs) increases on TCR and CD28 co-stimulation [23]. Upregulation of SNATs on T-cell activation suggests that they also modulate the rate or quantity of glutamine uptake.

Glutamine metabolism plays a crucial role in determining differentiation to T_{H1} and T_{H17} cells. Glutamine affects LAT1–CD98 activity, which promotes leucine uptake to induce the proliferation and differentiation of T_{H1} cells, T_{H17} cells, and effector $CD8^+$ T cells [23, 66]. Glutamine along with leucine activates mTORC1 and sustains metabolic flux during quiescence exit [28]. Further, utilization of glutamine to generate glutathione via glutaminolysis is essential for T-cell proliferation and differentiation into T_{H17} cells [27]. Glutaminolysis also generates α-ketoglutarate (α-KG), which promotes initial programming of T_{H1} cells. Glutaminolysis also affects IL-2 signaling, as it has been observed to suppress IL-2-induced mTORC1 activation during type 1 inflammation [27]. However, impaired glutaminolysis may promote abnormal leucine uptake to increase mTORC1 activation under such inflammatory conditions [23, 66]. Thus, glutamine and glutaminolysis have different roles during quiescence exit and upon T-cell activation.

During impaired glutaminolysis, the oxidation of pyruvate acts as a crucial checkpoint. The mitochondrial pyruvate carrier (MPC) transports pyruvate into the mitochondria to fuel the TCA cycle and OXPHOS and depletes it from the cytoplasm. The inhibition of MPC favors glycolysis over OXPHOS, particularly when glutaminolysis is also impaired. Downregulation of OXPHOS in T cells require inhibition of both MPC and glutaminase 1 (GLS 1) [67]. T_{H17} cells suffice their nutrient requirement using both glucose and glutamine, which otherwise is optional for other activated T cells. The plausible explanation for this phenomenon is the high-level expression of pyruvate dehydrogenase kinase 1 (PDK1) in T_{H17} cells, which prevents conversion of pyruvate to acetyl-CoA in mitochondria [53].

High expression of PDK1 diverts the pyruvate flux away from TCA in T_{H17} cells, and hence, the cell depends on glutamine to fuel the TCA cycle. The regulation of PDK1 is not well understood in T_{H17} cells; however, studies suggest that hypoxia-inducible factor 1α (HIF1α) might induce PDK1, promoting T_{H17} cell responses [53]. Also, lactate dehydrogenase A (LDHA), which catalyzes lactate formation from pyruvate, sustains glycolytic metabolism and promotes interferon-γ (IFNγ) expression in activated T cells [68].

Upon activation, amino acids play an important role in the functioning of activated T cells. Certain amino acids promote quiescence exit and proliferation of naïve T cells, whereas others might suppress proliferation and promote quiescence-like programs in naïve T cells. Majority of the biomass of activated T cells is made by amino acids. Uptake of essential amino acids such as leucine or conditionally essential amino acids such as glutamine are taken up by amino acid transporters, such as LAT1–CD98 or ASCT2 [23], but non-essential amino acids accumulate in T cells due to influx or de novo biosynthesis from glucose or glutamine. Accumulation of amino acid intermediates impact the functioning of activated T cells. Accumulation of kynurenine, an intermediate of tryptophan metabolism, suppresses T-cell proliferation [30]. Kynurenine accumulation might also result from its uptake through the LAT1-CD98 transporters [69]. Ornithine, an arginine intermediate, reduced glucose consumption via glycolysis. However, arginine supplementation increases serine biosynthesis and OXPHOS [31], which increases T-cell survival and promotes secondary effector responses.

Balanced redox reactions are one of the prerequisites for T-cell activation [70]. The NAD^+-NADH-dependent conversion of pyruvate to lactate is a major redox balancer of T cells. An accumulation of NAD^+ increases lysosome biogenesis, which can suppress T-cell activation. Mitochondrial reduction of NAD^+ levels is utilized to promote aspartate synthesis, which is necessary for T-cell proliferation [70]. Both NAD^+ and ATP cooperatively influence T-cell responses. Extracellular ATP augments quiescence exit and T-cell proliferation via the expression of purinergic receptor P2XY, which induces IL-2 production [71]. Conversely extracellular NAD^+ promotes T-cell death by increasing the ART2-dependent activation of P2XY [72].

Oxygen sensing by T cells also regulates their effector functioning [73]. OXPHOS, which requires oxygen, is essential for both T-cell quiescence and activation [70, 74]. OXPHOS generates ROS, which stimulates IL-2 production and promotes T-cell proliferation by activating nuclear factor of activated T-cell (NFAT) transcription factors [75]. Under pathological conditions, increased levels of mitochondria-derived ROS can have antagonizing T-cell responses, including T_{H17} cell differentiation [27, 53].

FOXP3 is an important determinant of Treg differentiation and the Treg cell responses and regulated via the metabolic regulation exerted by FOXP3 [76]. It promotes OXPHOS and inhibits glycolysis in Treg cells. Survival and function of these cells are reduced by excessive PI3K or mTOR activity as it decreases FOXP3 expression and increases glycolytic metabolism [77]. Treg cells, upon activation, upregulate mTOR signaling, which induces lipid synthesis, mevalonate metabolism, and mitochondrial function [78, 79]. These pathways influence activation programs to regulate Treg cell function.

Mitochondria-derived metabolites like acetyl-Coa, succinate, αKG, and 2-hydroxyglutarate (2-HG) alter epigenetic programs. Acetyl-CoA induces histone acetylation, which is permissive for transcription. α-KG promotes the activity of demethylases that target DNA or histones, whereas 2-HG antagonizes demethylases [80]. Demethylation in turn allows changes in gene transcription associated with specific T-cell effector programs. 2-HG accumulation downstream of the von Hippel–Lindau disease tumor suppressor (VHL)–HIF1α axis in T cells induces changes in DNA and histone methylation that increase CD8+ T-cell proliferation [80].

Thus, we observe metabolic regulation of T-cell activation and functioning at different levels. Mitochondria-derived metabolites affect the functioning and/or expression of various transcription factors through methylation-demethylation, acetylation processes or by mitochondria-derived ROS regulations. The effects of glucose metabolism in mTOR and c-Myc regulation have been implicated. Metabolites also regulate transcription factor activity. For example, transcriptional regulators BAZ1B, PSIP1 are activated by arginine and lipids or sterols regulate the activities of LXRs, PPARs, and SREBPs [81–83]. Further, metabolic processes also regulate processes at posttranscriptional and translational levels. For example, amino acid deprivation is sensed by GCN2 (or EIF2AK4) and leads to inhibition of protein translation by the EIF2α pathway, which supposedly leads to suppression of T-cell proliferation [84]. Also, GAPDH produced by the glycolytic pathway has been observed to suppress protein translational processes [85]. Metabolites also affect the activity of activated T cells by the regulation of transporter proteins and complexes. Amino acids like leucine, glutamine, tryptophan, and arginine and the intermediate metabolites generated during the biogenesis or catabolism of these amino acids like kynurenine, ornithine, etc., affect the functioning of T cells upon activation via the regulation of transporter proteins like LAT1-CD98 or ASCT2. To summarize, metabolism can influence the processes of T-cell differentiation, activation, and functioning by regulating molecular processes at different levels starting from gene and transcription regulation.

6.5   Methodologies to Unwind the Regulations of the Immune Response

A comprehensive understanding of the complex regulations underlying the immune responses under different environmental conditions, antigenic challenges, strength of stimulus, and metabolic demands have challenged the implementation of successful immunotherapy. A need to unveil these regulatory mechanisms has driven experimental researchers as well as computational biologists to implement different omic studies and model the immunome under different antigenic stimulus. In the following section, we have taken up examples of the studies of T-cell responses and differentiation during infectious diseases (e.g., Leishmaniasis) and cancer that will give a clear insight of how the immune responses are altered under specific antigenic challenges.

6.5.1 Immunomics and Enrichment Analysis

Transcriptomic analysis, e.g., microarray, RNAseq, have opened up new avenues of research that allows the analysis of gene expression profile of several patient cohorts under various disease conditions. While microarray involves detection and quantification of gene expression based on the pairing of an mRNA transcript with its probe on a chip, RNA-Seq involves direct sequencing of gene transcripts by high-throughput sequencing technologies. This enables the RNAseq technique to detect novel transcripts as it does not require transcript specific probes as well as confers higher specificity and sensitivity for the detection of a wider range of differentially expressed genes, allowing detection of genes even with low expression. Following the identification of differentially expressed genes, gene ontology (GO) and pathway enrichment tools enable the identification of the biological processes (BP), molecular functions (MF), cellular component (CC), and biochemical pathways that are significantly enriched or over-represented in a given scenario. Various online tools and web-servers such as DAVID, GeneCodis, Gene Set Enrichment Analysis, and Reactome are available freely for performing enrichment analysis [86–90].

Researchers have exploited these techniques to unearth the immunome landscape in the microenvironment where the spatio-temporal dynamics of 28 different immune cell-types (immunome) have been studied using 105 human colorectal cancer patient data. Here the immunome was made up of mRNA transcripts specific for most innate and adaptive immune cell subpopulations. Using an integrative analysis, it has been elucidated that the densities of T follicular helper (Tfh) cells and innate cells increased, whereas most other T-cell densities decreased along with tumor progression. However, the Tfh and B cell numbers are inversely correlated with the disease progression and recurrence, and CXCL13 and IL21 genes are essential for the Tfh/B cell axis that is correlated with higher chances of survival of the patient [91, 92].

RNAseq analyses in the case of Leishmaniasis have been performed, that has revealed *Leishmania* species–specific differences in the expression of mammalian macrophage genes due to infection [93]. Such analyses have helped in the understanding of the changes in immune response generated during infection by unveiling the notable changes induced in the cytokine expression profiles during the Leishmania invasion. Experiments using microarray techniques have been used to assess the host cell genes and pathways in human dendritic cells associated with early *Leishmania major* infection. The study revealed 728 genes were significantly differentially expressed in the infected cells, and molecular signaling pathway revealed that the type I IFN pathway was significantly enriched. Here it was elucidated that *L. major* induces expression of IRF2, IRF7, and IFIT5, which indicates that the regulation of type I IFN-associated signaling pathways is responsible for the production of IL-12. However, this is not observed in the case of *L.donovani* [94].

6.5.2 Computational Methods for the Study of Immune Responses

The understanding of intra-cellular and inter-cellular signaling pathways involved in the generation of immune responses requires the study of a complex network of biochemical pathways under different disease-affected micro-environmental conditions. This is an extremely challenging task that can rarely be achieved using in vitro or in vivo experimental techniques. In order to gain insight into the immune-regulatory modules involved in T-cell functioning as well as study the immune-modulatory mechanisms employed by pathogen and the tumor cells, computational tools and mathematical modeling approaches have been extremely useful in obtaining a systems-level understanding. These have also helped the researchers and medical practitioners in the prediction of immunotherapeutic strategies and design of treatment protocols. Here we will throw light onto some of the most popular tools and techniques used for such studies and also explore a few of the mathematical models that have helped us unravel some of the intriguing problems in immunology.

6.5.2.1 Signaling and Metabolic Pathway Databases

The signaling pathway databases are important sources of information that collate pathway data from experimental studies regarding the intracellular signaling pathways in different immune cells [95, 96]. The KEGG provides information regarding the core TCR-mediated pathway along with a few co-receptor signaling pathways. The database also contains the pathways responsible for the T_{H1}, T_{H2}, and T_{H17} differentiation. Another popular database called Reactome provides detailed biochemical reactions involved in each step of the protein–protein interactions involved in the T-cell signaling pathway. It also enlists the pathway information related to CD28 and PD-1 co-signaling pathways. Simultaneously, Reactome forms a very important source for cytokine signaling pathways that includes different interleukin families, interferons, tumor necrosis factor, and a few growth hormones. A list of few of the available databases and the available information in each has been listed down in Table 6.2. However, the information regarding the intercellular cross-talks in the immune system is lacking in most of these databases that can be extracted through a thorough literature survey.

Few databases also provide data regarding the changes in the pathway during disease condition. The KEGG database has a sufficient amount of pathway information regarding the endocytosis of the *Leishmania* pathogen as well as the signaling events that occurs inside the infected macrophage. BioLegend database contains the cancer immune-editing network that consists of the intercellular signaling cross-talks governing the immune responses generated during cancer.

For the analysis of these biochemical pathways, the BIOPYDB database also provides an integrated platform for performing network analysis, logical steady-state analysis, knock-out analysis, etc. It contains detailed information regarding each protein involved in the immunological pathways as well as links them to the specific diseases associated with them. Apart from the TCR co-receptor-mediated

Table 6.2 List of a few signaling and metabolic pathway databases containing T-cell-specific pathway data and related cytokine pathways

Database	T-cell activation/ differentiation pathways/network	Cytokine pathways	URL of database
Kyoto Encyclopedia of Genes and Genomes (KEGG)	T-cell receptor signaling pathway, T_{H1} and T_{H2} cell differentiation, T_{H17} cell differentiation	IL-17, TNF, calcium signaling pathway	http://www.genome.jp/kegg/
Reactome	TCR-mediated pathway, CD28 co-signaling pathway	IFN-α/β, IFN-γ, TNF-α, IL-1, IL-2, IL-3, IL-5,GM-CSF, IL-4, IL-13, IL-6, IL-7, IL-10, IL-12, IL-17, IL-20 family cytokines	https://reactome.org/
Wikipathways	TCR-mediated pathway, B7-CD28, B7-CTLA4, PDL- PD1 pathways	IL-2, IL4, IL-5, IL-7, IL-9, IL-11, Type-1 IFN, TNF-α pathways	http://www.wikipathways.org
NCI – Pathway Interaction Database (PID)	TCR signaling network in naïve CD4 cells, B7-CD28 signaling networks	IL-1, IL-2, IL-3, IL-4, IL-5, IL-6, IL-8, IL-12, IL-23, IL-27, TNF signaling networks	http://www.ndexbio.org
BioLegend	T-fh, T_{H1}, T_{H2}, T_{H17}, Treg, $\gamma\delta$–T-cell signaling pathways	IL-1, IL-2, IL-4, IL-6, IL-10, IFN, TNF pathways and inter- cellular cytokine signaling network of immune cells	https://www.biolegend.com/pathways/
BIOPYDB	TCR-mediated pathway, co-receptor-mediated T-cell activation pathway	IL-1 α, IL-β, IL2, IL-4, IL-6, IL-12, IL-18, IL-36 α, IL-36 β, IL-36 γ, TNF α, TNF β, IFN α, IFN β, IFN γ, TGF β	http://biopydb.ncl.res.in/biopydb/index.php
HumanCyc	T_{H1}, T_{H2}, T_{H17}, Treg- associated processes and pathways	Cytokine pathways are not available separately, but integrated with the other immune processes	https://biocyc.org/HUMAN/
Brenda	T_{H1}, T_{H2}-related processes	IL-1, IL-3, IL-5, IL-6, IL-8, IL-12, IL-17, IL-18, IL-21, IL-33, IFN-α, IFN-β, IFN-γ, TNF-α, TNF-β ligands	https://www.brenda-enzymes.org/

and cytokine pathways, BIOPYDB also contains detailed information about the toll-like receptor (TLR) pathways that has an important role in the regulation of immune response [97].

With the realization of the importance of immune-metabolism as a decisive factor in eliciting immune responses, metabolic databases have started to incorporate such details into the database structure. Although the advent is very recent and only a limited number of databases have included this information. Two of the popularly used metabolic databases, HumanCyc [98] and Brenda [99], include information about immune-metabolites that are linked to immune responses. HumanCyc is the *Homo sapiens*–specific repertoire of the metabolic database BioCyc, which enlists metabolism specific to human. The database enlists a range of "Biological Process" and "Proteins" related to immune system. The biological processes are linked to their "Gene Ontology" term. A few of the important immune processes listed are "leukocyte-mediated cytotoxicity," "adaptive immune response," "immune effector process," "regulation of immune response," and "immune system development." The GO IDs of these processes link them to pathways and processes to which are linked/cross-linked, which are enlisted as "Parent Classes" and metabolites/proteins which are involved in these processes are enlisted under "Instances". These metabolites/proteins are linked to their detailed descriptions along with reactions in which they are involved and the reaction mechanism [98]. Brenda also provides details of immune-metabolites. The database has a wide range of entries as search option. Upon search of immune processes, it provides a variety of immnune-metabolites and proteins whose "Enzyme Nomenclature," "Enzyme-Ligand Interactions," "Diseases," "Functional Parameters," "Organism-related Information," "General Information," "Enzyme Structure," "Molecular Properties," "Applications," and "References" are provided.

6.5.2.2 Graph Theoretical Analysis

The Graph Theory was initiated with Euler's famous publication from 1736 on the Seven Bridges of Königsberg problem [100]. However, it was applied to biochemical networks much later with the advent of the concepts of small-world and scale-free networks in 1999 that describes the global architecture of any complex real-world network such as the network of biochemical reactions in a cell [101, 102]. Computational biologists have modeled biochemical pathways as network where each protein or metabolite has been considered a node and the reaction between any two such species have been denoted as an edge, thereby translating the entire reaction network as an interconnected mesh of nodes and edges. Various network parameters such as Degree (k), Betweenness Centrality, Closeness Centrality, Eccentricity, Edge Betweenness, and Clustering Coefficient are used to describe the topological properties of the network. These parameters help in the identification of important hubs, i.e., a highly connected node, and shortest paths in the biochemical reaction network that may have significant contribution in the functioning of the signaling or metabolic pathways. Tools such as Cytoscape, Gephi, Pajek are freely available for performing network analysis of large reaction network [103–105]. Cytoscape further offers downloadable plugins for identifying important motifs, extracting sub-networks, and performing enrichment analysis and a host of other functions required for visualizing and analyzing large biochemical reaction networks. These biochemical networks mostly follow the small-world property of a network that indicates a relatively short distance from any one node to another and a relatively high level of clustering. This network property, termed as scale-free property of a network, denotes a connectivity distribution that fits a power law depicted in Eq. 6.1 where the value of γ lies in the range $2 < \gamma < 3$ [106]. It has been observed that networks following the scale-free property are generally resistant to perturbations and thus are highly robust:

$$P(k) = \alpha\, k^{-\gamma} \qquad (6.1)$$

Graph Theory has successfully been applied to signaling pathway networks where the concept of shortest path has been used to hypothesize potential signaling mechanisms in Neuro2A cells downstream of CB1R receptors. Here the cells were stimulated with a CB1R agonist for the assessment of activity of transcription factors. This experiment revealed CB1R activation modulates the activity of 23 transcription factors [107]. Such methods are useful in the identification of important novel signaling routes between a cell-surface receptor and downstream transcription. In a recent study, Graph theoretic network analysis has been used to identify protein pathways responsible for cell death after neurotropic viral infection by Chandipura Virus (CHPV) [108]. Another important application of network analysis is that it can be used to identify important hub proteins that can be used as potential drug or immunotherapeutic target [109, 110].

6.5.2.3 Logic-Based Models

Logical modeling is gradually being recognized as a simple yet powerful tool in systems biology for the study of large and complex reaction networks. Here the information flow from one node to another in a network is determined by a combination of input nodes and their relation is specified using logic gates – AND, OR, NOT. It was first explained by Kauffmann where he modeled the gene as a binary device that can be either in the 'ON' or 'OFF' states signifying whether a gene expression is upregulated or downregulated, respectively [111]. Here he elucidated that a distinct advantage in this choice of a binary model for gene activity lies in the fact that the number of different possible rules by which a finite number (K) of inputs may affect the output behavior of a binary element is finite, i.e., 2^{2^K}. Figure 6.3a shows a simple toy model of three nodes interacting with one another. The reaction network can be represented using Boolean rules or equations (Eqs. 6.2, 6.3 and 6.4). The truth tables and the state transitions graphs of the reaction network show the temporal evolution of the states (0 or 1) of the nodes starting from different input combinations (Fig. 6.3a). Here, in this example we observe under the different input conditions the system tends to reach certain point steady-state attractors, i.e. 1–0–0 and 1–1–1 or cyclic attractor, i.e. 1–0–1 $\longleftrightarrow$ 1–1–0:

$$v1 = v1\; OR\; (NOT\; v3) \qquad (6.2)$$

$$v2 = v1\; AND\; v3 \qquad (6.3)$$

$$v3 = v2 \qquad (6.4)$$

Several software packages such as BoolNet (R-based), BooleanNet (Python based), and CellNetAnalyzer (software with GUI) are available for performing logical steady-state analysis of large biochemical networks [112–114]. This concept

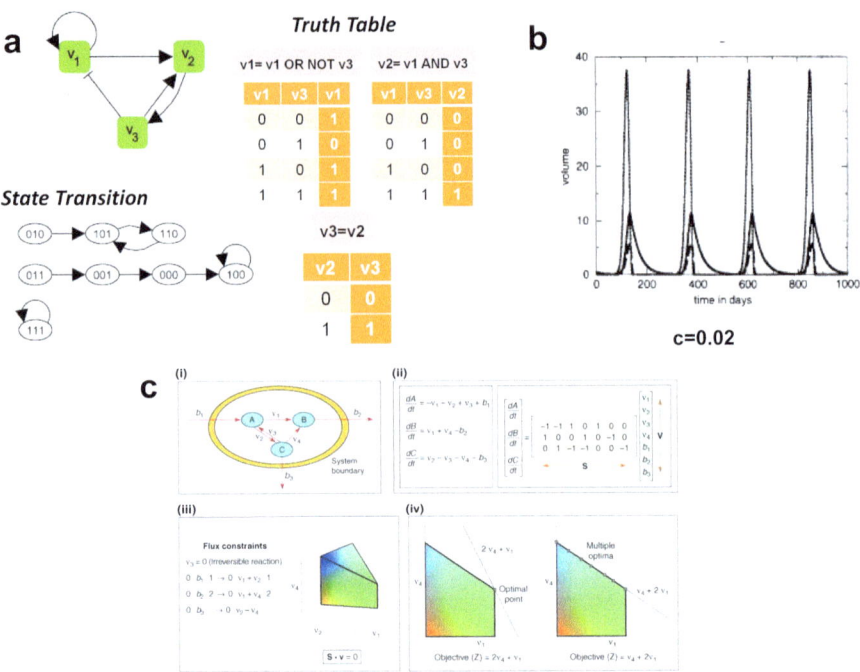

Fig. 6.3 Computational techniques used for study of large biochemical pathways. (**a**) Interaction Graph, Truth Table, and State Transition Graph for a Logic-Based Toy Model; (**b**) Temporal dynamics of Tumor, Effector cells, and IL-2 from an ODE-based model (adapted from Kirshner, et al. 1998 [150]); (**c**) A toy model describing (*i*) the flux distribution of metabolites A, B. and C through different reactions, (*ii*) the formation of stoichiometric matrix "*S*" and flux vector "*v*," (*iii*) defining constraints and (*iv*) defining objective and finding optimal solution within the solution space of linear optimization problem (Adapted from Kauffman et al. 2003 [157])

was later used by Huang and Ingber to model cell signaling networks for demonstrating that cellular phenotypes correspond to the dynamic steady states of the intracellular signaling molecules in a logic-based model. A key advantage of this strategy is that it does not require the knowledge of parameter values that is often not available for large biochemical networks. Later it has been extensively used for the study of cell signaling pathways and identification of drug targets for the treatment of cancer [109, 110]. Logical models have also been developed for the study of T-cell signaling pathways where the observations made from the *in silico* analysis were experimentally validated to establish the authenticity of their logic-based model. Using this model, the authors have predicted an alternative pathway of activation from CD28 to JNK that does not involve the canonical pathway involving LAT signalosome, nor does it involve the activation of PLCγ1 or calcium flux, but depends on the activation of the nucleotide exchange factor Vav1, which activates MEKK1 via the small G-protein Rac1 [115]. A logical steady-state model that captures the effect of the co-receptor signaling pathway cross-talks has been developed that shows that simultaneous activation of the TCR:CD3, CRAC, and OX40 pathways are important for sustained T-cell proliferation. At the same time, it has been shown that the co-receptor CD27 and LTBR pathways are important for regulating the cytokine production [116]. A further extension of this work for the study of immune responses during Leishmaniasis explains how the differentiation of T cell is altered during infection [117]. Another model employing Boolean formalism has been used in the study of differentiation of naive cells into T_{H1}, T_{H2}, T_{H17}, and Treg subtypes under different environmental conditions [118]. This model provides evidences that Foxp3$^+$ Treg cells and T_{H17} cells are highly plastic and labile, whereas the T_{H1} and T_{H2} subtypes remain steady under different environmental conditions. However, this model also predicts the existence of hybrid states and cyclic attractors expressing markers characteristic of two or more canonical cell types under certain environmental conditions that lays the foundation for the oscillatory behavior of T-cell differentiation. This study further elucidates that under proper polarizing environments, the Treg cells may differentiate into T_{H1} or T_{H2} subtypes [118]. Later another model based on the Boolean formalism was developed to study the molecular mechanisms controlling the cytokine-driven T_H cell differentiation and plasticity. This model explained the role for peroxisome proliferator–activated receptor gamma (PPARγ) in the regulation of T_{H17} to iTreg cell switching that gives promising cues for the prediction of therapeutic target for dysregulated immune responses and inflammation [119]. More recently, Probabilistic Boolean Control Network has also been employed for the study of T_H cell differentiation under varied environmental conditions. Here each input node is activated with a certain user-defined probability, which makes the system stochastic. Using this study, the authors have identified that the T-cell differentiation process is regulated by composition and dosage of signals that the cell receives from the environment. They have also predicted novel T-cell phenotypes using their model and have identified the specific environmental conditions that give rise to them [120].

6.5.2.4 Steady-State Metabolic Models

Immunometabolism has gained momentum in recent years as an emerging field of investigation at the interface between two highly discussed disciplines of immunology and metabolism [9, 10]. The idea of metabolism as a driver of the immune response [121] has been appreciated in recent years. However, capturing the bidirectional regulation of signaling and metabolism using a single computational platform is challenging. The mechanism of action of the two cascades is different, and the time scales in which the two processes occur also differ enormously. Mostly signaling cascades are faster than the metabolic reactions. This, along with the limitation of availability of information about how metabolism regulates immune cell responses and functioning, has limited the designing of immune-metabolic models to a small scale, mostly considering few parameters to design smaller dynamic models. An integrated systems-level computational model of immunometabolism is yet to be undertaken. Nevertheless, the currently employed computational approaches can be used to address immune-metabolism at a systems-level.

Genome-scale metabolic modeling (GSMM) is currently the most widely used systems-level modeling approach that accounts for whole-genome metabolism of biological systems. It is a constraint-based mathematical modeling approach that assimilates biochemical, genetic, and genomic information within a single computational platform [122–126]. It allows the study of the metabolic genotype-phenotype relationship of an organism. Genome-scale metabolic models have been used in in silico metabolic engineering for the design of studies like defining essentiality of the reaction/gene [127, 128], the relevance of distant pathways [129] and overexpression or knockout analyses of metabolites, reactions, and metabolic pathways [130]. These are efficient tools for the prediction of growth in living cells/tissues exposed to different external conditions [131]. They have been used to predict conditional and absolute essentiality of metabolites and reactions in metabolic networks.

Flux balance analysis (FBA) is the most popularly used constraint-based approach in systems-level metabolic modeling, which works on the basic principles of linear optimization [132]. The technique assumes a steady-state approach, where all the metabolites of the network are considered to be in steady state; i.e., the rate of change of metabolites over time remains zero (Fig. 6.3c). This ensures that the rate of formation of a metabolite in the network is always equal to the rate of its consumption and hence a net difference in the metabolite concentration over time always remains zero. All reactions of the network work as constraints to the optimization problem. The reactions are bounded between a lower and an upper bound, which creates the constraint. The metabolites are connected to respective reactions in the form of a stoichiometric matrix, "S," where the rows represent the metabolites (m) and the columns represent reactions (n). Thus, a "$m \times n$" matrix is generated in which the involvement of a metabolite in a reaction is represented by its respective stoichiometry in that reaction. A positive stoichiometric value represents the formation of the metabolite and a negative stoichiometric value represents consumption. The flux through the reactions is represented in a separate flux matrix "v," which is a "$n \times 1$" matrix. The outcome of the optimization is obtained by matrix multiplication of "$S.v = 0$." The matrix multiplication results in an optimized "v" matrix, which assigns an optimized flux to each of the reactions in the network. Generally, whole-genome models are large with a few hundreds of reactions and metabolites, which make it a multidimensional optimization problem. An objective is assigned to the model that depends on the biological question one wants to address. For example, if one wants to observe the behavior of the network when it tries to maximize ATP production, then one can assign ATP synthase (ATPS) reaction as the objective and try optimizing the model by maximizing the objective function. Thus, the model gets optimized a per the requirement of maximizing or minimizing the objective function.

A further extension of the modeling technique has been done to incorporate dynamic regulation of metabolic regulations by signaling pathways. This is popularly known as dynamic FBA (dFBA), where the initial activation of the metabolic FBA model depends on the output of signaling response generated by dynamic analysis. In yet another extension of FBA, the initial signaling response is analyzed using Boolean analysis. This is known as rFBA. The method that takes into account a combined FBA, Boolean regulatory, and ODE approach is known as integrative FBA (iFBA).

There are various tools available for performing these analyses. COBRA Toolbox is the most widely used platform for flux balance analysis [133]. This is a Matlab extension, which allows user-interface for ease in analysis. Other platforms are COBRApy [134], PSAMM [135], OptFlux [136], FBASimVis [137], FluxViz [138], FlexFlux [139], FAME [140], and Escher-FBA [141].

6.5.2.5 Dynamic ODE-Based Immune Models

Several dynamic models have been developed for the study of immune responses for several diseases [142–146]. The study of immune responses during tumor formation using mathematical ODE-based models has helped clinicians in the prediction of tumor evolution and the determination of dosage schedules and treatment protocols [147–149]. A seminal work by Kirschner and Panetta has led to the development of many such similar models with further improvisations [150]. The model developed by them represents an ODE-based model of the tumor-immune interaction and the production of IL-2 that has important roles in the regulation of immune response generated during tumor progression (Eqs. 6.5, 6.6, and 6.7). The model considers that the proliferation of the effector immune cells increases proportional to the antigenicity of the tumor. The model equations comprise three variables, viz. tumor (T), effector cells (E), and IL2 (I_L), that interact among themselves, and 12 parameters that describe the rate at which these interactions occur. In this model the antigenicity, denoted with c, of the tumor has been considered as an essential parameter that regulates the dynamics of the effector cell population:

$$\frac{dE}{dt} = cT - \mu_2 E + \frac{p_1 E I_L}{g_1 + I_L} + s_1 \tag{6.5}$$

$$\frac{dT}{dt} = r_2 \left(1 - bT\right)T - \frac{aET}{g_2 + T} \tag{6.6}$$

$$\frac{dI_L}{dt} = \frac{p_2 ET}{g_3 + T} - \mu_3 I_L + s_2 \tag{6.7}$$

Figure 6.3b (adapted from Kirshner et al. 1998 [150]) shows the temporal evolution of the system and the oscillating steady state behavior of the variables when antigenicity parameter $c = 0.02$. This model explains short-term oscillations in tumor sizes as well as long-term tumor relapse. This model has been further used to explore the effects of adoptive cellular immunotherapy for the tumor elimination [150].

A more recent tumor–immune interaction model developed for understanding the dynamics of immune-mediated tumor rejection focuses mainly on the role of natural killer (NK) and CD8+ T cells in tumor surveillance. Here the techniques of parameter estimation and sensitivity analysis have been exploited for the model calibration and validation with experimental results. This study has revealed the variable to which the model is most sensitive is patient specific and that there exists a direct positive correlation between the patient-specific efficacy of the CD8+ T-cell response and the likelihood of a patient favorably responding to immunotherapy treatments [151]. A more detailed model of immune responses during tumor progression has been developed using 13 variables and 71 parameters. The model considers cytokine feedbacks and five different immune cells present in the tumor microenvironment. This model is useful for optimizing combinatorial treatment dose and schedules for maximal tumor reduction using immunotherapy [152].

There is a range of ODE models that investigate various pathways involved in metabolism under different pathological conditions. Immune metabolic models are available for glucose metabolism [153], glutathione metabolism [154], folate-mediated one-carbon metabolism [154], and arsenic metabolism [155]. A composite review of these metabolic models is available in Nijhout et al.'s work [156]. The recent understanding from experimental research on the metabolic regulation of the immune response [9] will help to adapt these mathematical models to the reality of metabolic pathways inside immune cells.

6.6   Challenges and Future Directions

The immune-regulatory network forms a complex mesh of interacting cells and biochemical reactions that work in a coordinated fashion to eliminate the pathogen-infected cells and trigger the remission of any neoplastic growth inside the body. However, the intricacies of the immune signaling network are far from being completely understood, and the regulations governing the differential immune response of the T cells under varied antigenic challenges still remain elusive to immunologists. In this context, the knowledge regarding the signaling routes is essential to understand the mechanistic regulations such as the feedback and feed-forward loops and the alternative signaling pathways that govern the production of effector molecules from the lymphocytes. Hence, an in-depth study of the co-receptor signaling pathways and their cross-talks is essential that will provide valuable information regarding the pathways involved in the cytokine regulation and effector functions of the immune cells.

T-cell plasticity that determines their differentiation, de-differentiation, subtype specification, and T helper memory cell formation under different environmental conditions is yet another area that has remained very less explored. Although the recent developments in the field elucidate the process of T-cell differentiation with respect to changes in the cytokine milieu under in vitro conditions, the complex interactions in the human immunome needs to be studied using a holistic integrative approach in order to gain clear insights into the changes of immune responses due to changes in quality and quantity of the antigenic challenge, the strength of the stimulus, and the role of the other interacting immune cells. Such studies will throw light into the modulations of T-cell subtype ratios that has a substantial impact on the disease prognosis and response of a patient to an immunotherapeutic intervention.

Metabolic regulation of immune cell in determining T-cell activation, proliferation, and differentiation is a newer area of research; and studies are in progress to understand these processes. Many questions related to immune-metabolism still remain unanswered. How metabolism alters during transition from quiescent T cells to activated effector T cells remains poorly understood. Although mTORC1 activity has been observed to be central to signaling and metabolic cross-talk and the master kinase in guiding quiescence exit of T cells, how nutrients tune mTORC1 activity remains to be explored further. Redox metabolism and oxygen sensing have been implicated in T-cell proliferation and activation; however, the exact mechanism of how they regulate T-cell quiescence and activation in different tissues remains unaddressed. Also, the cross-talks between signaling and metabolic pathways are only partially explored. A clear understanding of these mechanisms will help augment immune responses and pave way for immunotherapy under different pathogenic conditions.

References

1. Cruse JM, Lewis RE, Wang H (eds) (2004) Chapter 1 – Molecules, cells, and tissues of immunity. Immunology guidebook. Academic, San Diego, pp 1–15
2. Rabb H (2002) The T cell as a bridge between innate and adaptive immune systems: implications for the kidney. Kidney Int 61(6):1935–1946
3. Moticka EJ (2016) Chapter 20 – Activation of T lymphocytes and MHC restriction. A historical perspective on evidence-based immunology. Elsevier, Amsterdam, pp 169–179
4. Moticka EJ (2016) Chapter 37 – Tumor immunology. A historical perspective on evidence-based immunology. Elsevier, Amsterdam, pp 329–339
5. Kara EE, Comerford I, Fenix KA, Bastow CR, Gregor CE, McKenzie DR et al (2014) Tailored immune responses: novel effector helper T cell subsets in protective immunity. PLoS Pathog 10(2):e1003905
6. Moticka EJ (2016) Chapter 23 – T lymphocyte subpopulations. A historical perspective on evidence-based immunology. Elsevier, Amsterdam, pp 197–205
7. Pearce EL, Pearce EJ (2013) Metabolic pathways in immune cell activation and quiescence. Immunity 38(4):633–643
8. Zeng H, Chi H (2017) mTOR signaling in the differentiation and function of regulatory and effector T cells. Curr Opin Immunol 46:103–111
9. Ganeshan K, Chawla A (2014) Metabolic regulation of immune responses. Annu Rev Immunol 32:609–634
10. Assmann N, Finlay DK (2016) Metabolic regulation of immune responses: therapeutic opportunities. J Clin Invest 126(6):2031–2039
11. Patel CH, Leone RD, Horton MR, Powell JD (2019) Targeting metabolism to regulate immune responses in autoimmunity and cancer. Nat Rev Drug Discov 18(9):669–688
12. Jung J, Zeng H, Horng T (2019) Metabolism as a guiding force for immunity. Nat Cell Biol 21(1):85–93
13. Brownlie RJ, Zamoyska R (2013) T cell receptor signalling networks: branched, diversified and bounded. Nat Rev Immunol 13:257
14. Ware CF (2008) Targeting lymphocyte activation through the lymphotoxin and LIGHT pathways. Immunol Rev 223:186–201
15. Elgueta R, Benson MJ, de Vries VC, Wasiuk A, Guo Y, Noelle RJ (2009) Molecular mechanism and function of CD40/CD40L engagement in the immune system. Immunol Rev 229(1):152–172
16. Munroe ME, Bishop GA (2007) A Costimulatory function for T cell CD40. J Immunol 178(2):671–682
17. Redmond WL, Ruby CE, Weinberg AD (2009) The role of OX40-mediated co-stimulation in T cell activation and survival. Crit Rev Immunol 29(3):187–201

18. Croft M, So T, Duan W, Soroosh P (2009) The significance of OX40 and OX40L to T-cell biology and immune disease. Immunol Rev 229(1):173–191
19. Ledbetter JA, Deans JP, Aruffo A, Grosmaire LS, Kanner SB, Bolen JB et al (1993) CD4, CD8 and the role of CD45 in T-cell activation. Curr Opin Immunol 5(3):334–340
20. Chen L, Flies DB (2013) Molecular mechanisms of T cell co-stimulation and co-inhibition. Nat Rev Immunol 13(4):227–242
21. Feske S (2007) Calcium signalling in lymphocyte activation and disease. Nat Rev Immunol 7:690
22. Baine I, Abe Brian T, Macian F (2009) Regulation of T-cell tolerance by calcium/NFAT signaling. Immunol Rev 231(1):225–240
23. Nakaya M, Xiao Y, Zhou X, Chang J-H, Chang M, Cheng X et al (2014) Inflammatory T cell responses rely on amino acid transporter ASCT2 facilitation of glutamine uptake and mTORC1 kinase activation. Immunity 40(5):692–705
24. Palmer CS, Ostrowski M, Balderson B, Christian N, Crowe SM (2015) Glucose metabolism regulates T cell activation, differentiation, and functions. Front Immunol 6:1
25. Chi H (2012) Regulation and function of mTOR signalling in T cell fate decisions. Nat Rev Immunol 12(5):325
26. Chapman NM, Boothby MR, Chi H (2020) Metabolic coordination of T cell quiescence and activation. Nat Rev Immunol 20:55–70
27. Johnson MO, Wolf MM, Madden MZ, Andrejeva G, Sugiura A, Contreras DC et al (2018) Distinct regulation of Th17 and Th1 cell differentiation by glutaminase-dependent metabolism. Cell 175(7):1780–95.e19
28. Dodd KM, Tee AR (2012) Leucine and mTORC1: a complex relationship. Am J Physiol Endocrinol Metab 302(11):E1329–42
29. Fallarino F, Grohmann U, Vacca C, Bianchi R, Orabona C, Spreca A et al (2002) T cell apoptosis by tryptophan catabolism. Cell Death Differ 9(10):1069
30. Fallarino F, Grohmann U, Hwang KW, Orabona C, Vacca C, Bianchi R et al (2003) Modulation of tryptophan catabolism by regulatory T cells. Nat Immunol 4(12):1206
31. Geiger R, Rieckmann JC, Wolf T, Basso C, Feng Y, Fuhrer T et al (2016) L-arginine modulates T cell metabolism and enhances survival and anti-tumor activity. Cell 167(3):829–42. e13
32. Zhao Y, Hu X, Liu Y, Dong S, Wen Z, He W et al (2017) ROS signaling under metabolic stress: cross-talk between AMPK and AKT pathway. Mol Cancer 16(1):79
33. Franchina DG, Dostert C, Brenner D (2018) Reactive oxygen species: involvement in T cell signaling and metabolism. Trends Immunol 39(6):489–502
34. Zhu J, Yamane H, Paul WE (2010) Differentiation of effector CD4 T cell populations. Annu Rev Immunol 28:445–489
35. Luckheeram RV, Zhou R, Verma AD, Xia B (2012) CD4+T cells: differentiation and functions. Clin Dev Immunol 2012:925135
36. Hawse WF, Cattley RT (2019) T cells transduce T-cell receptor signal strength by generating different phosphatidylinositols. J Biol Chem 294(13):4793–4805
37. Yamane H, Paul WE (2013) Early signaling events that underlie fate decisions of naive CD4(+) T cells towards distinct T-helper cell subsets. Immunol Rev 252(1):12–23
38. Peter C, Waldmann H, Cobbold SP (2010) mTOR signalling and metabolic regulation of T cell differentiation. Curr Opin Immunol 22(5):655–661
39. MacIver NJ, Michalek RD, Rathmell JC (2013) Metabolic regulation of T lymphocytes. Annu Rev Immunol 31:259–283
40. Klysz D, Tai X, Robert PA, Craveiro M, Cretenet G, Oburoglu L et al (2015) Glutamine-dependent α-ketoglutarate production regulates the balance between T helper 1 cell and regulatory T cell generation. Sci Signal 8(396):ra97
41. Mak TW, Grusdat M, Duncan GS, Dostert C, Nonnenmacher Y, Cox M et al (2017) Glutathione primes T cell metabolism for inflammation. Immunity 46(4):675–689
42. Stark JM, Tibbitt CA, Coquet JM (2019) The metabolic requirements of Th2 cell differentiation. Front Immunol 10:2318

43. Gerriets VA, Kishton RJ, Johnson MO, Cohen S, Siska PJ, Nichols AG et al (2016) Foxp3 and Toll-like receptor signaling balance T reg cell anabolic metabolism for suppression. Nat Immunol 17(12):1459

44. Angelin A, Gil-de-Gómez L, Dahiya S, Jiao J, Guo L, Levine MH et al (2017) Foxp3 reprograms T cell metabolism to function in low-glucose, high-lactate environments. Nat Immunol 25(6):1282–93.e7

45. Geginat J, Paroni M, Maglie S, Alfen JS, Kastirr I, Gruarin P et al (2014) Plasticity of human CD4 T cell subsets. Front Immunol 5:630

46. Pollizzi KN, Powell JD (2015) Regulation of T cells by mTOR: the known knowns and the known unknowns. Trends Immunol 36(1):13–20

47. Yang K, Neale G, Green DR, He W, Chi H (2011) The tumor suppressor Tsc1 enforces quiescence of naive T cells to promote immune homeostasis and function. Nat Immunol 12(9):888

48. Sauer S, Bruno L, Hertweck A, Finlay D, Leleu M, Spivakov M et al (2008) T cell receptor signaling controls Foxp3 expression via PI3K, Akt, and mTOR. Proc Natl Acad Sci U S A 105(22):7797–7802

49. Macintyre AN, Gerriets VA, Nichols AG, Michalek RD, Rudolph MC, Deoliveira D et al (2014) The glucose transporter Glut1 is selectively essential for CD4 T cell activation and effector function. Cell Metab 20(1):61–72

50. Wang R, Dillon CP, Shi LZ, Milasta S, Carter R, Finkelstein D et al (2011) The transcription factor Myc controls metabolic reprogramming upon T lymphocyte activation. Immunity 35(6):871–882

51. Yang K, Shrestha S, Zeng H, Karmaus PW, Neale G, Vogel P et al (2013) T cell exit from quiescence and differentiation into Th2 cells depend on Raptor-mTORC1-mediated metabolic reprogramming. Immunity 39(6):1043–1056

52. Tan H, Yang K, Li Y, Shaw TI, Wang Y, Blanco DB et al (2017) Integrative proteomics and phosphoproteomics profiling reveals dynamic signaling networks and bioenergetics pathways underlying T cell activation. Immunity 46(3):488–503

53. Gerriets VA, Kishton RJ, Nichols AG, Macintyre AN, Inoue M, Ilkayeva O et al (2015) Metabolic programming and PDHK1 control CD4+ T cell subsets and inflammation. J Clin Invest 125(1):194–207

54. Blagih J, Coulombe F, Vincent EE, Dupuy F, Galicia-Vázquez G, Yurchenko E et al (2015) The energy sensor AMPK regulates T cell metabolic adaptation and effector responses in vivo. Immunity 42(1):41–54

55. Mossmann D, Park S, Hall MN (2018) mTOR signalling and cellular metabolism are mutual determinants in cancer. Nat Rev Cancer 18(12):744–757

56. Buck MD, O'sullivan D, Pearce EL (2015) T cell metabolism drives immunity. J Exp Med 212(9):1345–1360

57. Mullen PJ, Yu R, Longo J, Archer MC, Penn LZ (2016) The interplay between cell signalling and the mevalonate pathway in cancer. Nat Rev Cancer 16(11):718

58. Doerig C, Rayner JC, Scherf A, Tobin AB (2015) Post-translational protein modifications in malaria parasites. Nat Rev Microbiol 13(3):160–172

59. Ho P-C, Bihuniak JD, Macintyre AN, Staron M, Liu X, Amezquita R et al (2015) Phosphoenolpyruvate is a metabolic checkpoint of anti-tumor T cell responses. Cell 162(6):1217–1228

60. Wang F, Beck-García K, Zorzin C, Schamel WW, Davis MM (2016) Inhibition of T cell receptor signaling by cholesterol sulfate, a naturally occurring derivative of membrane cholesterol. Nat Immunol 17(7):844–50

61. Demetriou M, Granovsky M, Quaggin S, Dennis JW (2001) Negative regulation of T-cell activation and autoimmunity by Mgat5 N-glycosylation. Nature 409(6821):733–9

62. Lee K, Gudapati P, Dragovic S, Spencer C, Joyce S, Killeen N et al (2010) Mammalian target of rapamycin protein complex 2 regulates differentiation of Th1 and Th2 cell subsets via distinct signaling pathways. Immunity 32(6):743–753

63. Frauwirth KA, Riley JL, Harris MH, Parry RV, Rathmell JC, Plas DR et al (2002) The CD28 signaling pathway regulates glucose metabolism. Immunity 16(6):769–777

64. Kimura MY, Pobezinsky LA, Guinter TI, Thomas J, Adams A, Park J-H et al (2013) IL-7 signaling must be intermittent, not continuous, during CD8+ T cell homeostasis to promote cell survival instead of cell death. Nat Immunol 14(2):143–51

65. Jacobs SR, Herman CE, MacIver NJ, Wofford JA, Wieman HL, Hammen JJ et al (2008) Glucose uptake is limiting in T cell activation and requires CD28-mediated Akt-dependent and independent pathways. J Immunol 180(7):4476–4486

66. Sinclair LV, Rolf J, Emslie E, Shi Y-B, Taylor PM, Cantrell DA (2013) Control of amino-acid transport by antigen receptors coordinates the metabolic reprogramming essential for T cell differentiation. Nat Immunol 14(5):500

67. Bricker DK, Taylor EB, Schell JC, Orsak T, Boutron A, Chen Y-C et al (2012) A mitochondrial pyruvate carrier required for pyruvate uptake in yeast, Drosophila, and humans. Science 337(6090):96–100

68. Shi LZ, Wang R, Huang G, Vogel P, Neale G, Green DR et al (2011) HIF1α–dependent glycolytic pathway orchestrates a metabolic checkpoint for the differentiation of TH17 and Treg cells. J Exp Med 208(7):1367–1376

69. Sinclair LV, Neyens D, Ramsay G, Taylor PM, Cantrell DA (2018) Single cell analysis of kynurenine and System L amino acid transport in T cells. Nat Commun 9(1):1981

70. Baixauli F, Acín-Pérez R, Villarroya-Beltrí C, Mazzeo C, Nuñez-Andrade N, Gabandé-Rodriguez E et al (2015) Mitochondrial respiration controls lysosomal function during inflammatory T cell responses. Cell Metab 22(3):485–498

71. Seman M, Adriouch S, Scheuplein F, Krebs C, Freese D, Glowacki G et al (2003) NAD-induced T cell death: ADP-ribosylation of cell surface proteins by ART2 activates the cytolytic P2X7 purinoceptor. Immunity 19(4):571–582

72. Adriouch S, Hubert S, Pechberty S, Koch-Nolte F, Haag F, Seman M (2007) NAD+ released during inflammation participates in T cell homeostasis by inducing ART2-mediated death of naive T cells in vivo. J Immunol 179(1):186–194

73. Clever D, Roychoudhuri R, Constantinides MG, Askenase MH, Sukumar M, Klebanoff CA et al (2016) Oxygen sensing by T cells establishes an immunologically tolerant metastatic niche. Cell 166(5):1117–31.e14

74. Tarasenko TN, Pacheco SE, Koenig MK, Gomez-Rodriguez J, Kapnick SM, Diaz F et al (2017) Cytochrome c oxidase activity is a metabolic checkpoint that regulates cell fate decisions during T cell activation and differentiation. Cell Metab 25(6):1254–68.e7

75. Sena LA, Li S, Jairaman A, Prakriya M, Ezponda T, Hildeman DA et al (2013) Mitochondria are required for antigen-specific T cell activation through reactive oxygen species signaling. Immunity 38(2):225–236

76. Chinen T, Kannan AK, Levine AG, Fan X, Klein U, Zheng Y et al (2016) An essential role for the IL-2 receptor in T reg cell function. Nat Immunol 17(11):1322

77. Yang K, Blanco DB, Neale G, Vogel P, Avila J, Clish CB et al (2017) Homeostatic control of metabolic and functional fitness of T reg cells by LKB1 signalling. Nature 548(7669):602

78. Zeiser R, Maas K, Youssef S, Dürr C, Steinman L, Negrin RSJI (2009) Regulation of different inflammatory diseases by impacting the mevalonate pathway. Immunology 127(1):18–25

79. Huynh A, DuPage M, Priyadharshini B, Sage PT, Quiros J, Borges CM et al (2015) Control of PI (3) kinase in T reg cells maintains homeostasis and lineage stability. Nat Immunol 16(2):188

80. Tyrakis PA, Palazon A, Macias D, Lee KL, Phan AT, Veliça P et al (2016) S-2-hydroxyglutarate regulates CD8+ T-lymphocyte fate. Nature 540(7632):236

81. Kidani Y, Elsaesser H, Hock MB, Vergnes L, Williams KJ, Argus JP et al (2013) Sterol regulatory element–binding proteins are essential for the metabolic programming of effector T cells and adaptive immunity. Nat Immunol 14(5):489

82. Bensinger SJ, Bradley MN, Joseph SB, Zelcer N, Janssen EM, Hausner MA et al (2008) LXR signaling couples sterol metabolism to proliferation in the acquired immune response. Cell 134(1):97–111

83. Angela M, Endo Y, Asou HK, Yamamoto T, Tumes DJ, Tokuyama H et al (2016) Fatty acid metabolic reprogramming via mTOR-mediated inductions of PPARγ directs early activation of T cells. Nat Commun 7:13683

84. Munn DH, Sharma MD, Baban B, Harding HP, Zhang Y, Ron D et al (2005) GCN2 kinase in T cells mediates proliferative arrest and anergy induction in response to indoleamine 2, 3-dioxygenase. Immunity 22(5):633–642

85. Chang C-H, Curtis JD, Maggi LB Jr, Faubert B, Villarino AV, O'Sullivan D et al (2013) Posttranscriptional control of T cell effector function by aerobic glycolysis. Cell 153(6):1239–1251

86. Sherman BT, Lempicki RA (2009) Systematic and integrative analysis of large gene lists using DAVID bioinformatics resources. Nat Protoc 4(1):44–57

87. Huang DW, Sherman BT, Lempicki RA (2008) Bioinformatics enrichment tools: paths toward the comprehensive functional analysis of large gene lists. Nucleic Acids Res 37(1):1–13

88. Carmona-Saez P, Chagoyen M, Tirado F, Carazo JM, Pascual-Montano A (2007) GENECODIS: a web-based tool for finding significant concurrent annotations in gene lists. Genome Biol 8(1):R3

89. Subramanian A, Tamayo P, Mootha VK, Mukherjee S, Ebert BL, Gillette MA et al (2005) Gene set enrichment analysis: a knowledge-based approach for interpreting genome-wide expression profiles. Proc Natl Acad Sci U S A 102(43):15545–15550

90. Croft D, Mundo AF, Haw R, Milacic M, Weiser J, Wu G et al (2013) The Reactome pathway knowledgebase. Nucleic Acids Res. 42(D1):D472–D477

91. Bindea G, Mlecnik B, Tosolini M, Kirilovsky A, Waldner M, Obenauf Anna C et al (2013) Spatiotemporal dynamics of Intratumoral immune cells reveal the immune landscape in human cancer. Immunity 39(4):782–795

92. Fridman WH, Pagès F, Sautès-Fridman C, Galon J (2012) The immune contexture in human tumours: impact on clinical outcome. Nat Rev Cancer 12:298

93. Dillon LAL, Suresh R, Okrah K, Corrada Bravo H, Mosser DM, El-Sayed NM (2015) Simultaneous transcriptional profiling of Leishmania major and its murine macrophage host cell reveals insights into host-pathogen interactions. BMC Genomics 16:1108

94. Favila MA, Geraci NS, Zeng E, Harker B, Condon D, Cotton RN et al (2014) Human dendritic cells exhibit a pronounced type I IFN signature following Leishmania major infection that is required for IL-12 induction. J Immunol 192(12):5863–5872

95. Chowdhury S, Sarkar RR (2015) Comparison of human cell signaling pathway databases—evolution, drawbacks and challenges. Database 2015:bau126

96. Sherriff MR, Sarkar RR (2008) Computational approaches and modelling of signaling processes in immune system. Proc Indian Natl Sci Acad 74:187–200

97. Chowdhury S, Sinha N, Ganguli P, Bhowmick R, Singh V, Nandi S et al BIOPYDB: a dynamic human cell specific biochemical pathway database with advanced computational analyses platform. J Integr Bioinform 15(3):20170072

98. Trupp M, Altman T, Fulcher CA, Caspi R, Krummenacker M, Paley S et al (2010) Beyond the genome (BTG) is a (PGDB) pathway genome database: HumanCyc. Genome Biol 11(1):O12

99. Schomburg I, Chang A, Schomburg D (2002) BRENDA, enzyme data and metabolic information. Nucleic Acids Res 30(1):47–49

100. Euler L (1736) Commentarii Academiae Scientiarum Imperialis Petropolitanae 8:128

101. Watts DJ, Strogatz SH (1998) Collective dynamics of 'small-world' networks. Nature 393(6684):440

102. Barabási A-L, Albert R (1999) Emergence of scaling in random networks. Science 286(5439):509–512

103. Smoot ME, Ono K, Ruscheinski J, Wang P-L, Ideker T (2010) Cytoscape 2.8: new features for data integration and network visualization. Bioinformatics 27(3):431–432

104. Bastian M, Heymann S, Jacomy M (2009) Gephi: an open source software for exploring and manipulating networks. Third international AAAI conference on weblogs and social media.

105. Batagelj V, Mrvar A (1998) Pajek-program for large network analysis. Connect 21(2):47–57

106. Pavlopoulos GA, Secrier M, Moschopoulos CN, Soldatos TG, Kossida S, Aerts J et al (2011) Using graph theory to analyze biological networks. BioData Min 4(1):10

107. Bromberg KD, Ma'ayan A, Neves SR, Iyengar R (2008) Design logic of a cannabinoid receptor signaling network that triggers neurite outgrowth. Science 320(5878):903–909

108. Ghosh S, Kumar GV, Basu A, Banerjee A (2015) Graph theoretic network analysis reveals protein pathways underlying cell death following neurotropic viral infection. Sci Rep 5:14438
109. Chowdhury S, Pradhan RN, Sarkar RR (2013) Structural and logical analysis of a comprehensive hedgehog signaling pathway to identify alternative drug targets for glioma, colon and pancreatic cancer. PLoS One 8(7):e69132
110. Chowdhury S, Sarkar R (2013) Drug targets and biomarker identification from computational study of human notch signaling pathway. Clin Exp Pharmacol 3(137):2161–1459
111. Kauffman SA (1969) Metabolic stability and epigenesis in randomly constructed genetic nets. J Theor Biol 22(3):437–467
112. Müssel C, Hopfensitz M, Kestler HA (2010) BoolNet—an R package for generation, reconstruction and analysis of Boolean networks. Bioinformatics 26(10):1378–1380
113. Albert I, Thakar J, Li S, Zhang R, Albert R (2008) Boolean network simulations for life scientists. Source code for biology and medicine. Source Code Biol Med 3(1):16
114. Klamt S, Saez-Rodriguez J, Gilles ED (2007) Structural and functional analysis of cellular networks with CellNetAnalyzer. BMC Syst Biol 1(1):2
115. Saez-Rodriguez J, Simeoni L, Lindquist JA, Hemenway R, Bommhardt U, Arndt B et al (2007) A logical model provides insights into T cell receptor signaling. PLOS Comp Biol 3(8):e163
116. Ganguli P, Chowdhury S, Bhowmick R, Sarkar RR (2015) Temporal protein expression pattern in intracellular signalling cascade during T-cell activation: a computational study. J Biosci 40(4):769–789
117. Ganguli P, Chowdhury S, Chowdhury S, Sarkar RR (2015) Identification of Th1/Th2 regulatory switch to promote healing response during leishmaniasis: a computational approach. EURASIP J Bioinform Syst Biol 2015(1):13
118. Naldi A, Carneiro J, Chaouiya C, Thieffry D (2010) Diversity and plasticity of Th cell types predicted from regulatory network modelling. PLOS Comp Biol 6(9):e1000912
119. Carbo A, Hontecillas R, Kronsteiner B, Viladomiu M, Pedragosa M, Lu P et al (2013) Systems modeling of molecular mechanisms controlling cytokine-driven CD4+ T cell differentiation and phenotype plasticity. PLOS Comp Biol 9(4):e1003027
120. Puniya BL, Todd RG, Mohammed A, Brown DM, Barberis M, Helikar T (2018) A mechanistic computational model reveals that plasticity of CD4+ T cell differentiation is a function of cytokine composition and dosage. Front Physiol 9:878
121. Kelly PN (2019) Metabolism as a driver of immune response. Science 363(6423):137–139
122. Price ND, Reed JL, Palsson BØ (2004) Genome-scale models of microbial cells: evaluating the consequences of constraints. Nat Rev Microbiol 2(11):886
123. Blazier AS, Papin JA (2012) Integration of expression data in genome-scale metabolic network reconstructions. Front Physiol 3:299
124. Orth JD, Palsson BØ (2010) Systematizing the generation of missing metabolic knowledge. Biotechnol Bioeng 107(3):403–412
125. O'Brien EJ, Monk JM, Palsson BO (2015) Using genome-scale models to predict biological capabilities. Cell 161(5):971–987
126. Bordbar A, Monk JM, King ZA, Palsson BO (2014) Constraint-based models predict metabolic and associated cellular functions. Nat Rev Genet 15(2):107–120
127. Patil KR, Nielsen J (2005) Uncovering transcriptional regulation of metabolism by using metabolic network topology. Proc Natl Acad Sci U S A 102(8):2685–2689
128. AbuOun M, Suthers PF, Jones GI, Carter BR, Saunders MP, Maranas CD et al (2009) Genome scale reconstruction of a salmonella metabolic model comparison of similarity and differences with a commensal Escherichia coli strain. J Biol Chem 284(43):29480–29488
129. Pharkya P, Burgard AP, Maranas CD (2004) OptStrain: a computational framework for redesign of microbial production systems. Genome Res 14(11):2367–2376
130. Pharkya P, Maranas CD (2006) An optimization framework for identifying reaction activation/inhibition or elimination candidates for overproduction in microbial systems. Metab Eng 8(1):1–13

131. Förster J, Famili I, Fu P, Palsson BØ, Nielsen J (2003) Genome-scale reconstruction of the Saccharomyces cerevisiae metabolic network. Genome Res 13(2):244–253
132. Orth JD, Thiele I, Palsson BØ (2010) What is flux balance analysis? Nat Biotechnol 28(3):245
133. Becker SA, Feist AM, Mo ML, Hannum G, Palsson BØ, Herrgard MJ (2007) Quantitative prediction of cellular metabolism with constraint-based models: the COBRA Toolbox. Nat Protoc 2(3):727
134. Ebrahim A, Lerman JA, Palsson BO, Hyduke DR (2013) COBRApy: constraints-based reconstruction and analysis for python. BMC Syst Biol 7(1):74
135. Steffensen JL, Dufault-Thompson K, Zhang Y (2016) PSAMM: a portable system for the analysis of metabolic models. PLoS Comput Biol 12(2):e1004732
136. Rocha I, Maia P, Evangelista P, Vilaça P, Soares S, Pinto JP et al (2010) OptFlux: an open-source software platform for in silico metabolic engineering. BMC Syst Biol 4(1):45
137. Grafahrend-Belau E, Klukas C, Junker BH, Schreiber F (2009) FBA-SimVis: interactive visualization of constraint-based metabolic models. Bioinformatics 25(20):2755–2757
138. König M, Holzhütter H-G (2010) Fluxviz—Cytoscape plug-in for visualization of flux distributions in networks. Genome Inform 24:96–103
139. Marmiesse L, Peyraud R, Cottret L (2015) FlexFlux: combining metabolic flux and regulatory network analyses. BMC Syst Biol 9(1):93
140. Kirchmair J, Williamson MJ, Afzal AM, Tyzack JD, Choy AP, Howlett A et al (2013) FAst MEtabolizer (FAME): a rapid and accurate predictor of sites of metabolism in multiple species by endogenous enzymes. J Chem Inf Model 53(11):2896–2907
141. Rowe E, Palsson BO, King ZA (2018) Escher-FBA: a web application for interactive flux balance analysis. BMC Syst Biol 12(1):84
142. Beerenwinkel N, Schwarz RF, Gerstung M, Markowetz F (2015) Cancer evolution: mathematical models and computational inference. Syst Biol 64(1):e1-25
143. Banerjee S, Sarkar RR (2008) Delay-induced model for tumor-immune interaction and control of malignant tumor growth. Biosystems 91(1):268–288
144. d'Onofrio A, Gatti F, Cerrai P, Freschi L (2010) Delay-induced oscillatory dynamics of tumour–immune system interaction. Math Comput Model 51(5–6):572–591
145. dePillis L, Caldwell T, Sarapata E, Williams H (2013) Mathematical modeling of regulatory T cell effects on renal cell carcinoma treatment. Discrete Cont Dyn-B 18(4):915–943
146. d'Onofrio A (2008) Metamodeling tumor–immune system interaction, tumor evasion and immunotherapy. Math Comput Model 47(5–6):614–637
147. Leder K, Pitter K, Laplant Q, Hambardzumyan D, Ross BD, Chan TA et al (2014) Mathematical modeling of PDGF-driven glioblastoma reveals optimized radiation dosing schedules. Cell 156(3):603–616
148. Robertson-Tessi M, El-Kareh A, Goriely A (2012) A mathematical model of tumor-immune interactions. J Theor Biol 294:56–73
149. Powathil GG, Adamson DJ, Chaplain MA (2013) Towards predicting the response of a solid tumour to chemotherapy and radiotherapy treatments: clinical insights from a computational model. PLoS Comput Biol 9(7):e1003120
150. Kirschner D, Panetta JC (1998) Modeling immunotherapy of the tumor–immune interaction. JMB 37(3):235–252
151. de Pillis LG, Radunskaya AE, Wiseman CL (2005) A validated mathematical model of cell-mediated immune response to tumor growth. Cancer Res 65(17):7950
152. Ganguli P, Sarkar RR (2018) Exploring immuno-regulatory mechanisms in the tumor microenvironment: model and design of protocols for cancer remission. PLoS One 13(9):e0203030
153. Chew YH, Shia YL, Lee CT, Majid FAA, Chua LS, Sarmidi MR et al (2009) Modeling of glucose regulation and insulin-signaling pathways. Mol Cell Endocrinol 303(1–2):13–24
154. Reed MC, Thomas RL, Pavisic J, James SJ, Ulrich CM, Nijhout HF et al (2008) A mathematical model of glutathione metabolism. J Theor Biol 5(1):8
155. Lawley SD, Yun J, Gamble MV, Hall MN, Reed MC, Nijhout HF et al (2014) Mathematical modeling of the effects of glutathione on arsenic methylation. J Theor Biol 11(1):20

156. Nijhout HF, Best JA, Reed MC (2015) Using mathematical models to understand metabolism, genes, and disease. BMC Biol 13(1):79
157. Kauffman KJ, Prakash P, Edwards JS (2003) Advances in flux balance analysis. J Coib 14(5):491–496

Asma Naseem and Hashim Ali

Abstract

Cardiovascular disease is the leading cause of death worldwide, despite the growing advances that have been made in the development of therapeutics. Almost all aspects of the pathogenesis underlying a cardiac injury are critically influenced by the inflammatory response. Over the past two decades, researchers have shown that the myocardium triggers an intense innate immune response that activates various immune effectors including the pattern recognition receptors.

In this chapter, we will give an overview of the innate immune cells involved in the cardiac homeostasis and their responses after cardiac injuries, focusing on the role of innate immune signaling pathways in the progression of various cardiovascular diseases.

Keywords

Atherosclerosis · Cardiomyocytes · Cardiac injury · DAMPs · Granulocytes · Innate sensing · Neutrophil extracellular traps · Plaque formation · Pattern-recognition receptors · Toll-like receptors

Asma Naseem and Hashim Ali have equally contributed to this chapter.

A. Naseem
Cellular Immunology Lab, ICGEB, Trieste, Italy

H. Ali (✉)
Department of Cardiology, King's College London, London, UK
e-mail: hashim.ali@kcl.ac.uk

© Springer Nature Singapore Pte Ltd. 2020
S. Singh (ed.), *Systems and Synthetic Immunology*,
https://doi.org/10.1007/978-981-15-3350-1_7

Abbreviations

Apo E	apolipoprotein E
CARD	caspase recruitment domains
CD	cluster of differentiation
CVD	cardiovascular disease
DAMP	danger-associated molecular pattern
ECM	extracellular matrix
HF	heart failure
HMGB1	high-mobility group box 1
HSP	heat shock protein
IFN	interferon
IKK	inhibitor of kappa B kinase
IL	interleukin
IRAK	IL-1 receptor-associated kinase
LRR	leucine-rich repeat
MDA5	melanoma differentiation-associated protein 5
MMP9	matrix metalloproteinase 9
MyD88	myeloid differentiation primary response protein 88
NETs	neutrophil extracellular traps
NF-κB	nuclear factor κ-light-chain-enhancer of activated B cells
NLR	NOD-like receptor
NLRP	NOD-, LRR-, and pyrin domain-containing protein 3
NOD	nucleotide-binding oligomerization domain
PAMP	pathogen-associated molecular pattern
PRR	pattern recognition receptor
RIG-I	retinoic acid-inducible gene I
RLR	RIG-I-like receptor
TAK	transforming growth factor-β–activated kinase
TLR	toll-like receptor
TNF-α	tumor necrosis factor-α
TRAF	tumor necrosis factor receptor-associated factor
TRAM	TRIF-related adaptor molecule
TRIF	toll/IL-1 receptor homology domain–containing adapter inducing IFN-β

7.1 Introduction

Cardiovascular disorders (CVDs) represent the most life-threatening disease, causing more deaths, disability, and economic costs than other diseases. CVDs alone are accountable for approximately 18 million deaths annually, which represent ~30% of all deaths. According to the 2015 WHO report, the overall burden of CVD and the associated heart failure (HF) continue to grow in the developed countries as well as in low- and middle-income countries. In India, these conditions have become the leading cause of mortality, being responsible for 24% of total deaths [1–3]. In particular, the frequency of coronary heart disease in India has increased at an extremely fast rate, rising from 2% in 1960 to 14% in 2013 in urban areas and from 1.7% to 7.4% in rural areas [2, 4]. A vast part of the CVD burden is due to the biological inability of the myocardium to repair the damaged cardiac tissue by regeneration of cardiomyocytes. Mammalian cardiomyocytes exhibit robust proliferative activity during embryonic and fetal development; this suddenly stops during the first weeks of postnatal life. However, it is now well established that adult zebrafish [5] and neonatal mice [6] can regenerate their hearts after injuries by the proliferation of existing cardiomyocytes, suggesting that latent regenerative capacity exists in the heart [6, 7]. Recent studies indicate that less than 1% of cardiomyocyte cycle every year in adult individuals and that this percentage increases after infarction [8]. Thus, from a clinical outlook, the heart is a postmitotic organ in which repair of damaged tissue occurs through the formation of fibrotic scar, and in this process, the immune system plays a key role.

According to epidemiological studies, ischemic heart disease (IHD) is the leading cause of various cardiac diseases. Other common causes include dilated cardiomyopathy, hypertension, atrial fibrillation, atherosclerosis, infections and myocarditis, and inflammation-related cardiomyopathy [9]. In 1990s, Levine et al. established the first link between heart failure (HF) and inflammation by reporting enhanced levels of tumor necrosis factor-α (TNF-α) in HF patients [10]. Numerous studies later showed that HF patients have elevated levels of circulating inflammatory cytokines such as interleukin-6 (IL-6) and interleukin-1β reflecting underlying pathogenic mechanisms [11].

To delineate self- from non-self-structures, the innate immune system evolved to delimit tissue injury as well as balance the homeostatic responses within the heart. A vast amount of literature suggests that intrinsic stress response is mediated by a family of pattern recognition receptors (PRRs). In this chapter, we review the roles of individual immune cell subsets and how the innate immune signaling pathways (involving the PRRs) contribute to both the initial insults and the chronic phase of cardiac injuries.

7.2   Cellular Composition of the Heart

The adult mammalian heart is composed of a diverse, symbiotic population of interstitial cells [12, 13]. Cardiomyocytes (CMs) are the most abundant cells and responsible for generating contractile force and control the rhythmic beating of the heart. While non-CMs occupy a comparatively small portion, these cells are essential for normal cardiac homeostasis, providing the extracellular matrix (ECM), intercellular communication, and vascular resource indispensable for CM function and survival. In the heart, both CMs and non-CMs respond to physiological and pathological stimuli. However, non-CMs play a pivotal role upon cardiac injuries such as inflammation, innate immune system activation, and fibrosis and also participate in various cardiac pathologies and HF.

7.3   Nonimmune Cells in the Heart

The adult mammalian heart contains ~20–35% mature CMs [14], most of which exit the cell cycle soon after birth. Therefore, in cardiac injuries such as myocardial infraction (MI), a significant portion of CM death occurs due to ischemia/reperfusion (I/R) and fast replicating fibroblasts replace the lost CM space, which leads to contractile dysfunction and scarring. CM death itself gives the primary signal for cardiac repair by circulating damage-associated molecules called alarmins or damage-associated molecular patterns (DAMPs) [15]. Fibroblasts represent the second largest population of cardiac resident cells (~10%) and are allocated throughout the heart [12]. Fibroblasts secrete collagen and different components of ECM and provide the support to neighboring cells to migrate, proliferate, and also control electrical functions, thus being involved in both cardiac regeneration and pathological conditions [16–18]. Fibroblasts also serve as sentinels to sense myocardial injury and trigger inflammation via PPR activation [17, 19]. Endothelial cells (ECs) are the most abundant cardiac resident cells and constitute >60% of the non-CMs in both mammalian and zebrafish heart [17, 20]. These cells play several essential roles in angiogenesis, heart development, CM organization, and immune cell trafficking, besides being prominent cells in the process of healing and regeneration in postischemic injuries [21, 22]. Further, during inflammation, leukocyte extravasation requires the activation of endothelial cells. In contrast to the adult mammalian heart, neonatal mouse possesses robust regenerative capacity shortly after birth, while fibrosis and scarring prevail later. One-day-old mice in response to cardiac injuries regenerate lost CM and form functional myocardium within 2–3 weeks postinjury without fibrosis [6, 23, 24]. More interestingly, cardiac regeneration is also noticed in newborn pigs and humans after cardiac injury [25, 26]. However, the regenerative potential of neonatal heart is lost on postnatal day 7 [6]. This is possibly due to the accumulation of various limiting factors and depletion of positive regulators of CM proliferation [27–30]. However, there are some evidences that support the notion that loss of neonatal regenerative potential soon after birth also overlaps with remodeling and maturation of the immune system (reviewed in [31]). Moreover, it has been shown that immune cells also play a pivotal role in CM proliferation after cardiac injuries [32–34].

7.4   Immune Anatomy of the Myocardium

Healthy mammalian hearts contain relatively small populations of immune cells [13, 35–37]; the amount and composition of these cells depend on the developmental stages, species, and cardiac pathologies. Cardiac immune cells include the network of residing or infiltrating cells of which ~25% are of the lymphoid lineage (B- and T-cells and NK cells), while ~75% are of the myeloid lineage (macrophages, monocytes, neutrophils, dendritic cells, mast cells, and eosinophils) [38–41]. By exploiting various genetic tools and cellular markers expressed on the surface of immune cells, scientists made successful attempts to identify and quantify the immune cell population in mammalian heart (Fig. 7.1); this quantification is based

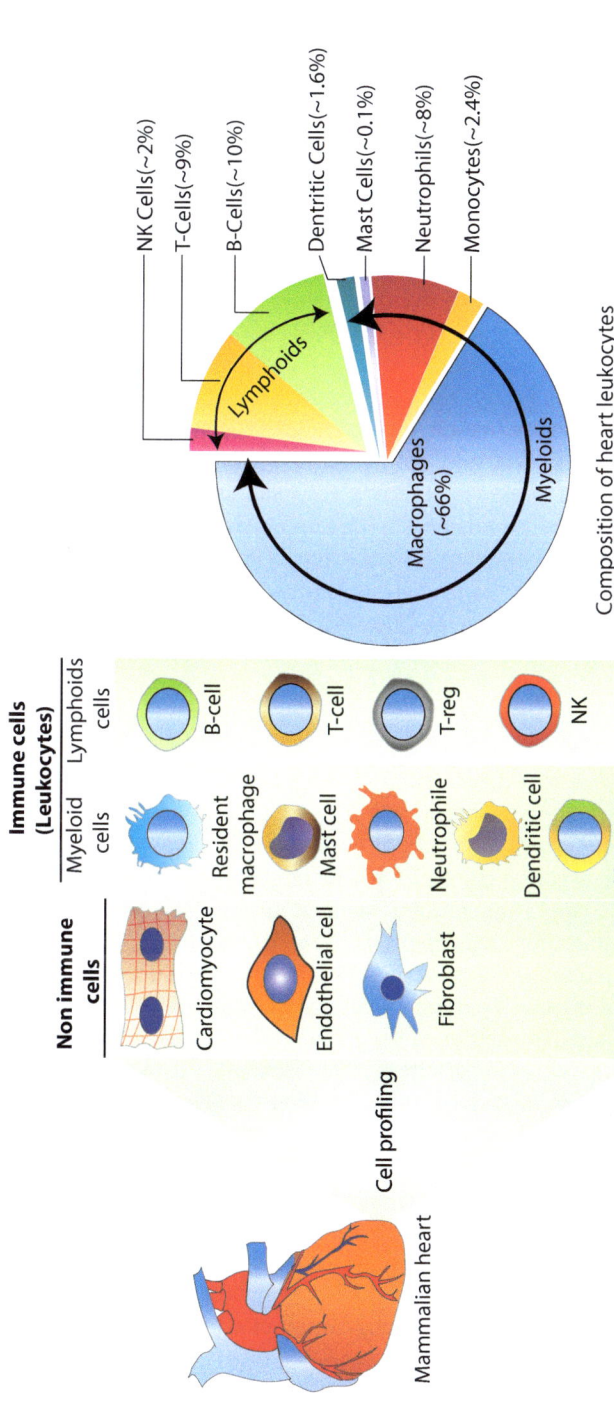

Fig. 7.1 Cellular composition of the myocardium. Among the immune cells, a major proportion is comprised of the macrophages found mainly surrounding the epithelial cells or within the interstitial space. Dendritic cells (DCs), although scanty, are found in the cardiac valves, possibly to sample antigens. Mast cells, B cells, and regulatory T cells (T_regs) are also found in a resting heart, though in sparse populations, while neutrophils and monocytes are recruited to the myocardium in large numbers only upon cardiac injury. Among the nonimmune cells, a vast population is composed of cardiomyocytes, followed by fibroblasts (FBs), endothelial cells (ECs), and perivascular cells

on the quality of antibodies used in the sorting assay [12, 42]. However, cardiovascular pathologies and aging change the subpopulation and composition of the immune cells in the heart, in order to promote tissue growth and repair.

Mast cells (MCs) are conventional granular resident cardiac immune cells that act as key effectors of innate immune responses. Because of their perivascular location and capacity to store and release proinflammatory signaling molecules, they act as cellular effectors in inflammatory responses post cardiac injuries. In the injured heart, MCs quickly degranulate and release proinflammatory signaling molecules such as TNF-α and histamine, and TNF-α subsequently triggers the signaling cascade and recruits the proinflammatory leukocyte in the infarct myocardium [35, 43].

Neutrophils (polymorphonuclear granulocytes or PMNs) are leukocytes that play a key role in innate immunity by removing foreign pathogens through different mechanisms including degranulation and oxidative. PMNs are the immune cells which are promptly employed into the injured myocardium by DAMPs and other immune modulators (cytokines, chemokines, and histamine) [44, 45]. Following a cardiac injury, neutrophils peak within 1–3 days and then drop down to their steady-state levels approximately a week later. In comparison to neonates, the adult heart is inadequate to remove infiltrating neutrophils, which subsequently decrease macrophage recruitment following a cardiac injury. This leads to increased matrix degradation, delayed collagen deposition, and increased susceptibility to heart rupture [32, 33, 46]. Neutrophils also play an important role in the resolution of inflammation by secreting myeloperoxidase (MPO) during neutrophil extracellular trap (NET) formation [47]. NETs are extruded from activated neutrophils as extracellular weblike structures in a process known as NETosis [48] and are composed primarily of chromatin (DNA and histones) along with a milieu of inflammatory mediators such as neutrophil elastase (NE), myeloperoxidases (MPO), reactive oxygen species (ROS), cathelicidin or LL-37, TNF-α, cathepsin G, and several cytoplasmic proteins such as annexin I to name a few. In CVDs including atherosclerosis, neutrophils activate leukocytes, platelets, and endothelial cells in the lumen creating a proinflammatory setting that leads to endothelial loss of function and paves the way to plaque formation. Progressing plaque lesions may eventually rupture, thus inducing intraluminal thrombosis leading to acute events of cardiac stress and ischemic stroke [49, 50]. Pharmacological inhibition of peptidylarginine deiminase 4 (PAD4), the enzyme that converts arginine into citrulline on histone tails, promoting chromatin decondensation in the nucleus, blocked NET formation accompanied by a reduced recruitment of neutrophils and macrophages to arteries. This resulted in reduced atherosclerosis burden in a murine model and strongly suggests a causative role for NET formation in atherosclerosis [51, 52]. Further, neutrophil depletion has also been shown to protect against atherosclerosis. Importantly, the protective role of PAD4 in atherosclerotic disease confirms the importance of NET formation in murine atherosclerosis [53–55].

7.5   Macrophage-Resident and Monocyte-Derived Cells

In the mammalian heart, resident macrophages are the most abundant immune cell subset, which respond to cardiac tissue damage by producing proinflammatory cytokines and admitting recruitment of neutrophils [56, 57]. These cells are present throughout the myocardium, maintained by local proliferation and intermingling directly with CMs and endothelial cells [35, 56]. A combination of studies including genetic fate mapping, parabiosis, and adoptive transfer concluded that rather than being a homogeneous population, cardiac macrophages comprise three subsets based on the differential expression of MHC class II and chemokine receptor 2 (CCR2). Of these, the dominating class is of the embryonic yolk sac origin and expresses MHC class II^{hi}. The other subset is also derived from embryonic precursors and is MHC class II^{lo}. These two subsets are CCR2$^-$ and they renew themselves through in situ proliferation without the need of circulatory monocytic input. The third subset expresses CCR2, is MHC class II$^-$, and in contrast to the other two subsets is dependent on circulating monocytic input [35, 56]. These macrophages are bona fide activators of the inflammasome as they express high levels of proinflammatory genes including those affiliated with the NLRP3 inflammasome, where they contribute pooling IL-1β to the heart under cardiac stress [35, 58]. Recent reports have confirmed the presence of CCR2$^-$ and CCR2$^+$ macrophage populations in human heart as well where the former plays a more reparative function, while the latter are inflammatory in nature [59]. Further, similar to mouse CCR2$^-$ macrophages, human CCR2$^-$ macrophages exist independent of monocyte input. However, transcriptomic analysis provided evidence that monocytes contribute to maintenance of CCR2$^+$ macrophages [59]. In this context, recent studies have provided fascinating new insights into the regulatory mechanisms of monocytosis relevant to atherosclerosis. It has been shown that two subsets of tissue-resident macrophages, CCR2$^-$ and CCR2$^+$, differentially regulate monocyte recruitment upon cardiac injury [59]. Moreover, depletion of CCR2$^+$ cardiac macrophages significantly reduces inflammation and fibrosis, which subsequently improves heart function and repairs the injured myocardium [59]. Tissue-resident embryonically derived macrophages are likely to have critical roles in the tissue-repair response. This is evident as cardiac injuries lead to gradual replacement of resident macrophages with infiltrating monocyte-derived macrophages taking over which contribute to the worsening of the cardiac regenerative potential because these monocyte-derived macrophages are proinflammatory, pro-fibrotic, and less pro-proliferative in nature [33, 60]. Conversely, it has also been reported that inflammation or injury are not necessarily required for replacement of embryonically derived cardiac tissue macrophages by monocytes [61]. In the injured myocardium, two distinct Ly6C^{hi} and Ly6C^{lo} monocyte-derived macrophage populations show sequential dominance [62]. Ly6C^{hi} macrophages peak during the early proinflammatory phase and have mainly phagocytic, proteolytic, and inflammatory functions, while Ly6C^{lo} macrophages come 3 days post injury, exhibit attenuated inflammatory properties, and express mainly endothelial cell growth factor (VEGF) and matrix metalloproteinase 9 (MMP 9) [60]. In humans, these two populations of monocyte-derived macrophages are related to circulating $CD14^+CD16^-$ and $CD14^+CD16^+$ macrophage, which also intrude infarcted myocardium in early and late stages, respectively [63–65], similar to rodent macrophages [60, 62].

7.6   Dendritic Cells

Dendritic cells (DCs) are professional antigen-presenting cells (APCs), which connect innate and adaptive immune system by presenting antigens to the T-cells. In the myocardium, a small number of DCs are present in localized region such as cardiac valves and aortic sinus, presumably to sample non-self-antigens [35, 66]. It has been observed that DCs infiltrate into injured cardiac tissue, and therefore depletion of these cells impaired cardiac remodeling after myocardial infarction. This function of DCs might be associated with enhanced inflammatory cytokine production, MMP9 protease activation, and infiltration of proinflammatory monocytes/macrophages into the infarcted myocardium [67]. Similarly, in humans, a smaller fraction of DCs in cardiac tissue is associated with macrophage infiltration, impaired reparative fibrosis, and eventually heart rupture after myocardial infraction [68].

7.7   Adaptive Immune Cells

In addition to innate immune cells, mammalian heart also contains some lymphocytes (B and T cells) and NK cells, of which 45% are B-cell population [41]. It has been shown that this percentage of lymphocytes increases up to 5–10 times more during cardiac injuries [62]. Depletion of B cells using anti-CD20 antibody in heart significantly reduces postischemic injury, prevents adverse ventricular remodeling, and improves cardiac function after myocardial infarction [69]. Regarding the molecular mechanisms involved in B-cell recruitment, secretion of cytokines, which induces the deployment of other immune cells in heart diseases, much remains unknown and requires detailed investigation.

7.8   Pattern Recognition Receptors

Recent advances in the field of cardiac injury have revealed that mammalian hearts use both innate and adaptive immune components to respond to cardiac insults such as ischemia or hemodynamic overloading. Immune cells residing in the heart are triggered by pathogen-associated molecular patterns (PAMPs) or danger-associated molecular patterns (DAMPS) and induce an appropriate inflammatory response through their binding to innate immune receptors, known as pattern recognition receptors (PRRs). Classic examples of PRRs include toll-like receptors (TLRs), C-type lectin receptors (CLR), nucleotide-binding and oligomerization domain (NOD)-like receptors (NLRs), retinoic acid–inducible gene (RIG)-I-like receptors (RLRs), absent in melanoma (AIM) 2-like receptors (ALRs), and advanced glycation end-product-specific receptors (AGER/RAGE) [70, 71].

TLRs are type I transmembrane glycoproteins comprising a leucine-rich repeat (LRR) extracellular motif and an intracellular signaling motif that is similar to interleukin (IL-1) [72, 73]. TLRs have been classified into two main groups depending on their subcellular localization; TLR1, TLR2, TLR4, TLR5, TLR6, and TLR11 are expressed on the plasma membrane, whereas TLR3, TLR7, TLR8, and TLR9 are found in endosomes [74, 75]. TLRs 1–10 have been identified in the human heart, of which TLR4 and TLR2 have been reported to be the most abundant [76]. TLRs need to dimerize for ligand binding [77, 78]. Each TLR recruits a member of a set of toll/IL-1 receptor (TIR) domain-containing adaptors differentially such as myeloid differentiation factor 88 (Myd88), Myd88 adaptor-like protein (Mal), or TIR domain-containing adaptor protein (TIRAP), TIR domain-containing adaptor protein inducing interferon (IFN)-β-mediated transcription factor (TRIF), and TRIF-related adaptor molecule (TRAM) [74, 75, 79]. Based on the type of adaptor recruited, TLR signaling can be divided into two general pathways, namely, the Myd88-dependent and Myd88-independent pathways. All TLRs except TLR3 use MyD88 as an adaptor protein. TLR3 uses TRIF as the adaptor protein belonging to Myd88-independent pathways, whereas TLR4 employs both the Myd88-dependent and Myd88-independent pathways [11, 74, 75, 80].

7.9   MyD88-Dependent Signaling

MyD88 is the canonical adaptor that can induce signaling from several TLRs, located either at the plasma membrane or in endosomes [81]. Moreover, MyD88 signaling can lead to the production of pro- or anti-inflammatory cytokines as well as type I IFNs [82]. The Myd88-dependent pathway is initiated via Myd88 after TLR activation in which the death domain (DD) of Myd88 recruits IL-1 receptor-associated kinase 4 (IRAK4) and activates one of other IRAK family members, that is, IRAK1 or IRAK2 in a large oligomeric complex known as the myddosome [83]. These IRAKs then dissociate from the Myd88-IRAK complex and activate the RING domain E3 ubiquitin ligase TNF receptor-associated factor 6 (TRAF6), so that it can interact with transforming growth factor-β-activated kinase 1 (TAK1), TAK1-binding protein 1 (TAB1), and TAB2. TAK1 then activates the complex of inhibitory κB (IκB) kinase α (IKKα)/IKKβ/IKKγ and induces IκB phosphorylation. After phosphorylation, the IκB undergoes proteasome degradation, allowing NF-κB to translocate into the nucleus and induce the expression of various proinflammatory cytokines (illustrated in Fig. 7.2) [74, 75].

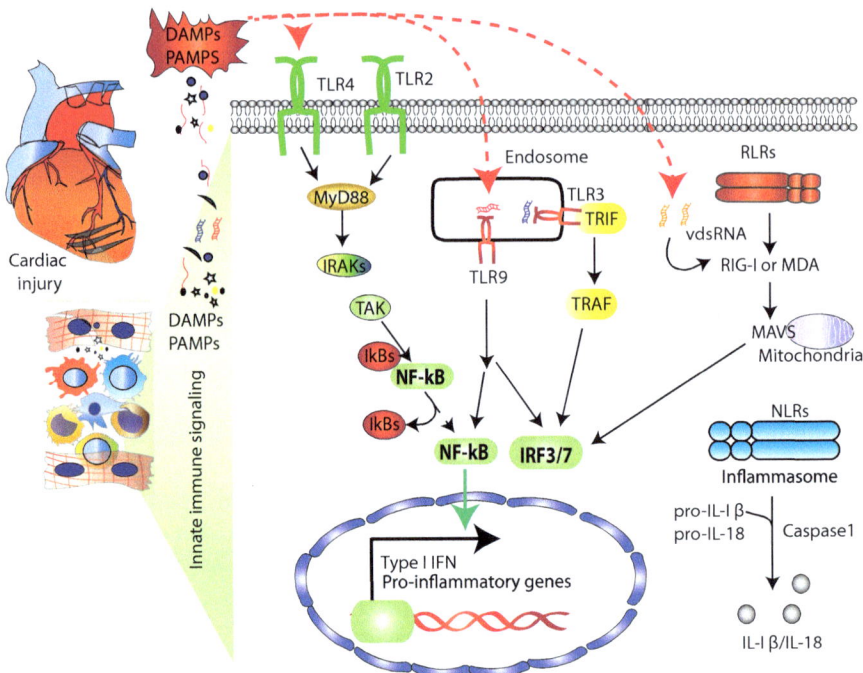

Fig. 7.2 PRR activation in cardiac cells by PAMPs and DAMPs during cardiac injury. Heart cells express a variety of PRRs including TLRs (mainly TLR2, TLR3, TLR4, and TLR9), NLRs, and RLRs. DAMPs and PAMPs, which include endotoxin, HSP60, HMGB1, ROS, lipoproteins, viral RNA, mtDNA, pore-forming toxins, crystalline substances, peptide aggregates, etc., are involved in CVD. All these PRRs induce the innate immune responses resulting in the expression of proinflammatory cytokines and interferons. Activation of NF-kB also increases the expression of NLRP3, which, in subsequent steps, activates inflammasome leading to the production of IL-1B and IL-18 (see text for details)

7.10 Myd88-Independent Signaling

The Myd88-independent pathway (also known as TRIF-dependent pathway) leads to the activation of both interferon regulatory factors (IRFs) and NF-κB [75]. This pathway is initiated by TRIF (at the endosome by TLR3) and TRAM (TRAM is a particular adaptor for TLR4). After recruitment by the TLR, TRIF interacts with TRAF6, which activates TRAF family member-associated NF-κB activator-binding kinase 1 (TBK1) and IKK-ε for phosphorylation of IRFs. Activated IRFs translocate into the nucleus and induce the production of IFNs. In another signaling cascade, TRIF interacts with TRAF6, and the latter recruits the receptor-interacting serine/threonine protein kinase 1 (RIP-1), which in turn interacts with and activates the TAK1 complex, resulting in the activation of NF-κB and MAPKs for the induction of inflammatory cytokines. Thus, activation of NF-κB contributes to the expression of proinflammatory cytokines, whereas the activation of IRF3 is dispensable for the expression of IFNs [11, 75].

7.11   Role of TLR Signaling in Cardiac Disease

Cardiac tissue injury can occur through a variety of pathophysiological processes, which can be of either ischemic or nonischemic etiology. In regard to the global disease burden, ischemic injury is the main pathophysiological mechanism of injury and is accompanied by the generation of endogenous signals that are potent activators of the innate immune system [70, 72]. The damaged ECM of the infarcted heart and the intracellular constituents released after tissue necrosis promote activation of TLRs [84]. In cardiac monocytes derived from neonatal rats, expression levels of TLRs 2, 3, 4 and 6 have been identified [85]. The mRNA expression levels for various TLR in the human heart are as follows: TLR4 > TLR2 > TLR3 > TLR5 > TLR 1 > TLR6 > TLR7 > TLR8 > TLR9 > TLR10 [71, 76]. Regarding the regulation of TLR expression in the heart, studies have been scarce, although some studies have implicated the importance of TLRs in atherosclerosis-related inflammation. For example, the expression of TLRs 1, 2, 4, and 5 has been shown in atherosclerotic plaques by resident cells and migrating leukocytes into the arterial wall. Moreover, TLR4 expression is upregulated and is concentrated in the shoulder region of the plaque, which is known to be the most sensitive area to undergo plaque rupture [71, 86]. Furthermore, genetic loss-of-function studies have emphasized the importance of TLR2 and TLR4, both located at the cell surface, as important mediators of postinfarction inflammation [87, 88]. Mann DL and colleagues have shown in mouse and rat experimental heart failure models that sustained TLR activation is maladaptive and can lead to left ventricular (LV) dysfunction and adverse cardiac remodeling [89]. Mice with a missense mutation of TLR4 or targeted disruption of TLR2, TLR4, or MyD88 have reduced infarct sizes as compared to the wild-type controls [90–94]. Moreover, treatment with eritoran, a TLR4 antagonist, led to reduced nuclear translocation of NF-κB, decreased expression of proinflammatory cytokines such as IL-6 and TNF-α, and reduction in infarct sizes when compared to vehicle-treated animals [94]. Further, targeted disruption of TLR2/4 in mice resulted in reduced mortality, preservation of cardiac function, increased survival rate, and attenuation of myocardial fibrosis after MI [94, 95]. In an early report, it has been shown that the decrease in the size of an infarct in a TLR2-deficient mouse ensuing an I/R injury was revoked in chimeric TLR2-deficient mice that underwent bone marrow transplantation (BMT) with WT bone marrow cells [96]. TLR4 is known to recognize some endogenous ligands, such as high-mobility group box 1 (HMGB1) and HSP [80], whose association with cardiac injuries and HF is very well known. Plasma concentration of HMGB1 was found to be elevated in HF and correlated with disease severity in patients with HF [97]. The study of Maqbool A et al. showed that tenascins can stimulate TLR4 to upregulate the expression of IL-6, further aggravating the worsening and progression of HF [19].

Besides the inevitable roles of TLR2 and TLR4, some reports have also shown the involvement of TLR9 in the progression of cardiac diseases. TLR9 is an endosomal TLR that recognizes cytosine-phosphate-guanine (CpG) repeats which are present within bacterial DNA [98, 99]. In one study using TLR9 KO mice, significant reduction in cardiac inflammation with sustained heart function was observed, suggesting an important role of TLR9 in promoting cardiac inflammation and associated HF [100]. In a similar study, ApoE$^{-/-}$/TLR9$^{-/-}$ double-knockout mice showed a further worsening of atherosclerotic lesions with an accumulation of inflammatory cells. Moreover, CD4+ T-cell depletion in these DKO mice or treatment of ApoE−/− mice with a TLR9 agonist resulted in a significant reduction in the size of atherosclerotic lesions [101]. Similar to bacterial DNA, mitochondrial DNA also contains CpG and is sensed by TLR9 as potent DAMP. In the setting of hemodynamic stress, mitochondria are typically damaged; however, if degradation of mitochondrial DNA (mtDNA) is inhibited, a TLR9-dependent inflammation-induced cardiomyopathy develops [59, 102, 103].

7.12 Other Pattern Recognition Receptors in Cardiac Diseases

C-type lectin receptors. CLRs are calcium-dependent carbohydrate-binding receptors, such as dectin-1 and dectin-2, that specifically recognize major carbohydrate structures in fungal cell walls [104]. Although their expression has been reported in human and murine heart tissue, very little is known about their role in cardiac injury and future studies will be needed to fully define the functions of this class of receptors [105].

NOD-like receptors. NLRs are the cytosolic innate sensors that sense intracellular DAMPs and PAMPs. The human NLR family is composed of 22 intracellular pattern recognition molecules that share a conserved central NACHT domain and a carboxy-terminal leucine-rich repeat (LRR) region [106, 107]. Upon activation, some of the NLRs assemble macromolecular protein complexes called inflammasomes. NLR family pyrin domain (PYD)-containing 1 (NLRP1) was the first member of the NLR family able to assemble into inflammasomes [108], which convert the inactive pro-caspase-1 into the catalytically active caspase-1 in the canonical pathway. The canonical inflammasome activation is complemented by a noncanonical pathway, which promotes activation of caspase-11 (in mice) and caspases-4 and 5 (in humans). These caspases in turn activate NLRP3 inflammasomes or caspase-1 [109, 110]. Caspase-1 then converts its substrates (pro-IL-1β, pro-IL-18, and gasdermin-D) into their bioactive and secreted forms upon inflammasome activation (Fig. 7.2) [111].

Analysis of human heart tissue has shown that NOD1, NOD2, NLRP2, and NLRP3 are expressed in the cytosol and activate canonical inflammasomes in the heart. They play important roles in adverse cardiac remodeling following I/R injury and myocardial infarction; however, the cell types involved remain to be investigated [112]. Inhibition of NLRP3 has been shown to be cardioprotective after ischemic as well as nonischemic injury (doxorubicin treatment) in rodents [113]. The proinflammatory cytokine IL-18 downstream of NLRP3 inflammasome is being considered as a therapeutic target in acute MI and heart failure [114]. The Canakinumab Anti-inflammatory Thrombosis Outcomes Study (CANTOS) trial, using canakinumab, a human monoclonal antibody that potently inhibits IL-1β, has shown good results for the anti-inflammatory therapies in recurrent vascular events and acute MI [115]. Besides, some other drugs targeting the NLRP3 inflammasome have also been evaluated in clinical trials. For instance, colchicine, which is generally effective in gout treatment, has been reported in a recent study to significantly reduce cardiovascular events in patients with stable coronary artery disease [116]. Collectively, these reports suggest that NLRP3 inflammasome plays an important role in modulating cardiac inflammation that progresses to heart failure.

RIG-I-like receptors. The RLR family is composed primarily of the helicases RIG-I and melanoma differentiation-associated gene 5 (MDA5). RLRs are localized in the cytoplasm and the structure is composed of the caspase activation and recruitment domain (CARD), RNA helicase domain, and a C-terminal domain. They are specialized in the recognition of genomic RNA of double-stranded (ds) RNA viruses and dsRNA generated as the replication intermediate of ssRNA viruses. RIG-I is expressed by macrophages, endothelial cells, DCs, and fibroblasts in human atherosclerotic lesions.

Following activation, RLRs recruit the adapter molecule mitochondrial antiviral signaling protein (MAVS) and CARD adapter inducing interferon beta (Cardif), followed by the activation of IRF-3 and NF-κB and ultimately leading to the production of proinflammatory responses (Fig. 7.2) [112, 117]. It has been reported that RNA stimulation of endothelial cells leads to an increased RIG-I expression, impaired vasodilation (endothelial cell-dependent), and augmented production of ROS [118]. Involvement of RIG-I has also been shown in the 25-hydroxycholesterol-induced IL-8 production in atherosclerosis [119].

7.13 Conclusion

Myocardial inflammation including myocarditis, MI, I/R injury, and HF has been critically involved to play an important role in the physiological and pathological mechanisms of cardiac injury and repair. Inflammation is required for host defense against damage and tissue repair and timely repression of this inflammatory process is critical for effective healing. This chapter elaborates on the emerging roles of various innate immune signaling pathways in excessive chronic myocardial inflammation leading to HF. In particular, TLR signaling pathway regulates a much broader regulation of inflammatory mediators. Therapeutic strategies targeting specific components of the inflammatory responses emanating from the various innate sensing pathways especially TLRs are promising for patients with myocardial infarction. Besides, biomarker and imaging-based approaches identifying patient groups with overactive proinflammatory signaling might contribute to rational design of therapies to prevent HF.

Acknowledgment The authors would like to thank Drs. M Rahman and MK Zakaria for helpful comments and suggestions.

Conflicts of Interest The authors have no conflicts to declare.

References

1. Gupta R (2004) Trends in hypertension epidemiology in India. J Hum Hypertens 18(2):73–78
2. Gupta R, Mohan I, Narula J (2016) Trends in coronary heart disease epidemiology in India. Ann Glob Health 82(2):307–315
3. Ndrepepa G (2017) Atherosclerosis & ischaemic heart disease: Here to stay or gone tomorrow. Indian J Med Res 146(3):293–297
4. Prabhakaran D, Jeemon P, Roy A (2016) Cardiovascular diseases in India: current epidemiology and future directions. Circulation 133(16):1605–1620
5. Poss KD, Wilson LG, Keating MT (2002) Heart regeneration in zebrafish. Science 298(5601):2188–2190
6. Porrello ER et al (2011) Transient regenerative potential of the neonatal mouse heart. Science 331(6020):1078–1080
7. Kikuchi K et al (2010) Primary contribution to zebrafish heart regeneration by gata4(+) cardiomyocytes. Nature 464(7288):601–605
8. Senyo SE et al (2013) Mammalian heart renewal by pre-existing cardiomyocytes. Nature 493(7432):433–436
9. Yancy CW et al (2013) ACCF/AHA guideline for the management of heart failure: executive summary: a report of the American College of Cardiology Foundation/American Heart Association Task Force on practice guidelines. Circulation 128(16):1810–1852
10. Levine B et al (1990) Elevated circulating levels of tumor necrosis factor in severe chronic heart failure. N Engl J Med 323(4):236–241
11. Yu L, Feng Z (2018) The role of toll-like receptor signaling in the progression of heart failure. Mediat Inflamm 2018:9874109
12. Pinto AR et al (2016) Revisiting cardiac cellular composition. Circ Res 118(3):400–409
13. Zhou P, Pu WT (2016) Recounting cardiac cellular composition. Circ Res 118(3):368–370
14. Bergmann O et al (2015) Dynamics of cell generation and turnover in the human heart. Cell 161(7):1566–1575
15. Zhang W et al (2015) Necrotic myocardial cells release damage-associated molecular patterns that provoke fibroblast activation in vitro and trigger myocardial inflammation and fibrosis in vivo. J Am Heart Assoc 4(6):e001993
16. Furtado MB et al (2016) View from the heart: cardiac fibroblasts in development, scarring and regeneration. Development 143(3):387–397
17. Shinde AV, Frangogiannis NG (2014) Fibroblasts in myocardial infarction: a role in inflammation and repair. J Mol Cell Cardiol 70:74–82
18. Tallquist MD, Molkentin JD (2017) Redefining the identity of cardiac fibroblasts. Nat Rev Cardiol 14(8):484–491
19. Maqbool A et al (2016) Tenascin C upregulates interleukin-6 expression in human cardiac myofibroblasts via toll-like receptor 4. World J Cardiol 8(5):340–350
20. Turner NA (2016) Inflammatory and fibrotic responses of cardiac fibroblasts to myocardial damage associated molecular patterns (DAMPs). J Mol Cell Cardiol 94:189–200
21. He L et al (2017) Preexisting endothelial cells mediate cardiac neovascularization after injury. J Clin Invest 127(8):2968–2981
22. Klotz L et al (2015) Cardiac lymphatics are heterogeneous in origin and respond to injury. Nature 522(7554):62–67
23. Haubner BJ et al (2012) Complete cardiac regeneration in a mouse model of myocardial infarction. Aging (Albany NY) 4(12):966–977
24. Porrello ER, Olson EN (2014) A neonatal blueprint for cardiac regeneration. Stem Cell Res 13(3 Pt B):556–570
25. Haubner BJ et al (2016) Functional recovery of a human neonatal heart after severe myocardial infarction. Circ Res 118(2):216–221
26. Ye L et al (2018) Early regenerative capacity in the porcine heart. Circulation 138(24):2798–2808

27. Ali H, Braga L, Giacca M (2019) Cardiac regeneration and remodelling of the cardiomyocyte cytoarchitecture. FEBS J
28. Heallen TR et al (2019) Stimulating Cardiogenesis as a treatment for heart failure. Circ Res 124(11):1647–1657
29. Tzahor E, Poss KD (2017) Cardiac regeneration strategies: staying young at heart. Science 356(6342):1035–1039
30. Uygur A, Lee RT (2016) Mechanisms of cardiac regeneration. Dev Cell 36(4):362–374
31. Sattler S, Rosenthal N (2016) The neonate versus adult mammalian immune system in cardiac repair and regeneration. Biochim Biophys Acta 1863(7 Pt B):1813–1821
32. Aurora AB et al (2014) Macrophages are required for neonatal heart regeneration. J Clin Invest 124(3):1382–1392
33. Lavine KJ et al (2014) Distinct macrophage lineages contribute to disparate patterns of cardiac recovery and remodeling in the neonatal and adult heart. Proc Natl Acad Sci U S A 111(45):16029–16034
34. Leid J et al (2016) Primitive embryonic macrophages are required for coronary development and maturation. Circ Res 118(10):1498–1511
35. Epelman S et al (2014) Embryonic and adult-derived resident cardiac macrophages are maintained through distinct mechanisms at steady state and during inflammation. Immunity 40(1):91–104
36. Frangogiannis NG et al (1999) Histochemical and morphological characteristics of canine cardiac mast cells. Histochem J 31(4):221–229
37. Gersch C et al (2002) Mast cells and macrophages in normal C57/BL/6 mice. Histochem Cell Biol 118(1):41–49
38. Bonner F et al (2012) Resident cardiac immune cells and expression of the ectonucleotidase enzymes CD39 and CD73 after ischemic injury. PLoS One 7(4):e34730
39. Janicki JS, Brower GL, Levick SP (2015) The emerging prominence of the cardiac mast cell as a potent mediator of adverse myocardial remodeling. Methods Mol Biol 1220:121–139
40. Swirski FK, Nahrendorf M (2018) Cardioimmunology: the immune system in cardiac homeostasis and disease. Nat Rev Immunol 18(12):733–744
41. Yu YR et al (2016) A protocol for the comprehensive flow cytometric analysis of immune cells in normal and inflamed murine non-lymphoid tissues. PLoS One 11(3):e0150606
42. Farbehi N et al (2019) Single-cell expression profiling reveals dynamic flux of cardiac stromal, vascular and immune cells in health and injury. Elife:8
43. Frangogiannis NG et al (1998) Resident cardiac mast cells degranulate and release preformed TNF-alpha, initiating the cytokine cascade in experimental canine myocardial ischemia/reperfusion. Circulation 98(7):699–710
44. McDonald B et al (2010) Intravascular danger signals guide neutrophils to sites of sterile inflammation. Science 330(6002):362–366
45. Soehnlein O, Lindbom L (2010) Phagocyte partnership during the onset and resolution of inflammation. Nat Rev Immunol 10(6):427–439
46. Lorchner H et al (2015) Myocardial healing requires Reg3beta-dependent accumulation of macrophages in the ischemic heart. Nat Med 21(4):353–362
47. Pase L et al (2012) Neutrophil-delivered myeloperoxidase dampens the hydrogen peroxide burst after tissue wounding in zebrafish. Curr Biol 22(19):1818–1824
48. Brinkmann V et al (2004) Neutrophil extracellular traps kill bacteria. Science 303(5663):1532–1535
49. O'Neil LJ, Kaplan MJ, Carmona-Rivera C (2019) The role of neutrophils and neutrophil extracellular traps in vascular damage in systemic lupus erythematosus. J Clin Med 8(9)
50. Sorensen OE et al (2014) Papillon-Lefevre syndrome patient reveals species-dependent requirements for neutrophil defenses. J Clin Invest 124(10):4539–4548
51. Knight JS et al (2014) Peptidylarginine deiminase inhibition reduces vascular damage and modulates innate immune responses in murine models of atherosclerosis. Circ Res 114(6):947–956
52. Rohrbach AS et al (2012) Activation of PAD4 in NET formation. Front Immunol 3:360

53. Franck G et al (2018) Roles of PAD4 and NETosis in experimental atherosclerosis and arterial injury: implications for superficial erosion. Circ Res 123(1):33–42
54. Li Y et al (2018) Neutrophil extracellular traps formation and aggregation orchestrate induction and resolution of sterile crystal-mediated inflammation. Front Immunol 9:1559
55. Wang H et al (2018) Obesity-induced endothelial dysfunction is prevented by neutrophil extracellular trap inhibition. Sci Rep 8(1):4881
56. Pinto AR, Godwin JW, Rosenthal NA (2014) Macrophages in cardiac homeostasis, injury responses and progenitor cell mobilisation. Stem Cell Res 13(3 Pt B):705–714
57. Yamasaki S et al (2008) Mincle is an ITAM-coupled activating receptor that senses damaged cells. Nat Immunol 9(10):1179–1188
58. Hulsmans M et al (2017) Macrophages facilitate electrical conduction in the heart. Cell 169(3):510–522.e20
59. Bajpai G et al (2019) Tissue resident CCR2- and CCR2+ cardiac macrophages differentially orchestrate monocyte recruitment and fate specification following myocardial injury. Circ Res 124(2):263–278
60. Nahrendorf M et al (2007) The healing myocardium sequentially mobilizes two monocyte subsets with divergent and complementary functions. J Exp Med 204(12):3037–3047
61. Molawi K et al (2014) Progressive replacement of embryo-derived cardiac macrophages with age. J Exp Med 211(11):2151–2158
62. Yan X et al (2013) Temporal dynamics of cardiac immune cell accumulation following acute myocardial infarction. J Mol Cell Cardiol 62:24–35
63. Cheng B et al (2017) Harnessing the early post-injury inflammatory responses for cardiac regeneration. J Biomed Sci 24(1):7
64. Nahrendorf M, Swirski FK (2013) Monocyte and macrophage heterogeneity in the heart. Circ Res 112(12):1624–1633
65. van der Laan AM et al (2014) Monocyte subset accumulation in the human heart following acute myocardial infarction and the role of the spleen as monocyte reservoir. Eur Heart J 35(6):376–385
66. Choi JH et al (2009) Identification of antigen-presenting dendritic cells in mouse aorta and cardiac valves. J Exp Med 206(3):497–505
67. Anzai A et al (2012) Regulatory role of dendritic cells in postinfarction healing and left ventricular remodeling. Circulation 125(10):1234–1245
68. Nagai T et al (2014) Decreased myocardial dendritic cells is associated with impaired reparative fibrosis and development of cardiac rupture after myocardial infarction in humans. J Am Heart Assoc 3(3):e000839
69. Zouggari Y et al (2013) B lymphocytes trigger monocyte mobilization and impair heart function after acute myocardial infarction. Nat Med 19(10):1273–1280
70. Epelman S, Liu PP, Mann DL (2015) Role of innate and adaptive immune mechanisms in cardiac injury and repair. Nat Rev Immunol 15(2):117–129
71. Mann DL (2011) The emerging role of innate immunity in the heart and vascular system: for whom the cell tolls. Circ Res 108(9):1133–1145
72. Mann DL (2015) Innate immunity and the failing heart: the cytokine hypothesis revisited. Circ Res 116(7):1254–1268
73. Ohto U et al (2018) Toll-like receptor 9 contains two DNA binding sites that function cooperatively to promote receptor dimerization and activation. Immunity 48(4):649–658.e4
74. Kawai T, Akira S (2010) The role of pattern-recognition receptors in innate immunity: update on Toll-like receptors. Nat Immunol 11(5):373–384
75. Kawasaki T, Kawai T (2014) Toll-like receptor signaling pathways. Front Immunol 5:461
76. Nishimura M, Naito S (2005) Tissue-specific mRNA expression profiles of human toll-like receptors and related genes. Biol Pharm Bull 28(5):886–892
77. Jin MS et al (2007) Crystal structure of the TLR1-TLR2 heterodimer induced by binding of a tri-acylated lipopeptide. Cell 130(6):1071–1082
78. Latz E et al (2007) Ligand-induced conformational changes allosterically activate Toll-like receptor 9. Nat Immunol 8(7):772–779

79. Xu Y et al (2000) Structural basis for signal transduction by the Toll/interleukin-1 receptor domains. Nature 408(6808):111–115
80. Kawai T, Akira S (2009) The roles of TLRs, RLRs and NLRs in pathogen recognition. Int Immunol 21(4):317–337
81. Medzhitov R et al (1998) MyD88 is an adaptor protein in the hToll/IL-1 receptor family signaling pathways. Mol Cell 2(2):253–258
82. Reparaz L et al (1992) The epidemiology and cost/effectiveness analysis of diabetic angiopathy in vascular surgery. Angiologia 44(6):225–233
83. Lin SC, Lo YC, Wu H (2010) Helical assembly in the MyD88-IRAK4-IRAK2 complex in TLR/IL-1R signalling. Nature 465(7300):885–890
84. Dobaczewski M, Chen W, Frangogiannis NG (2011) Transforming growth factor (TGF)-beta signaling in cardiac remodeling. J Mol Cell Cardiol 51(4):600–606
85. Frantz S, Kelly RA, Bourcier T (2001) Role of TLR-2 in the activation of nuclear factor kappaB by oxidative stress in cardiac myocytes. J Biol Chem 276(7):5197–5203
86. Holloway JW, Yang IA, Ye S (2005) Variation in the toll-like receptor 4 gene and susceptibility to myocardial infarction. Pharmacogenet Genomics 15(1):15–21
87. Michelsen KS et al (2004) Lack of Toll-like receptor 4 or myeloid differentiation factor 88 reduces atherosclerosis and alters plaque phenotype in mice deficient in apolipoprotein E. Proc Natl Acad Sci U S A 101(29):10679–10684
88. Mullick AE, Tobias PS, Curtiss LK (2005) Modulation of atherosclerosis in mice by Toll-like receptor 2. J Clin Invest 115(11):3149–3156
89. Sakata S et al (2007) Transcoronary gene transfer of SERCA2a increases coronary blood flow and decreases cardiomyocyte size in a type 2 diabetic rat model. Am J Physiol Heart Circ Physiol 292(2):H1204–H1207
90. Chong AJ et al (2004) Toll-like receptor 4 mediates ischemia/reperfusion injury of the heart. J Thorac Cardiovasc Surg 128(2):170–179
91. Feng Y et al (2008) Innate immune adaptor MyD88 mediates neutrophil recruitment and myocardial injury after ischemia-reperfusion in mice. Am J Physiol Heart Circ Physiol 295(3):H1311–H1318
92. Kim SC et al (2007) Toll-like receptor 4 deficiency: smaller infarcts, but no gain in function. BMC Physiol 7:5
93. Oyama J et al (2004) Reduced myocardial ischemia-reperfusion injury in toll-like receptor 4-deficient mice. Circulation 109(6):784–789
94. Shishido T et al (2003) Toll-like receptor-2 modulates ventricular remodeling after myocardial infarction. Circulation 108(23):2905–2910
95. Riad A et al (2008) Toll-like receptor-4 modulates survival by induction of left ventricular remodeling after myocardial infarction in mice. J Immunol 180(10):6954–6961
96. Arslan F et al (2010) Myocardial ischemia/reperfusion injury is mediated by leukocytic toll-like receptor-2 and reduced by systemic administration of a novel anti-toll-like receptor-2 antibody. Circulation 121(1):80–90
97. Volz HC et al (2012) HMGB1 is an independent predictor of death and heart transplantation in heart failure. Clin Res Cardiol 101(6):427–435
98. Hemmi H et al (2000) A Toll-like receptor recognizes bacterial DNA. Nature 408(6813):740–745
99. Latz E et al (2004) TLR9 signals after translocating from the ER to CpG DNA in the lysosome. Nat Immunol 5(2):190–198
100. Lohner R et al (2013) Toll-like receptor 9 promotes cardiac inflammation and heart failure during polymicrobial sepsis. Mediators Inflamm 2013:261049
101. Koulis C et al (2014) Protective role for Toll-like receptor-9 in the development of atherosclerosis in apolipoprotein E-deficient mice. Arterioscler Thromb Vasc Biol 34(3):516–525
102. Bliksoen M et al (2016) Extracellular mtDNA activates NF-kappaB via toll-like receptor 9 and induces cell death in cardiomyocytes. Basic Res Cardiol 111(4):42
103. Oka T et al (2012) Mitochondrial DNA that escapes from autophagy causes inflammation and heart failure. Nature 485(7397):251–255

104. Hardison SE, Brown GD (2012) C-type lectin receptors orchestrate antifungal immunity. Nat Immunol 13(9):817–822
105. Lech M et al (2012) Quantitative expression of C-type lectin receptors in humans and mice. Int J Mol Sci 13(8):10113–10131
106. Latz E, Xiao TS, Stutz A (2013) Activation and regulation of the inflammasomes. Nat Rev Immunol 13(6):397–411
107. Lu A et al (2014) Unified polymerization mechanism for the assembly of ASC-dependent inflammasomes. Cell 156(6):1193–1206
108. Bertin J et al (2000) CARD9 is a novel caspase recruitment domain-containing protein that interacts with BCL10/CLAP and activates NF-kappa B. J Biol Chem 275(52):41082–41086
109. Kayagaki N et al (2011) Non-canonical inflammasome activation targets caspase-11. Nature 479(7371):117–121
110. Kesavardhana S, Kanneganti TD (2017) Mechanisms governing inflammasome activation, assembly and pyroptosis induction. Int Immunol 29(5):201–210
111. Man SM, Kanneganti TD (2016) Converging roles of caspases in inflammasome activation, cell death and innate immunity. Nat Rev Immunol 16(1):7–21
112. Zimmer S, Grebe A, Latz E (2015) Danger signaling in atherosclerosis. Circ Res 116(2):323–340
113. Stachon P et al (2015) Two-year survival of patients screened for transcatheter aortic valve replacement with potentially malignant incidental findings in initial body computed tomography. Eur Heart J Cardiovasc Imaging 16(7):731–737
114. O'Brien LC et al (2014) Interleukin-18 as a therapeutic target in acute myocardial infarction and heart failure. Mol Med 20:221–229
115. Huet F et al (2017) Anti-inflammatory drugs as promising cardiovascular treatments. Expert Rev Cardiovasc Ther 15(2):109–125
116. Nidorf SM et al (2013) Low-dose colchicine for secondary prevention of cardiovascular disease. J Am Coll Cardiol 61(4):404–410
117. Zimmer A et al (2019) Innate immune response in the pathogenesis of heart failure in survivors of myocardial infarction. Am J Physiol Heart Circ Physiol 316(3):H435–H445
118. Asdonk T et al (2012) Endothelial RIG-I activation impairs endothelial function. Biochem Biophys Res Commun 420(1):66–71
119. Wang F et al (2012) Interferon regulator factor 1/retinoic inducible gene I (IRF1/RIG-I) axis mediates 25-hydroxycholesterol-induced interleukin-8 production in atherosclerosis. Cardiovasc Res 93(1):190–199

Role of Regulatory T Lymphocytes in Health and Disease

Niti Shokeen, Chaman Saini, Leena Sapra, Zaffar Azam,
Asha Bhardwaj, Ayaan Ahmad, and Rupesh K. Srivastava

8.1   Introduction

T cells are conventionally categorized into two basic types, viz., CD4+ helper and CD8+ cytotoxic T cells. CD4+ T cells were known to "help" in the activation and differentiation of various immune cells such as NK cells, macrophages, and dendritic cells, whereas CD8+ T cells were known to kill foreign antigens. In 1970s, it was reported that functions exhibited by T cells were not merely restricted to augmenting an immune response but also to dampen it [1]. These T supressor cells were famously named as regulatory T cells or Tregs. Suppression caused by Tregs on various T cells was believed to mediate immunological tolerance by discriminating between self- and non-self-antigen [2, 3]. Tregs are believed to play an important role in maintaining homeostasis of the immune system by restricting the enormity of effector responses and permitting the initiation of immunological tolerance [4–6]. Treg populations are majorly divided into two major types: nTregs (natural Tregs) originating from the thymus and iTregs (induced Tregs) arising extrathymically, i.e., from secondary lymphoid organs or inflamed tissues [7]. Tregs are further differentiated into five subtypes based upon their origin, phenotypes, and expression of markers.

Niti Shokeen and Chaman Saini have equally contributed to this chapter.

N. Shokeen · C. Saini · L. Sapra · Z. Azam · A. Bhardwaj · A. Ahmad · R. K. Srivastava (✉)
Department of Biotechnology, All India Institute of Medical Sciences (AIIMS),
New Delhi, India

© Springer Nature Singapore Pte Ltd. 2020
S. Singh (ed.), *Systems and Synthetic Immunology*,
https://doi.org/10.1007/978-981-15-3350-1_8

8.2   Timeline of Treg Discovery

With the gain of knowledge about the importance of cell-mediated immune (CMI) responses in a diseased condition, it was validated that T lymphocytes were mediators of the CMI. The primary function of CD4$^+$ T cells was to regulate immune response against foreign antigens. But with the effort of Gershon and Kondo in 1970s, it was found that CD4$^+$ T cells are capable of suppressing immune response and were termed as "suppressor T cells (T$_s$ cells)." These cells were assessed by expression of Lyt-1 (CD5 in mice) and Lyt-2 (CD8 in mice). Existence of T$_s$ cells as distinct subset was deserted by the end of 1980s due to the poor characterization of the cells and lack of peculiar markers [8, 9]. Advancements such as immunological tolerance regulated via clonal deletion and anergy questioned the immunosuppression triggered by suppressor cells [10, 11]. Also, molecular characterization of varied cytokines such as the IL-10 disclosed their redundancy, pleiotropic, and cross-regulatory functions [12]. All of these discoveries led to the conclusion that immunosuppression was attributed to the immunosuppressive or cross-regulatory cytokines secreted by T cells, where suppressor T cells played no significant role [13]. Investigation of T cell suppression was done by examining how autoimmune diseases develop by breaching natural tolerance and how it can be inhibited, rather than inspecting tolerance particularly towards an exogenous antigen. This approach convicted that under normal conditions, the immune system fosters T cells with autoimmune suppressive activity [6]. Autoimmune suppressive activity of CD4$^+$ T cells was validated by systematic examination done by Nishizuka and Sakakura in 1969. They showed that thymectomized mice underwent destruction of ovaries which was earlier connected with ovarian dysgenesis. But with subsequent studies, this ovarian lesion was found to be autoimmune in nature [14]. Their results also suggested about the coexistence of two different CD4$^+$ T cells in peripheral circulation, one likely mediating autoimmunity and the other authoritatively suppressing autoimmunity. Both of these populations can be distinguished based on the expression of the CD5 marker.

CD5^{low}CD4$^+$ T cells produced autoimmune disorders when transferred to Balb/c, a thymic nude mice congenitally deficient in T cell population [15]. Scientists were in need to find more markers that could differentiate between autoimmune-inducing and inflammation-inhibiting T cells. In 1995, Sagakuchi's group identified CD25 molecule specific for operational identification of CD25$^+$ CD4$^+$ T cells as distinct subtype of T cells with suppressive functions (17), and in the mid-1990s, the proposition of a new T cell population was made and euphemistically called as regulatory T (T$_R$) cells [16].

FOXP3 gene, a member of the forkhead/winged-helix family of transcription regulators encoded in X-chromosome, was identified as a disease-causing gene in scurfy mice in 2001 and a single gene mutation in X chromosome resulted in the development of severe autoimmune and inflammatory conditions [4, 17]. In case of humans also, it has been reported that mutation in the *FOXP3* gene was the major cause of IPEX (immune dysregulation, polyendocrinopathy, enteropathy, X-linked syndrome) [18]. Similarities in disease conditions between IPEX and autoimmune disease in humans that resulted from T$_R$ cell-depleted conditions convinced several

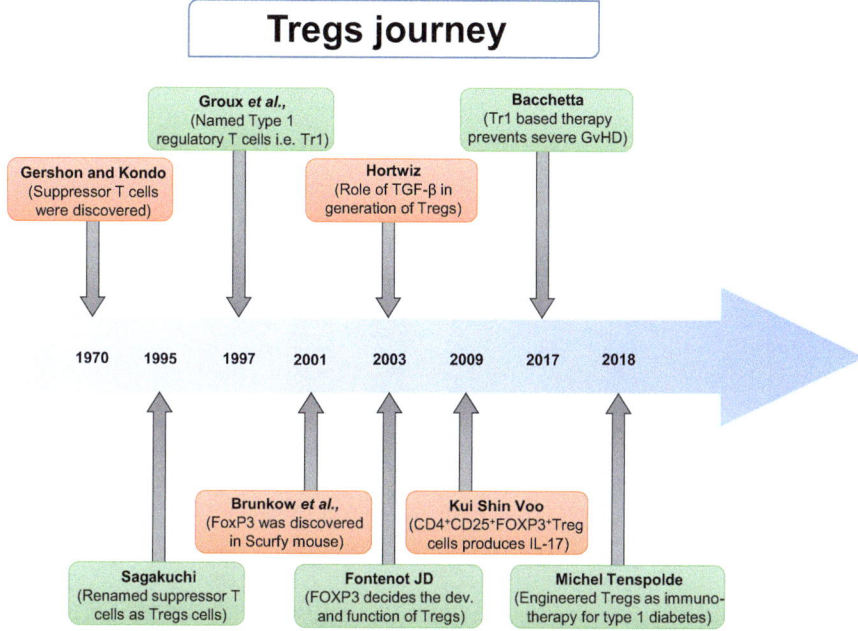

Fig. 8.1 Chronological journey of Tregs. A timeline representing important events in the discovery of Tregs and their establishment as a functionally distinct lineage

groups to investigate the possible role of FOXP3 in natural T_R cell development. In 2003, a study reported that FOXP3 was a key molecule involved in T_R development and functions. Two studies demonstrated that retroviral transduction of FOXP3 to CD25^{neg}CD4$^+$ T cells transformed these T cells into phenotypically and functionally T_R-like cells [5, 19]. These transduced cells showed suppressive functions in vivo and in vitro. These findings collectively suggest that FOXP3 (transcription factor) could be a master gene regulator that controls the development and functions of T_R cells (Fig. 8.1).

8.3   Types of Tregs

Broadly five major subsets of Tregs have been identified based on the markers present and their location of origin and maturation, namely, thymic, peripheral, Tr1 cells, CD8$^+$ Tregs, and IL-17-producing Tregs (Fig. 8.2 and Table 8.1).

8.3.1   Thymic Tregs (tTregs)

This subset is termed as custodians of tissue-specific and systematic immunity. CD4$^+$FOXP3$^+$ Tregs are named as natural or thymic Tregs because of their evident origin from the naïve CD4$^+$ T cells in the thymus itself. These Tregs arise in the

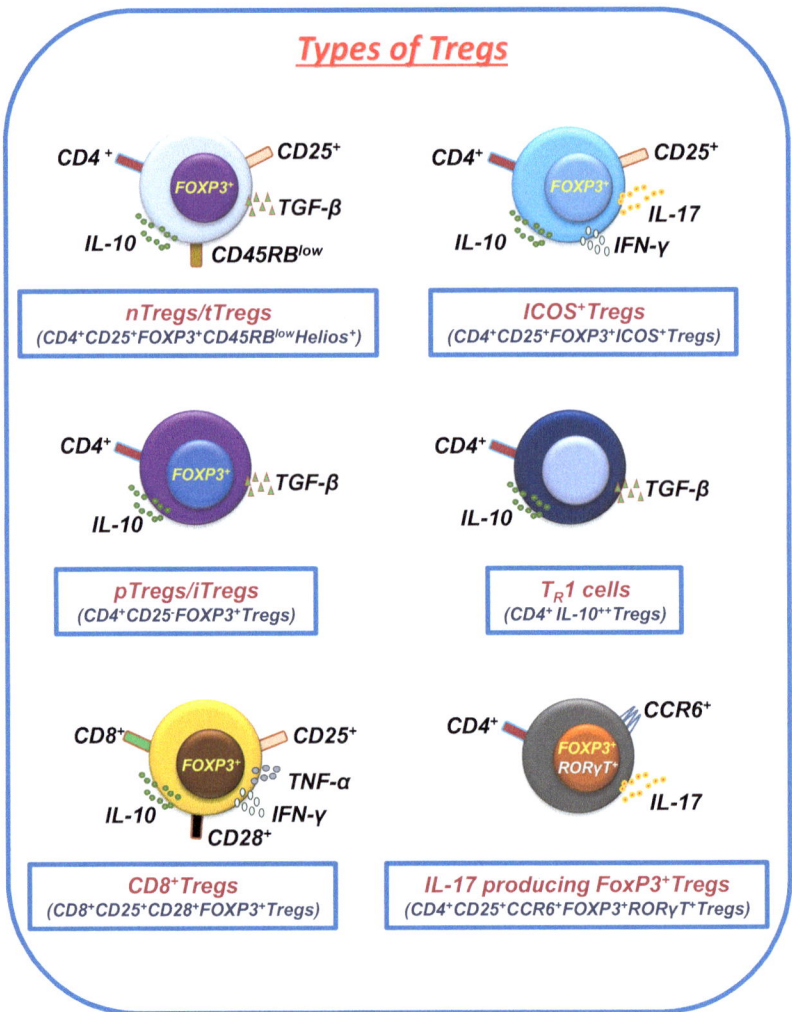

Fig. 8.2 **Tregs subsets.** Tregs have been classified into six different subtypes based on differential expression of surface markers

thymus in response to self-antigens and were termed as natural or naïve Tregs (nTregs). nTregs are believed to migrate from the thymus to the periphery and comprise only 5–10% of the peripheral CD4+ T cell population [31]. Thymic Tregs (also called as tTregs) are generated in the thymus itself through positive selection by MHC-II-restricted self-peptides with greater affinity presented to the CD4+ thymocytes. The critical compartment of tTreg development is the thymic medulla [32]. The direction of CD4+ thymocytes in the thymic medulla towards the tTreg lineage is driven by the signal strength of TCR stimulation. TCR stimulation should be higher than that required for positive selection and lesser than that

Table 8.1 Types of Tregs showing their phenotypes, mechanism of action, and functions

S. no.	Types of Tregs	Abbreviated form	Phenotype	Mechanism of action	Function	References
1.	Natural or thymic regulatory T cells	nTregs/ tTregs	CD4+ CD25+ FOXP3+ CD45RB^{low} Helios+	TGF-β, IL-10	Control allergy and allograft rejection, suppress antigen-specific autoimmune responses	[5, 20, 21]
2.	ICOS+ natural regulatory T cells	ICOS+ Tregs	CD4+ CD25+ FOXP3+ ICOS+	IL-10, IL-17, and IFN-γ	Involved in antitumor, allogenic graft rejection, antiviral response, wound healing	[22–26]
3.	Peripheral or induced regulatory T- cells	pTregs/ iTregs	CD4+ CD25⁻ FOXP3+	TGF-β, IL-10	Involved in immunological response at inflammatory sites especially mucosal surfaces	[27, 62]
4.	Type 1 regulatory T cells	T_R1 cells	CD4+ CD25+	IL-10	Inhibit migration and functions of effector T_H cells, suppress eosinophils, basophils, and mast cells	[28–30]
5.	CD8+ regulatory T cells	CD8+ Tregs	CD8+ CD25+ CD28+ FoxP3+	TNF-α, IL-10, and IFN-γ	Suppress activation of naive and effector T cells, inhibit IgA/IgE responses	[31]
6.	IL-17-producing FOXP3+ regulatory T cells	IL-17 FOXP3+-producing Tregs	CD4+ ROR-γt+ FOXP3+ CCR6+	IL-17	Suppress formation of CD4+ effector T cells	[119]

required for negative selection. In Rag2⁻/⁻ mice, expression of MHC-II-restricted transgenic TCRs resulted in positive selection and development of CD4+ thymocytes rather than tTregs [33]. An experiment where IL2R⁻/⁻ and CD28⁻/⁻ knockouts failed to produce tTregs indeed led to the development of lethal autoimmunity disorders early in life [34, 35]. This study confirmed the significance of IL-2 and CD28 in tTreg development. IL-2 was considered important but not entirely necessary for the development of tTregs [36]. CD28 stimulation was believed as the most essential factor for tTreg development [37]. In contrast to this, a recent study reported the generation of normal numbers of tTregs in CD28 conditional knockout mice [38]. However, these knockout mice developed critical autoimmunity due to dysfunction of tTregs. Another important factor involved in tTreg development is TGF-β. It is not directly driving the tTreg lineage commitment and development but might be providing useful signals needed for survival during initial tTreg development [39]. Ultimately, APCs are considered as the key regulators behind Treg development. It was proposed that plasmacytoid dendritic cells (pDCs) in the human thymus could initiate progression of CD4+CD25+FOXP3+ tTregs after activation with IL-3 and CD40 ligand (CD40L) [40]. Also, IL-3 expands the population of Tregs in mice [41]. It was revealed that the CD27-CD70 co-stimulatory pathway was important for the development of tTregs by liberating them from apoptosis, following induction of FOXP3 by CD28 and TCR signals [42]. Expression of CD70 on mTECs (medullary thymic epithelial cells) and DCs in the medulla region of the thymus stimulates the CD27 signal on tTregs to encourage their survival chances by suppressing apoptosis in the mitochondria [43]. Conclusively, microenvironment, APCs, co-stimulatory signals, and cytokine milieu all cooperate in generating and maintaining tTregs. Tregs are known to develop in the thymus by a two-step process [44, 45]. The first step involves TCR-dependent strong signals which upregulate CD25 expression, as CD25 is the key element of the IL-2 receptor as well as TNF receptor superfamily members TNFR2, OX40, and GITR [44, 46]. The second step, which is TCR independent, involves the conversion of the progenitor CD25+ Treg population to mature CD25+ FOXP3+ Tregs which is dependent on IL-2 and STAT5 (signal transduction and activator of transcription 5), a transcription factor [44, 45, 47, 48].

Natural Tregs, in mice, make up for 5–10% of the total peripheral CD4+ T cell population. Characteristic features of natural Tregs involve lowered expression of CD45RB and constitutive expression of CD25 [49, 50]. In humans, it comprises 1–2% of CD4+ T cells, especially the ones with the highest expression of CD25 [5]. But CD25 is not unique to Tregs as it is also present on activated T cells and expressed by effector T cells such as Th1 and Th2. It has also been found that activated T cells in humans are also capable of expressing FOXP3 without having suppressive activities [51, 52]. Various markers were studied such as CTLA-4, GITR, CD26L^high, CD103, neuropilin1, CD5, CD38, CD39, CD27, CD73, CD122, CD134(OX-40), CCR4, CCR7, and CCR8, but none was found to be exclusive for Tregs [53]. Presently, to distinguish Tregs from conventional/activated CD4+ T cells, low expression of CD127 and modulated expression of CD45RB are used as co-markers along with expression of CD25 and FOXP3 [54, 55].

A distinctive marker employed for identification of tTregs is Helios, which is a zinc finger transcription factor [56]. About 70% of Tregs circulating in the peripheral blood of humans and peripheral lymphoid tissue present in mice are Helios+. As reported, over 95% of Treg population residing in the thymus of mice are Helios+. More than 90% of Treg population were found to be Helios+ when analyzed from the specimens of human thymus and umbilical cord [57]. An interesting study by

Dhamen and McClymont groups in 2011 and 2012 demonstrated a methylation pattern of Tregs in a Treg-specific demethylation region (TSDR) of the FOXP3 promoter. It was found that in humans, Helios+ FOXP3+ Tregs have less than 10% CpG methylation within the TSDR. On the other hand, Helios^{neg} Foxp3+ subset was reported to be more than 40% methylated [58, 59]. Furthermore, thymus-derived FOXP3+ Tregs are classified into two subtypes based on the differential expression of a co-stimulatory molecule known as inducible T cell co-stimulator (ICOS). A subset, which is ICOS+FOXP3+, is endowed with increased IL-10 generating capacity and ICOS^{neg}FoxP3+ subset is provided with increased TGF-β production [60]. Both of these subsets use contact-dependent and contact-independent mechanisms for suppression in periphery. Since there are no specific cell surface markers for distinguishing nTregs, a number of cell surface proteins expressed by nTregs such as CD25 can help in the selective study of this Treg subtype. nTregs are CD25+ CD4+ FoxP3+ cells which secrete TGF-β and IL-10 and represent one of the largest subsets of the Treg population. But an nTreg population with CD4+CD25+CD127$^{(low/neg)}$FoxP3+ expression was detected in the thymus of neonates which acts by suppressing the proliferative response to allogenic stimulation of CD25^{neg} and CD4+ T cells. It has also been reported that Treg turnover and suppressive activity increase with advancing age. It was also found that there is an inverse relationship between CD127 and FOXP3 expression suggesting that cell surface expression of CD127 can be used along with FOXP3 for functional analysis of Tregs [61]. ICOS+ Tregs are a subtype of nTregs that arise by expansion of nTregs in response to its allied antigen. ICOS+ Tregs are distinguished from all other FOXP3+ Tregs by the expression of IL-10, IL-17, and IFN-γ.

8.3.2 Peripheral Tregs (pTregs)

pTregs differentiate from naïve CD4+CD25^{neg} T cells particularly in peripheral lymphoid tissues. It has been shown that upon antigenic interaction, adoptive transfer of CD4+CD25^{neg} T cell into antigen-expressing transgenic mice from Rag$^{-/-}$ TCR mice leads to their conversion to CD4+CD25+ Treg population [62, 63]. CD103+ DCs found in the lamina propria and mesenteric lymph nodes of the small intestine can also trigger conversion of pTregs [43]. It has been found that CD8+CD205+ splenic DCs in the peripheral lymphoid tissue are involved in pTreg development [64]. Various studies showed the importance of antigenic challenge in governing the polarization of Tregs into pTregs. In 2010, Gottschalk et al. reported that induction of pTregs in vivo could be done by providing low dosage of high-affinity TCR ligand [65]. Another study demonstrated the role of a high peptide dose or increased polyclonal TCR stimuli in preventing the induction of FOXP3 via NF-κB-mediated cytokine production [66, 67]. Therefore, interpretation drawn from the above studies is that tTregs arise as a result of moderate/strong affinity interaction with self-antigens in the thymus, whereas induction of pTregs in the periphery occurs as a response to suboptimal/low dosage of strong affinity alloantigen. Commensal microbiota of the colon can also serve as an antigenic source for peripheral induction of pTregs [68, 69]. For example, *Clostridium* species inhabiting the intestine can encourage induction of pTregs which is correlated with enhanced bioavailability of cytokines such as TGF-β [69]. *Lactobacillus acidophilus* and *Bacillus clausii* have also been reported by our group in the induction of pTregs [70, 71]. Polysaccharide A present in *Bacteroides fragilis* was also able to stimulate the proliferation of pTregs via TLR2 signaling [72]. Conclusively, an interaction between various signaling pathways like TGF-β, retinoic acid, IL-2, TLRs, and a milieu of cytokines is required for differentiation of naïve T cells towards pTregs or other effector subsets of T cells. The role of CNS (conserved noncoding sequence 1) in driving differentiation of pTregs in gut-associated lymphoid tissue (GALT) was investigated by Rudensky's group. CNS3 is crucial for the development of both tTregs and pTregs. The CNS role has been further demonstrated by the same group that selectively blocks the differentiation of pTregs in CNS1$^{-/-}$ mice that did not cause aggravation of pathologies related to induced tissue-specific autoimmunity, enhanced proinflammatory responses to Th17 and Th1 cells, or unprovoked multiorgan autoimmunity [73] But these mice impromptu developed Th2 pathologies like asthma and allergic inflammation at the mucosal sites in the GI tract and lungs. It was further reported that mice had altered microbiota indicating the importance of pTregs in maintaining a balance between intestinal immunity and gut microbiome. pTregs serve an essential and distinct function in directing the adaptive immune response to restrict inflammation at mucosal surfaces due to allergic reactions [74]. Following the removal of invaded pathogens, induction of pTregs can function as mediators to repress antigen-specific immune response and avert genesis of cross-reactive T cell. Consequently, failure of any of the mentioned mechanisms can lead to emergence of immune-mediated disorders. pTregs can be distinguished from tTregs based on the expression of Helios. From the studies performed, it was concluded that Helios can be used as a distinctive marker for tTregs and the Helios^{neg} subset constitutes pTregs [57]. But some controversies prevail whether Helios can be used to accurately define pTregs. A CNS discovery by Zheng et al. (2011) revealed the essentiality of CNS1 in the development of pTregs. He developed a model where CNS1$^{-/-}$ FOXP3 GFP^{-} T cells possessed the ability to transform into pTregs in vivo whilst smaller than wild-type controls [75]. Unfortunately, it was not discussed whether Tregs present in CNS1$^{-/-}$ mice were predominantly Helios^{+} or Helios^{neg}. Several groups have challenged the claim of Helios as a marker for distinguishing pTregs from tTregs. The first study in 5C.C7 Rag2$^{-/-}$ transgenic mice showed that Helios could be expressed in vivo in pTregs and in vitro in iTregs [76]. A study conducted using human experiments has reported that tTregs can be Helios^{neg} [77]. Hence, it becomes unclear how "naïve" Helios^{neg} Tregs were stimulated to become pTregs without changing their naïve markers. Also known as effector or induced Tregs, iTregs are derived from naïve CD4^{+} T cells in the periphery which upon encounter with a foreign antigen begins to express FOXP3, exhibiting a suppressive function as that of nTregs. Th3 cells are a subtype of iTregs which secrete TGF-β and IL-10. However, the nomenclature of Tregs as natural and peripheral is ambiguous and to some extent inaccurate as it may indicate the existence of peripheral Tregs as unnatural [78].

8.3.3 T_R1 Cells

A new subset of Tregs was discovered in 1997 by Roncarolo et al. which can sup-press antigen-specific T cell response and prevent colitis. These are the CD4$^+$ sub-set of Tregs which do not express FOXP3 but secrete IL-10 and have a suppressive action on effector T cells. Tr1 cells induced by IL-27 are known to play a role in suppressing immune responses by producing IFN-γ [79]. It has been found that T_R1 cells play a protective role during colitis by supporting the immune homeosta-sis to the intestinal microbiome [80]. T_R1 cells are demarcated from conventional Tregs in FoxP3 and CD25 expression, where T_R1 cells are CD25^{neg} FoxP3^{neg}IL10$^+$ [81]. T_R1 cells are usually differentiated from the alternate CD4$^+$ T cell population by the expression of unique cytokines such as IL10$^+$, IFN-γ, IL5$^+$, TGF-β^+, IL2$^{low/-}$, and IL4 [82, 83]. T_R1 cells are known to have a mediocre expression of CD49b, LAG3, CD69, CD40L, CD28, CD152/CTLA-4, PD-1 (programmed cell death pro-tein), and HLA-DR (human leukocyte antigen-DR) and a higher expression of regulatory factors like GITR, OX40/CD134, and TNFRSF9 [83]. It has been known that T_R1 cells have a substantial expression of ICOS and overexpression of CD18 integrin [84, 85].

8.3.4   CD8$^+$ Tregs

This subset of Treg was discovered by Gershon and Kondo in 1970 [1]. CD8$^+$ Tregs were known to have dual effects in immune responses. Primarily, they suppress the immunological response against pathogens and also the host's inflammation caused by pathogen infection [86–88]. These CD8$^+$ Tregs are mainly characterized by FOXP3 expression and IL-10 secretion and this subtype of Tregs originates from OT-1 CD8 T cells in the presence of IL-12 and IL-4. CD8$^+$ Tregs can be generated in vitro from naïve CD8$^+$ T cells through polyclonal stimulation which are predominantly CD25^{high} CD28 ^{high} and secrete increased level of granzyme B, TNF-α, and IFN-γ. Common markers for Tregs are CD25, CD39, CD127, CD73, and FOXP3 [17] and markers that can differentiate CD8$^+$ Tregs from the conventional CD8$^+$ T cells are CD25, HLA-DR, CD28, CD122, LAG-3, CD38, CD27, CD103, CD8$\alpha\alpha$, and GITR [89]. In mice, FoxP3 is predominantly expressed in CD4$^+$CD25$^+$ T cells but has limited expression in CD8$^+$ Tregs. However, in humans, FOXP3 expression is significantly higher in CD4$^+$ Tregs when compared with the CD8$^+$ Tregs [5, 90]. In humans, majority of CD8$^+$ Tregs are predominantly CD8$^+$CD28^{neg} Tregs, but two subtypes of CD8$^+$ Tregs were produced in vitro by induction, specifically CD8$^+$CD28^{neg} Tregs and CD8$^+$CD28$^+$ Tregs [91, 92]. Three different types of CD8$^+$ CD28^{neg} Tregs are recognized till now, viz., types I, II, and III. Type I cells directly interact with DCs and negatively regulate the expression of CD80 and CD86 (co-stimulatory molecules). Type II cells exert an inhibitory role by secreting IFN-γ and IL-6 without directly involving with APCs (antigen-presenting cells), whereas type III acts by secreting IL-10 [93–95]. Varied classes of CD8$^+$ Tregs function by producing inhibitory chemokines and cytokines such as TGF-β, IL-10, IFN-γ, IL-16, and CCL4 (chemokine C-C ligand 4). Certain subtypes of CD8$^+$ Tregs exert their inhibitory role via a contact-dependent manner, where TGF-β and CTLA-4 (cytotoxic T lymphocyte–associated protein) present on the cell play pivotal roles [96, 97]. Ovulation is considered as an inflammatory process but there is little understanding regarding the participation of the immune system [98, 99]. Unconventional CD8$\alpha\alpha^+$ Tregs were recognized in the thecal region of antral follicles [100]. Existence of CD8$\alpha\alpha^+$ Tregs was validated with the observation that ovaries of nude mice (lacking thymus) and anovulatory C31F$_1$ mice undergoing estradiol treatment had low fertility and lacked CD8$\alpha\alpha$+ Tregs [101]. TECK (thymus-expressed chemokine) present in the ovaries reportedly attract the CD8$\alpha\alpha^+$ Tregs to ovaries. Nevertheless, TECK expression in anovulatory mice was found to be normal, indicating deprivation in migration of CD8$\alpha\alpha^+$ Tregs to the ovaries led to infertility. Ultimately, the origin of CD8$\alpha\alpha$+ Tregs residing in the ovaries was traced back to the thymus [100, 101]. An interesting finding suggests that a subset of CD8$^+$ Tregs is essential for maintaining self-tolerance and preventing autoimmunity in mice [102–104]. Any disruption in the interaction between Qa-1$^+$ follicular helper T cells and CD8$^+$ T cells can give rise to SLE (systematic lupus erythematosus), specifying the importance of this subtype of CD8$^+$ T cell in regulating immune response and monitoring the immunological tolerance [105, 106]. CD8$^+$ Tregs have also been reported by our group to be involved in bone remodeling by inhibiting bone loss in ovx mice model [107].

8.3.5   IL-17-Producing FOXP3$^+$ Tregs

These cells are characterized by co-expression of both RORγt and FOXP3 transcription factors. This subset was observed in peripheral blood along with lymphoid tissue but not in the thymus. The CD4$^+$FOXP3$^+$ T cells expressing CCR6 can produce IL-17 upon activation. IL-17-producing CCR6$^+$ FOXP3$^+$ Tregs are known to greatly inhibit the expansion of CD4$^+$ responder T cells [108]. IL-17-producing FOXP3$^+$ Tregs are considered as a population of immune cells which can have a novel crossover from Tregs into Th17 and are related with decreased suppressive function of CD4$^+$ FOXP3 T lymphocytes [54]. Conventional Tregs perform an immunosuppressive function via the production of anti-inflammatory cytokines like TGF-β, IL-35, and IL-10. But there have been studies directed towards a type of Tregs with the property of secreting proinflammatory cytokines [109–112]. IL-17A-producing FOXP3$^+$ Tregs originated from the induction of naïve CD25^{neg} Tregs either through the ectopic action of FOXP3 or TGF-β signaling [113]. Culturing murine CD4$^+$FOXP3$^+$ Tregs in an environment capable of inducing Th17 differentiation leads to induction of IL-17A from these CD4$^+$FOXP3$^+$ Tregs [114]. When naïve CD4$^+$ T cells were cultured in the same conditions, they too produced IL-17A and transiently expressed FOXP3 [114, 115]. Tregs producing proinflammatory cytokines are demarcated into two types based on their FOXP3 expression. The first subset termed as ex Treg cells [116, 117] are reprogrammed cells which have lost their *FOXP3* expression and attain the properties of T helper cells, i.e., releasing proinflammatory cytokines in immunocompromised conditions [118]. The second subset is the Th-like Tregs which substantially express FOXP3, secrete proinflammatory cytokines, and express lineage-specified transcription factors [119]. Th1-like Tregs express T-bet (transcription factor specific to T_h1 cells) and secrete IFN-γ, which is a type of Th1 cytokine. Th17-like Tregs express RORγt along with FOXP3 and secrete IL-17A [119]. During autoimmune diseases, inflammatory Tregs act like pathogen-eliminating effector T cells and in turn play a major role in inflammation and tissue injury [101]. In the course of pathogenic infections, inflammatory Tregs may eliminate pathogens by producing proinflammatory cytokines. A distinctive feature of IL-17A-producing Tregs is the expression of RORγt along with transcription factor specific to the T_h17 lineage [119]. Even though all FOXP3+ RORγt+ Tregs are not IL-17A-producing Tregs, the presence of CD49d, CCR6, CD161, and IL-1Rβ and absence of HLA-DR have been reported as selective markers for IL-17A-producing Tregs [108, 110, 120, 121]. Broadly, IL-17A-producing Tregs are CD4+, CD49d+CD25^{hi}, CD161+CCR6+, RORγt+, HLA-DR^{neg}CD45RA^{neg}, CD127^{lo}, Foxp3^{lo}, and Helios^{neg} T cells.

8.4 Proposed Mechanisms of Action

Regulatory mechanisms are activated by Tregs to perform the functions needed for maintaining immunological homeostasis especially under conditions of pathogen encounter or any other external stimuli inducing inflammation. Tregs receive help from a variety of immune components, viz., IL-10, IL-35, TGF-β, granzyme, perforin, CTLA-4, and many more, to maintain equilibrium. Herein, we will discuss various routes through which Tregs retain the balance of the immune system. Mechanisms involved are broadly divided into four different categories based on their modes of action: anti-inflammatory cytokines, viz., IL-10, TGF-β, and IL-35, cytolysis by the granzyme-/perforin-mediated pathway, and immune attenuation by CTLA-4 (Fig. 8.3).

8.4.1 Anti-inflammatory Cytokines

8.4.1.1 IL-10

IL-10, a pleiotropic cytokine, suppresses immune response at different levels by modulating the APCs [122] or inhibiting the T cell expansion [123] and most interestingly by sustaining the function of the Treg population [124, 125]. Along with TGF-β, IL-10 is also involved in the differentiation and function of iTregs. A justifying explanation is: how significant reduction in IL-10 production leads to the failure of IL-10Rβ-deficit M2 macrophages to form functional Tregs in the gut [126]. Coexistence of IL-10 with TGF-β is the main determinant of tolerance [127]. It is an established fact that IL-10 mainly functions by STAT3 phosphorylation [128]. Regulation of iTregs through STAT3 was validated in a study where IL-10-induced iTregs when cultured for 7 days showed upregulated STAT3 phosphorylation. The role of STAT3 phosphorylation was in turn demonstrated by treating the IL-10-induced iTregs with

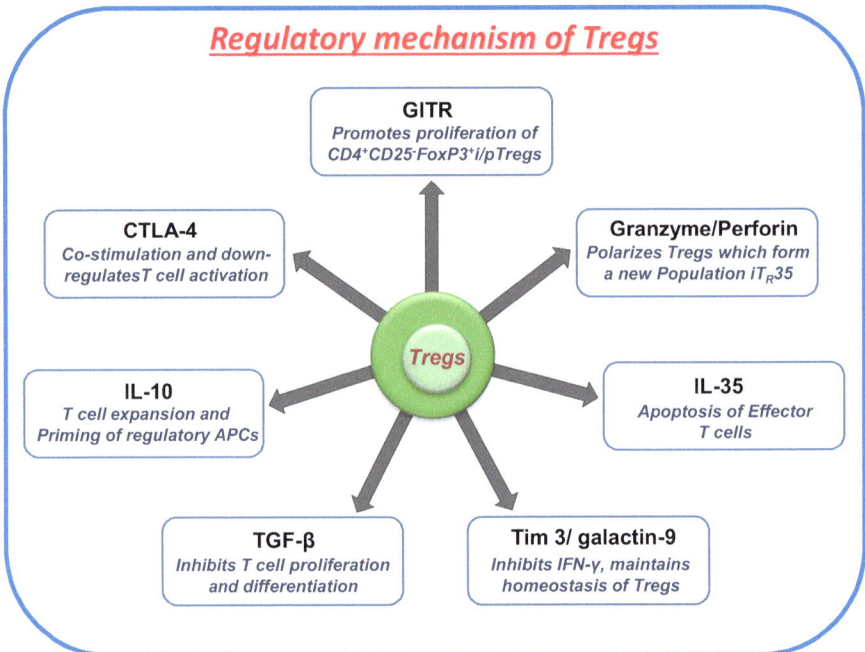

Fig. 8.3 **Regulatory mechanisms of Tregs**. Different mechanisms have been employed by Tregs to suppress immunological responses and maintain a state of homeostasis

Stattic V, a STAT3 inhibitor, at a concentration of 50 ng/ml [129]. In cell cultures where Stattic V was added along with IL-10, no significant increase in expression FOXP3 or CTLA-4 was observed when compared to cells cultured in the absence of IL-10. Inarguably, iTregs cultured with Stattic V and IL-10 were unsuccessful in gaining prominent suppressive activity. Along with STAT3 phosphorylation, IL-10 also regulates the Treg suppression by inhibiting the PI3K/Akt signaling pathway in effector T cells [130], since phosphorylated Akt regulates the expression of Foxo1 [131], important for Treg function [132]. Keeping in mind the divergent role of IL-10, it can be suggested that IL-10 may hold some therapeutic importance in the possible treatment of various immunological disorders such as allergy, autoimmunity, etc. via human iTregs generated in vitro, although some questions still need to be addressed on the stability of in vitro cultured iTregs.

8.4.1.2 TGF-β

It is found to be synthesized by different cell types and belongs to a superfamily of growth factors. Due to the diversity in functions performed by TGF-β, multiple responses are observed based on differentiation state and type of responder cell [133]. Several immune responses have been found to be affected by TGF-β such as T cell proliferation [134], differentiation [135–137], and apoptosis [138, 139]. A possible justification to differing effects displayed by TGF-β is that it acts at various levels for activation and maturation of lymphoid cell. TGF-β binds to its respective responsive cells via three different types of receptors, viz., TRI (TGF receptor Type 1), TRII (Type 2), and TRIII (Type 3) [140]. TRI and TRII have three main regions: an extracellular domain, a transmembrane segment, and a serine-threonine kinase domain found in the cytoplasmic region. TRII is capable of binding a free ligand; on the other hand, TRI only recognizes a ligand when it is bound to TRII. A heterotetramer formed as a result ligand interaction with TRI and TRII is crucial for signaling. The Smad protein family is known to mediate the signaling pathway by TGF-β receptors [141, 142]. Earlier studies have suggested that IL-10 is capable of inducing anergy [143] and driving differentiation of Tr1 (Type 1 regulatory T cells) Tregs [82]. An experiment by Zeller et al. showed that TGF-β enhances the function of IL-10, suggesting a synergistic relationship between TGF-β and IL-10 [144].

8.4.1.3 IL-35

IL-35 is a newly found cytokine involved in the suppression mediated by Tregs and have the potential to directly suppress proliferation of conventional T cells [145]. Belonging to the IL-12 family, IL-35 is made of the IL-12α chain p35 and another IL-27β chain EBI3 (Epstein-Barr virus–induced gene), joined by a disulfide bond [146]. EBI3 and IL-12 p40 are homologous [147]. The IL-12 p35 subunit has ubiquitous expression, whereas IL-12 p40 has inducible expression. Both of these subunits can dissociate and interact with some subunits to give rise to new cytokine profiles [148]. p40 may interact with a p19 subunit to form IL-23, which is considered to be important for T_H1 responses [148, 149], whereas EBI3 may associate with p28 to give rise to IL-27, which possess both pro- and anti-inflammatory roles [150, 151]. Deficiency in any of the IL-35 chains reported leads to alteration in the suppressive ability of Tregs under both in vivo and in vitro conditions in IBD mice model, but it did not show occurrence of autoimmune disease. In comparison to mice, humans lack a constitutive expression of IL-35 [152]. The role of IL-35 in humans was reported by culturing mouse T cells or naïve human T cells in the presence of IL-35. This treatment polarized a new population of Tregs called iT_R35 regulatory cells which act by producing IL-35 and did not require FOXP3, TGF-β, or IL-10 for suppression [153]. In in vivo mice models, iT_R35 cells have been reported to be considered as "strongly suppressive." Human Tregs are not known to have a higher expression of IL-35 but enhanced IL-35 production was indeed reported after long-term activation of Tregs for 3 days [154]. Suppression mediated via these mentioned long-term activated Tregs was contact independent and thus depended upon IL-35. Based on the above observations, IL-35 is also believed to play a role in infectious tolerance [155].

8.4.2 Cytolysis by the Granzyme-/Perforin-Mediated Pathway

Granzyme, a well-known serine protease, is found to be present in particularized cytotoxic vesicles of cytotoxic T lymphocytes (CTLs) and natural killer (NK) cells. Expression of granzymes and perforin is restricted to NK cells and CD8+ T cells

[156]. There have been reports of granzyme and perforin expression in CD4+ T cells but functional importance of granular exocytic pathway in CD4+ T cells has not yet been elucidated [157, 158]. Effector lymphocytes such as CTLs and NK cells destroy their cellular targets by employing two different mechanisms of action. In the first mechanism, perforin (protein-disrupting membrane) and granzyme (serine proteases) are produced through exocytosis and collectively stimulate cell death pathways in targeted cell [159]. Apoptotic pathways activated through exocytosis of granules act by stimulating apoptotic cysteine proteases (called caspases), but it can also cause cell death when activated caspases are absent [160, 161]. The second mechanism involves assembly of cell death receptors (such as FAS) and its respective ligands (like FASL) on the cell membrane which ultimately leads to classical caspase-dependent apoptotic pathway [162]. The main function employed by FAS-FASL pathway is to destroy lymphoid cells that have become self-reactive [163]. Granzymes and perforin induce apoptosis of the targeted cell in a cooperative manner. The apoptosis-inducting potential of granzymes has been found to be associated with perforin for their delivery into the targeted cell. Granzyme B is the most vigorous activator of caspase-mediated and independent cell death. It cleaves at specific aspartate residues of target cell proteins. Mannose-6-phosphate receptor is believed to be the mediator for the entry of granzyme into the target cells through endocytosis [164]. Granzyme A is incapable of activating caspases but can directly destroy cells by cleaving the nuclear proteins and in turn induces the formation of ssDNA breaks [165]. Activated CD4+ CD25+ natural Tregs principally express granzyme A, whereas adaptive Tregs express granzyme B. Both the Treg populations act on their autologous target, i.e., CD4+ and CD8+T cells and DCs, via the perforin-dependent pathway [166]. The mechanism by which Tregs recognize their targets is still not well understood, but several evidence hint that it is a TCR/MHC-independent mechanism, somewhat related to target recognition by NK cells. Some key points to understand the Treg mechanism are as follows: Adaptive Tregs are capable of killing K562 (an allogeneic tumor cell line) [166] which lacks the expression of MHC-I and MHC-II [166]. Effector Tregs did not interact with their target cells preceding the killing assay, and both subsets of Tregs could effectively kill their autologous targets. The different expressions of granzyme A and B in various subsets of Tregs enable them to kill their respective target cells. After stimulation, natural Tregs predominantly express granzyme A as nTregs target or suppress autologous activated target cells expressing endogenous factors capable of inhibiting granzyme B, like proteinase inhibitor-9 (PI-9) [167].

8.4.3   CTLA-4-Mediated Mechanism

CTLA-4 is a homologue of CD28 and a well-known immune attenuator; CTLA-4 interacts with B7.1 (CD80) and B7.2 (CD86) expressed on APCs with a much higher affinity than CD28 [168, 169]. The competitive binding of CTLA-4 to B7 with respect to CD28 prevents the co-stimulation by secreting inhibitory signals and downregulating T cell activation [168, 170, 171]. Thus, the interaction between CTLA-4 and B7 does not trigger a stimulatory signal in effector T cells. In fact, the relative proportion of CTLA-B7 binding and CD28-B7 binding is the deterministic factor that influences the fate of effector T cells, whether it enters into anergic state (functionally unresponsive) or undergoes activation [172]. TCR and CD28-B7 binding leads to stimulation of CTLA-4 and its expression on the cell surface is dependent on exocytosis of vesicles containing CTLA-4 [173]. Upregulation of CTLA on the cell surface occurs in a graded feedback mechanism where higher TCR signaling stimulates greater translocation of CTLA-4 to the cell surface. Net negative signal delivered through binding of CTLA-4 with B7 prevents activation of T cells by inhibiting the production of IL-2 and progression of cell cycle [172]. The discussion so far indicates that the main biological role of CTLA-4 is to negatively regulate CD28 signaling mechanism. Several studies performed on *Ctla* knockouts have shown that immune dysregulation was prevented by blockade of CD86 and CD80 with the CTLA-4 antibody [174]. Similar experiments conducted with triple knockout mice, lacking CTLA-4 along with CD80 and CD86, showed no symptoms of immune impairment linked to CTLA-4 deficiency [175]. From various studies, we can conclude that CTLA-4 plays a crucial role in inhibiting autoimmunity. Thus, an intersection exists between Tregs and CTLA-4 for mediating tolerance. Phenotypic similarities between FoxP3-deficient and CTLA-4-deficient mice attracted a lot of interest in whether both of these are connected through a common pathway. Tregs are known to maintain peripheral tolerance by regulating activity of effector T cells [176]. Constitutive expression of CTLA-4 in Tregs is believed to play an important role in its immune-suppressive actions [177]. An impairment in suppressive functions was observed in Tregs lacking CTLA-4 expression [177, 178]. CTLA-4 gene–deficient mice showed impaired T cell immunity leading to tissue infiltration and early death at the age of 3 weeks [179, 180]. The CTLA-4 pathway was considered as a deciding factor between tolerance and immunity. Antibodies against CTLA-4 aggravated autoimmunity in various mice models [181–183] and induced autoimmune expression such as oophoritis, gastritis, and sialoadenitis in normal mice [177]. Polymorphisms found in the *CTLA4* locus are believed to be involved in autoimmunity [184–186]. An interesting example showing the interdependency of Tregs and CTLA4 pathway is how the presence of wild-type cells can correct CTLA4$^{-/-}$ and Foxp3$^{-/-}$phenotypes. It was shown that injecting CD4$^+$ CD25$^+$ wild-type cells directly into scurfy mice could lead to the restoration of immune homeostasis [5]. Restoring the Treg deficiency in scurfy phenotype by administering CD4$^+$ CD25$^+$ wild-type cells could be easily anticipated as the role of Tregs in regulating cell extrinsic properties has been known. In 1999, Bachmann et al.'s group suggested that CTLA-4 deficiency could be corrected by combining CTLA-4$^{-/-}$ bone marrow along with wild-type bone marrow in chimeric mice. Their observation also included that CTLA-4$^{-/-}$ bone marrow reconstituted to Rag$^{-/-}$ mice resulted in death of mice roughly after 10 weeks, but those mice that received wild-type bone marrow were completely healthy [187]. Tregs lacking CTLA-4 can easily ignite an autoimmune response indicating a fundamental role played by CTLA-4 in regulating the Tregs lineage [178]. However, there have been reports that Tregs deficient in CTLA-4 were functional enough to cause suppression. A uniformity is absent in defining the role of CTLA-4 with respect to Treg functions.

8.5 Role of Human Regulatory T Cells in Infection

Bacterial and viral infections like leprosy, TB, and HIV in common wealthy countries show more morbidity and mortality. Treg cells are an active area of investigation over the last two decades in human chronic infections. Treg cell immune responses have now been implicated in a large range of pathogens like Mycobacterium leprae, M. tuberculosis (M.tb), human immunodeficiency virus (HIV), and malarial parasites that cause chronic infections [188–191]. During chronic infection, Treg cells secrete immunosuppressive cytokines such as TGF-β and IL-10 that play pivotal role in preventing tissue damage that occurs due to inflammation mediated by Th17, neutrophils, NK, and monocytes, but these immunosuppressive cytokines are also involved in preventing pathogen clearance from the host [192].

8.5.1 Bacterial Diseases

8.5.1.1 Leprosy

Treg, Tr1, and Th3 cells are the primary mediators of anti-inflammatory responses against exogenous antigen (*M. leprae*) such as those associated with mucosal immunity. High TGF-β promotes the development of a microenvironment required for differentiation of Th3 cells, *M. leprae* progresses with the TGF-β and IL-10 cytokine milieu, and increased production of TGF-β and CTLA-4 leads to T cell anergy [193, 194]. Some seminal studies also reported that acetylating FOXP3 leads to induction of Th3 environment via increased production of TGF-β from cholesterol-deprived *M. leprae*–infected macrophages [193, 195]. Reports on FOXP3+ cells are varied in leprosy patients where higher association with tuberculoid and ENL subjects was observed [196]. In contrast to this, other studies found that FOXP3+ cells were increased in leprosy patients [194]. A subset of CD4+CD25+IL10+ Treg cells was also observed in leprosy patients [197]. Also a genetic study described IL-10 and TNF-α cytokine gene polymorphism for determining predisposition to leprosy progression [198]. A study by Saini et al. reported increased TGFβ+FoxP3+ naïve and memory cells in these patients [194]. Saini et al. further confirmed the presence of natural Treg (nTreg) and induced Treg (iTreg) phenotypes, with the help of CD25^high, CD25^low, and CD25^neg FOXP3+Treg cells and observed that *p*STAT5A signaling activates TGF-β production [194]. Subsequently, unstimulated basal levels of the CD8+CD25+FoxP3+Treg phenotype were significantly higher in the leprosy group, but they lacked expression of intracellular TGF-β [194, 199]. Similar results were also observed in 15-year leprosy patients. All these results showed an increase in antigen-specific induced Tregs in leprosy [200]. Some studies showed the molecular mechanism of class II (HDAC7 and HDAC9) activated FOXP3-mediated immunosuppression of Treg cells in leprosy [201]. Subsequently, they silenced FOXP3 gene expression and showed downregulation of CTLA-4 and CD25 in lepromatous patients [201]. Hence, these results suggest that FOXP3 directly regulates the promotion of IL-2R and CTLA-4 genes and is involved in immune suppression in leprosy patients. Moreover, Tariqe et al. showed that IL-35+ Treg and Breg cells are associated with PD1-PD-L1 contact-dependent mechanism for immune suppression in leprosy [202]. Consequently, it indicates that IL-10-producing Breg cells promote CD4+CD25[neg] cells to CD4+CD25+cells in leprosy disease [203]. Importantly, nonconventional T cells (γδ) also expressed FOXP3 and TGF-β in stable leprosy patients associated with severity of leprosy [204]. Moreover, Saini et al. in 2018 showed that γδ T cells produce IL-17 and IFN-γ and also express FOXP3 in inflammatory leprosy reactions [205]. It has been reported earlier that because of hyperimmunization of mice with *M. leprae* sonicated antigens (MLSA), the frequency of Treg cell drops. It thus proves that *M. leprae* is capable of inducing homeostatic imbalance in the immune system of the host and is a major factor for the development of auto-reaction [206–208].

8.5.1.2 Tuberculosis

The immune response to *Mycobacterium tuberculosis (M.tb)* regulates various types of cells, cell surface markers, and cytokines. But recent studies have exposed that Treg cells also showed immunopathology in tuberculosis [209, 210]. Although the primary studies of Treg cells in tuberculosis (TB) give many proofs for their presence in *M.tb* infection, Guyot-Revol et al. and Ribeiro-Rodrigues et al. in 2006 suggested that Treg cells showed anti-inflammatory immune responses to prevent damage to host tissues during TB [211, 212]. A study by Shafiani et al. in 2010 showed the initial stages of immune T cell responses against *M.tb* infection [213]. *M.tb* infection develops Treg cell-mediated immune suppression and allows it to replicate inexhaustibly in the lungs until T helper cells finally reach the infection site. TB patients showed a high percentage of Tregs at the site of granuloma and in the blood that compromise protective Th1 response and interfere with stasis of *M.tb* bacterial growth in macrophages. *M.tb* causes pulmonary and extrapulmonary tuberculosis and manipulates immune response against immune tolerance and pathogen persistence. The involvement of Treg and Th17 cells in pulmonary TB has also been observed. In 2018, Saini et al. observed an increased TGF-β-producing FOXP3+Treg population in cutaneous tuberculosis (CTB) patients as compared to healthy individuals, suggesting that Treg cells play a pivotal role in negatively regulating T cell immune responses in CTB. In addition, the balance of Tregs and Th17 cells in terms of high TGF-β may downregulate IFN-γ and IL-17 responses leading to downregulation of antigen-specific immune responses associated with CTB patients [214].

8.5.1.3 Leishmaniasis

Leishmania is an intracellular protozoan parasite causing leishmaniasis. The role of Treg cell is also important in leishmania infection. The first study showed that there is reduction in immune response in mice infected with leishmania parasite. Negative selection of CD4+CD25+FoxP3+ cells during bacterial diseases resulted in improved cell-mediated immunity and rapid bacterial clearance [215]. Subsequent studies by Suffia et al. in 2006 showed that Treg cells mediated immunity-induced proliferation of antigen-presenting cells, suggesting that FOXP3+ cells bind to a leishmania-derived antigen [216]. Subsequently, Katar et al. in 2011 and 2013 supported the above findings and showed that gene expression of FOXP3, CD25, and IL-10 directly correlated with parasite load in an in situ study [217, 218]. Moreover, these results proved the positive correlation between frequencies of Treg cells with parasite burden. Taken together, Tregs showed immunopathologies in disease severity in dermal leishmaniasis.

The homeostasis between Tregs and T helper cells can be changed in cases of infection, controlling the recognition of antigen-specific effector T cells and reinfection of pathogens [219]. However, in case of human visceral leishmaniasis (VL), no evidence has been found to support the idea behind the role of FOXP3+cell–mediated immune suppression [220]. In humans, both IFN-γ- and IL-10-producing T helper cells showed a significantly higher percentage in leishmania antigen-specific stimulated PBMC cultures of VL patients [221]. IL-10 showed pathogenesis in cutaneous leishmaniasis (CL) produced by all FOXP3+ and non-FOXP3 cells in the chronic lesion of CL [217]. In chronic phase of the infection, both IFN-γ-producing CD4+FOXP3^{neg} and IL-10-producing FOXP3+ Treg cells migrate to the site of infection. In human VL, high level of IFN-γ gene expression in lymphoid organs is correlated by high expression of IL-10 [222, 223], where the predominant source of IL-10 is the T helper (CD3+FOXP3^{neg}) cells [220]. In accord to this, a type of regulatory dendritic cells in *L. donovani*–infected spleen produces IL-10 that induces the development of IL-10-producing regulatory T cells, inhibiting the antimicrobial potential by reactive oxygen (RO) and nitrogen intermediates produced by macrophages and other phagocytic cells. IL-27-producing regulatory APCs and IL-21-producing T cells together drive the differentiation of Th1-like cells to Tregs, along with inhibiting Th17 cell development and IL-17 production. In conclusion, acquiring better knowledge about leishmania species-specific Treg cell phenotypes and functions, their network of interaction and regulation with other subsets of T cells could further help in finding a novel immunological target for the cure and management of leishmaniasis.

8.5.2 Viral Diseases

Tregs play a pivotal role in viral infections and a balance between useful and harmful effects of Tregs can be changed in case of acute and chronic phases of virus infection. Tregs have been reported in RNA, DNA, and retrovirus viral infections in human as well as in mice models [224]. In chronic viral infection, CD4+FOXP3+ and CD8+FOXP3+Treg subsets have been identified but not in acute infections [225]. Although, in hepatitis A virus, infection showed acute inflammatory conditions, hepatitis A virus and its HAV cellular receptor (HAVCR1) suppress Treg function [226]. In acute dengue virus cases, the ratio of CD4+CD25^{high}FoxP3+Treg cells/T effector cells increases, indicating that the rise in this ratio is beneficial for the disease outcome [227]. In contrast, blockade of Treg functions in acute viral infection may help in viral clearance, at the cost of temporarily high inflammation, which can be due to effector immune responses. Higher inflammation is related with low activity of Tregs. On the other hand, TGF-β-producing Treg cells also assist the host during acute infection: First, negative selection of Treg cells in murine herpes simplex infection improved lymph node levels of interferon-α and interferon-γ. But due to downregulation of IFN-γ in infection site, the influx of antigen-presenting cells, natural killer cells, and T helper cells at the infected lesion is delayed [228], resulting in the role for Treg promotion in the lymph node and efflux of Th17 cells [229]. Second, FOXP3$^+$Treg cells showed a protective nature in early HIV infection and inhibited the proliferation of infected cells. Because of this, infection did not establish at the mucosal entry level [230, 231]. One study on West Nile virus infection model showed that Tregs play a vital role in memory T cell formation through activating antigen persistence [232].

The role of human Tregs in chronic viral infection showed high CD4$^+$Treg population in chronic hepatitis B virus (HBV) infection as compared to acute HBV infection and noninfected individuals. This study supports a positive correlation of Tregs with disease progression and viral load [233]. The higher percentage of FOXP3$^+$Treg cells seen in chronic HCV infection, on the other hand, lessened the inflammatory T cell immune responses [234, 235]. Subsequently, Riezu-Boj et al. described the recruitment of Tregs with the help of CCL17and CCL22 migratory molecules in the liver [236], promoting pathogen persistence. However, Tregs may also be involved in HCV-induced liver damage by chronic inflammation [235]. Although CD4$^+$FOXP3$^+$Treg cells remain also high in chronic HIV infection as compared to healthy individuals, Treg-mediated immune homeostasis on anti-HIV immune responses always remained a matter for debate [237]. Moreno-Fernandez et al. showed that CD39 mediated ectonucleotide shifts to block HIV replication in T cells in vitro via CD4$^+$FoxP3$^+$Tregs [238]. Subsequently, Treg cells showed transfer of cAMP via gap junctions formed with conventional (αβ) T cells [238]. During the antigen presentation of the virus from dendritic cells to T cells, FOXP3$^+$ cells inhibited immunological synapse and contained virus spreading [238]. In 2011, Nikolova M et al. performed a CD39 experiment to show the maintenance of cytokine productions by HIV-1 gag protein–stimulated cytotoxic T cells [239] resulting in disease progression and HIV viral load correlating with the percentage of CD4$^+$CD39$^+$ Tregs [240]. These mechanisms of Tregs may be explained by viral load and control of viral replication by CD4$^+$CD39$^+$ Tregs. It may also be important for early and late HIV infection with a partial number of infected cells. Taken together during chronic HIV infection, Tregs are unable to suppress proliferation of proinflammatory immune response and potentially become more harmful due to decreasing anti-HIV immune responses. This points to the need for more detailed analyses of Treg functions in acute vs. chronic inflammation.

8.5.3 Autoimmunity

Autoimmune diseases are estimated to affect 3–5% of individuals in western countries [241]. Autoimmunity cannot be permanently diagnosed, which adversely affects the health-related value of life of patients and is a leading cause for morbidity and mortality. This is unclear what triggers the original event that breaks down immune tolerance to autoimmunity against self-antigens and allows for the activation of autoreactive immune cells [242]. This is found to be associated with specific human leukocyte antigen (HLA) haplotypes and the presentation of specific autoantigens via the major histocompatibility complex (MHC) [243]. Additionally, specific T lymphocytes also play key roles in autoimmune reactions [244]. Treg cells suppress autoinflammatory episodes in patients through various mechanisms. Treg cells secrete immunosuppressive cytokines such as TGF-β, IL-10, and IL-35 [52]. These suppressive cytokines can suppress multiple cell types at the site of inflammation. One mechanism by which Tregs are able to target autoreactive CD4$^+$T effector memory cells is through the generation of tolerogenic APCs. When APCs contact TGF-β and IL-10, they express a tolerogenic phenotype that promotes an anergic state of memory T cells that bind to their MHC molecules [245, 246]. APCs also induce IL-10-producing Treg cells, but it is unclear whether these T cells were naive or memory cells, when they are communicating with the tolerogenic APCs. This phenomenon allows for the targeting of antigen-specific memory T cells when cells become reactivated by tolerogenic APCs at the site of inflammation. Subsequently, IL-10-secreting Tregs, which cause anergy in CD4$^+$CD45RO$^+$ T cells and moreover activation of tolerogenic APCs, upregulate the programmed death ligand 1 (PD-L1) signaling pathway that is important for the suppression of memory T cells post activation in autoimmunity [247, 248]. After antigen activation, exhausted cytotoxic CD8$^+$ T cells upregulate the expression of certain cell surface markers, such as PD-1 [249]. This upregulation of PD-1 leaves CD8$^+$ T cells susceptible to PD-L1-dependent anergy. The generation of exhausted PD-1$^+$CD8$^+$ T cells involves the blockade of IL-2 in the cytokine milieu. Treg shows high amounts of the high-affinity IL-2R (CD25) and are capable of depleting local inflammatory cytokines [250]. In autoimmune diseases, Treg cells are expected to saturate IL-2 at the site of inflammation leading to exhaustion of CD8$^+$ T cells and leaving them prone to PD-1-PD-L1-mediated cell death. Moreover, B cells also play a significant role in the immunopathology of autoimmune disease via the secretion of autoantibodies. These antibodies target endogenous proteins and allow for direct binding of specific cell types by the Fc receptor and complement system [251]. The effector mechanism of autoantibodies has been verified via adoptive transfer of autoantibodies into animal models, whereby they exacerbate tissue pathology in a similar manner as in the human disease [252, 253]. Subsequently, Treg cells are also able to suppress autoantibody secretion from B cells via cell-to-cell contact-dependent manner [254]. Animal model studies have shown that negative selection of Treg cells leads to increase autoantibody production [255]. Patients which do not express the functional *FOXP3* gene, responsible for a disease condition called immune dysregulation, polyendocrinopathy, enteropathy, and X-linked (IPEX) syndrome, lack natural Treg cells and suffer from a number of acute autoimmune and inflammation disorders that are dangerous if not treated by bone marrow transplantation [256]. Immunotherapy for autoimmune disorders aims to inhibit the proinflammatory immune response by depleting specific adaptive immune cell populations like Th1 and Th17 or inhibiting the activation of these cells in target organs [257]. These immunotherapies are helpful in preventing proinflammatory immune response against organ-specific autoimmunity, but they also inhibit protective immunity and can leave the patient's immune system compromised and susceptible to infections. Newer therapies are thus being designed to utilize the suppressive capabilities of human Treg cells to suppress autoimmune cells in an antigen-specific manner [258].

8.5.4   Allergy

Tregs can change and modulate the progress of allergic diseases shifting the ongoing hypersensitivity and effector T cell development via many important pathways. Tregs dampen the proinflammatory immune response. Moreover, Tregs have the ability to promote dendritic cells to prime effector responses via IFN-γ, IL-4, and IL-17, along with initiating the expansion of tolerogenic dendritic cell phenotypes [178]. Subsequently, Tregs directly suppress humoral immune response and its cytokine milieu (via IL-4, IL-5, IL-13, and IL-9) for inactivation of allergen-specific immunity during an allergy [259–261]. Tregs also play an important role in suppression of mast cells, basophils, and eosinophils in allergic inflammation. TGF-β-producing Tregs are also involved in tissue remodeling with the help of resident tissue cells [239, 262]. Moreover, FOXP3+ regulatory T cells use suppressive cytokine in a contact-dependent manner to suppress hypersensitivity reactions by blocking entry of effector T cells into inflamed tissues due to allergic reaction [263]. Moreover, Treg cells also stop polarization of effector cells, to abrogate apoptosis of keratinocytes and bronchial epithelial cells, thereby preventing tissue injury [264]. Importantly, Tregs also suppress B cells and stop the production of allergen-specific IgE and IgG4 antibodies [265].

Numerous studies on healthy humans have showed predominantly IL-10-producing Treg cells against common environment allergy-specific immune response [259–261]. A phenotype of Tregs showed no difference between nonallergic healthy and allergic individuals as allergen-specific Th1, Th2, and TR1 (IL-10 producing) cells all recognize the same T cell epitopes. Accordingly, depending on the predominance of Th2 and TR1 subsets and their balance, allergic people may develop allergy with high Th2 immune response or recovery with TR1 predominance. A study in human models for the last decades established the fact that high-dose exposure of allergens leads to Treg induction [266, 267]. Taking into consideration the beekeepers that are generally exposed to bee venom allergens, there is reduction in T-cell-associated cutaneous late-phase response. In response, the allergen-specific T cells proliferate and release Th1 and Th2 cytokines. The above mechanism correlates with a clonal switch of venom antigen-specific CMI response towards IL-10-producing Tr1 cells [266]. Another study showed that high-dose exposure to cat allergens activate Tr1 and IgG4 antibody responses without following the development of a new hypersensitivity or asthma development [267]. The above study indirectly establishes the fact that Treg cells have a protective effect in allergy reaction. A study by Verhasselt et al. 2008 in mice has shown that breast milk mediated transfer of antigens to the neonates for the development of antigen-specific FOXP3+Treg cells and stop allergic airway inflammation [268].

This mechanism is dependent on TGF-β^+FOXP3$^+$Tregs and also depends on TGF-β signaling. Same is the case with children who develop milk allergy; these children possess a higher percentage of Tregs with reduced in vitro proliferative response than their counterparts with no tolerance to milk [269].

8.5.5 Cancer

T cells with suppressive function were reported in patients with cancer back in the 1990s [270–272]. However, these studies were unclear until the identification of CD4$^+$CD25$^+$FOXP3$^+$Treg cells in 1995 [273]. Various studies demonstrated the presence of FOXP3$^+$ regulatory T cells in patients with lung and ovarian carcinoma [274, 275]. Treg cells blocked antitumor immune responses and the higher frequency of Treg cells in peripheral blood of human cancers has been found to be increased [276–278]. From these studies we can conclude that the presence of Treg cells inhibits the development of antitumor immune responses; thus, methods of preventing the activity of FOXP3$^+$ Tregs may be crucial for the successful immunotherapeutic treatment in humans [279]. It has been reported that patients with gastrointestinal cancer had a significantly higher percentage of Treg cells in peripheral blood [280, 281]. Patients who had gastric carcinoma with higher percentages of FOXP3$^+$Treg cells had a poorer prognosis than those with lower percentages. Interestingly, FOXP3$^+$Treg cell proportions were also found to be enhanced in ascites from patients who had advanced-stage disease with peritoneal dissemination [280, 281]. Moreover, another study showed that the percentages of high CD25$^+$Treg cells in peripheral blood mononuclear cells (PBMCs) from patients with gastric and esophageal cancer were significantly higher as compared to healthy donors [282]. Ichihara et al. in 2003 showed that the percentage of Treg cells in the TILs of gastric cancer patients in the later stage was significantly higher as compared to patients with early-stage disease [283]. Moreover, it has been shown that prevalence of Treg cells in the peripheral blood of gastrointestinal cancer patients is significantly higher than that in early-stage patients and healthy controls [284]. Since Treg cell population is significantly reduced after curative surgery, it is possible that tumor cells may have induced and expanded the Treg cell pool [284]. Shen et al. in 2009 have characterized CD4$^+$CD25$^+$CD127^{neg} as the surface marker of Treg cells in gastric cancer and found that the frequency of Treg cells in the PBMCs of gastric cancer patients was significantly higher as compared to healthy controls [285]. They proposed that CD4$^+$CD25$^+$FOXP3$^+$CD127^{neg} can be used as a selective biomarker to enrich human Treg cells and also to perform functional in vitro assay in gastric cancer. Furthermore, a study by Xu et al. in 2009 has also shown that the prevalence of Treg cells in the peripheral blood of gastrointestinal cancer patients is significantly higher than that in healthy donors, but it also increased in parallel with tumor progression [286].

A study by Mizukami et al. in 2008 investigated the frequency of Treg cells in TILs, tumor-draining regional lymph nodes, and PBMCs of patients of gastric cancer and evaluated the relationship between the CCL17- and CCL22-producing cells with such an observation occurring in early-stage gastric cancer [287]. Some studies demonstrated that CCL22 chemokines derived from tumors induce the migration of Treg cells through CCR4, which is a chemokine receptor for CCL22, and impairs antitumor immunity in primary breast cancer and lung cancer [288, 289]. Moreover, in 2009, it has been found that Treg frequency is significantly higher in the peripheral blood of patients with IL-2-treated melanoma and in formalin-fixed tissue from patients with lung and colon cancer [290]. In addition, they also demonstrated that Treg cell numbers are predictively elevated in the peripheral blood of patients with various solid tumors. Patients with squamous cell carcinoma of the head and neck have increased number of Treg cells in their peripheral circulation compared with normal controls and have a depressed antitumor immunity [291, 292]. Surprisingly, higher frequency of Tregs and levels of suppression were observed in patients with no clinically defined disease than in untreated patients with active disease [291]. Furthermore, a study showed that patients with hepatocellular carcinoma also have increased numbers of FOXP3$^+$Treg cells in their peripheral blood, suggesting that the increased number of FOXP3$^+$Treg cells might play a role in the modulation of the immune responses against hepatocellular carcinoma and could be important in designing novel immunotherapeutic approaches [273]. Moreover, Treg cells are associated with hepatocellular carcinoma invasiveness and intratumoral balance of Tregs and cytotoxic T (CD8) cells are a promising independent biomarker for recurrence and survival in hepatocellular carcinoma [293]. It has also been showed that primary hepatic carcinoma develops in the liver that is immunosuppressed by a marked infiltration of CD4$^+$CD25$^+$FOXP3$^+$Treg cells. A high prevalence of Treg cells infiltrating hepatocellular carcinoma cells is thought to be an adverse prognostic indicator [294]. Prostate carcinoma patients showed significantly a higher frequency of CD4$^+$CD25^{high}Treg cells inside the prostate compared with benign tissue from the same prostate [295]. Moreover, Treg cells from blood and supernatants from cultured prostate tumor tissue samples exhibited immunosuppressive function in vitro. These studies point out that Treg cells are important for the development of early-stage prostate tumors, and thus new therapeutic strategies aimed at negative selection of Treg cells may improve prostate cancer immunotherapy [296]. Additionally, it has been reported that more than four hundred prostate cancer patients have elevated numbers of circulating and tumor-infiltrating Treg cells and increase tumor growth in vivo and these Treg cells potently inhibit tumor-specific T cells [297].

Interestingly, administration of high-dose IL-2 in patients with renal cell carcinoma increased the percentage of circulating Treg cells [298]. These studies suggest that selective inhibition of IL-2-mediated proliferation of Treg cells may improve the therapeutic values. Jensen et al. in 2009 reported that infiltration of FOXP3$^+$Treg cells significantly increased during IL-2-based immunotherapy, and after treatment, high FOXP3$^+$ cells were correlated with poor prognosis in patients with metastatic renal cell carcinoma [299]. In patients with ovarian cancer, tumor-associated T cells from patients with advance-stage ovarian cancer contain increased CD4$^+$CD25$^+$ T cells and were involved in T cell immune suppression [274]. In addition, higher percentage of CD4$^+$CD25$^+$Treg cells in PBMC, TIL, and tumor-associated lymphocytes in ovarian carcinoma patients has also been reported [300]. Tumor cells and macrophages produce the chemokine CCL22, which mediates trafficking of Tregs to the tumor. These studies suggested that this specific recruitment of Treg cells represents a mechanism by which tumors may promote immune privilege and block Treg cell migration [272, 301]. One line of evidence showed that FOXP3⁺Treg cells were not influenced by ovarian cancer tissue, but median disease-specific survival of patients with a high CD8⁺/FOXP3⁺ ratio in ovarian-derived tumor tissue was twice as high as in patients with a low CD8⁺/FOXP3⁺ratio [302].

In patients with breast and pancreatic cancer, the frequency of Treg cells in the peripheral blood is enhanced when compared with normal individuals. Similarly, Treg cells are present in TILs and tumor-draining lymph nodes (TDLNs) infiltrated by tumor. These cells secrete IL-10 and TGF-β and prevent activation of T helper cells [303]. Quantification of FOXP3⁺Treg cells in breast cancer for monitoring the disease prognosis and progression is an important therapeutic approach in breast cancer. Thus, FOXP3⁺Treg cells represent a novel marker for identifying late-relapse patients (Bates GJ 2006). In patients with acute myeloid leukemia (AML), the population of CD4⁺CD25^high^Treg cells in peripheral blood is significantly higher as compared to healthy individuals. Notably, Treg cells in AML presented significantly higher apoptosis and proliferation than healthy individuals [304]. It has been reported that Treg cells accumulate in the peripheral circulation of acute myeloid leukemia patients via contact-dependent and contact-independent mechanisms [305]. However, most of these studies have been performed on carcinomas, with the role of Treg cells in hematologic malignancies such as non-Hodgkin lymphoma being still unestablished. Such studies suggest that the role of Treg cells in the pathogenesis of these B cell lymphomas may be different than carcinomas. The majority of non-Hodgkin lymphomas are B cell dependent, but the tumor tissue can be variably infiltrated with T cells. A recent study showed that a subset of FOXP3⁺ Tregs with high level of CTLA-4 is identified in biopsy specimens of B cell non-Hodgkin lymphoma and these cells suppressed the production of IFN-γ and IL-4 by infiltrating T helper cells in response to phytohemagglutinin (PHA) stimulation [306].

8.5.6 Osteoporosis and Bone Health

Osteoporosis is a well-known systemic skeletal disease that in general leads to abnormal bone remodeling resulting in dysregulated bone resorption and bone formation process. Age associated decline in bone health is being observed in both men and women. In postmenopausal women, progression of osteoporosis is accelerated due to declining levels of estrogen hormone, known to have osteoprotective role [307]. A study by Tai's group in 2008, demonstrated that osteoprotective hormone estrogen can stimulate the proliferation of Tregs cells that have been shown to inhibit osteoclast function [308]. Numerous studiesin mice (including our group) and humans suggested that immune cells of both innate and adaptive arm of immune response plays an important role in dynamic regulation of bone homeostasis, a field coined by our group as "Immunoporosis" i.e. Immunology of osteoporosis. Among various immune cells FOXP3⁺ Treg cells play indispensable roles in immune homeostasis, differentiation of HSCs and functions of osteoblasts (bone forming cells) and osteoclasts (bone resorbing cells). Several in vitro studies have reported that Tregs exhibits the potential to inhibit osteoclastogenesis either by secreting inhibitory cytokines (TGF-β, IL-10 and IL-4) or in a cell-cell contact dependent manner [309, 310]. Various in vivo studies also have suggested that Tregs directly inhibit osteoclastogenesis by downregulating production of RANKL and MCSF and hence enhancing bone health [71, 311, 312]. Furthermore, it has been found that adoptive transfer of Treg cells ameliorated the disease in autoimmune arthritis animal model, whereas depletion of Treg cells induced the more severe form of arthritis [313–316]. Recently, a study in 2019 reported that Tregs via expressing CTLA-4 may interact with osteoclast precursors expressing CD80/CD86 and thus inhibit the differentiation of osteoclasts precursors into to mature osteoclasts [317]. Moreover, in Ovx mice, it has been found by our group that oral supplementation of probiotics viz. Lactobacillus rhamnosus and Bacillus clausii enhance Tregs population in lymphoid organs such as bone marrow, spleen etc. which in turn regulatesbone health by secreting immune suppressive cytokines such as IL-10 and IL-4 as compared to control groups [70, 71]. There are evidences which also suggest that Tregs play a role in bone formation by promoting differentiation of osteoblasts. A study in 2018, showed that supplementation of probiotic Lactobacillus rhamnosus GG enhances the Tregs population which further upregulates the expression of osteogenic factor Wnt10b by osteoblasts [318]. These observations raise the question that whether the accumulation of Foxp3+ Tregs within bone marrow is due to the recruitment of pre-formed FOXP3+ Tregs into the bone marrow microenvironment or due to the de novo induction of FOXP3- T cells to Foxp3+Treg cells. Altogether these studies indicate that any dysregulation in the population or functioning of Tregs would result in enhanced bone loss. In fact, this has been proposed and demonstrated in mice, although human studies are still lacking. Thus, exploring novel mechanisms regulating the correlation between Tregs and bone cells is highly anticipated for future clinical implications.

8.6 Conclusion

The past decades have provided outstanding insights into the diverse phenotypic and functional types of Tregs. A wealth of studies has demonstrated that Tregs are crucial in maintenance of immune tolerance. This chapter has focused specifically on the discovery of Tregs, its specific markers and how the various regulators control the development and functions of Tregs. Here, we have discussed the various proposed mechanisms of actions that are displayed by immune suppressive Tregs. The present global scenario arising from various studies using experimental models and human disorders validate the vital role of Tregs in several diseases including bone health. Together, these studies indicate that Tregs have the potential to modulate a number of immune pathologies. In the context of immunological conditions such as autoimmunity and transplantation, long-term usage of immunosuppressive drugs increases the likelihood of life-threatening infections. In certain conditions such as during graft transplantation, autoimmune diseases and so on, expansion of the immunosuppressive Tregs population is needed. Thus, elucidation of mechanismsthat govern the amplification and attenuation of theTreg lineage will have important implications for therapy. Thus, strategies can be exploited by therapeutically targeting Tregs can open new avenues in treating various immune-mediated diseases.

Acknowledgement This work was financially supported by projects: DST-SERB (EMR/2016/007158), Govt. of India and intramural project from All India Institute of Medical Sciences (AIIMS), New Delhi-India sanctioned to RKS. NS, CS, LS, ZA, AB, AA and RKS acknowledge the Department of Biotechnology, AIIMS, New Delhi-India for providing infrastructural facilities. NS and AA thanks DBT for fellowship. LS and ZA thank the UGC for their respective research fellowships. AB thanks DST for research fellowship.

Author Contributions RKS suggested the focus and outline of the chapter and wrote the review. NS, CS, LS, ZA, AB and AA participated in writing of the chapter. RKS suggested, and ZA created the illustrations.

Conflicts of Interest The authors declare no conflicts of interest.

References

1. Gershon RK, Kondo K (1970) Cell interactions in the induction of tolerance: the role of thymic lymphocytes. Immunology 18:723–737
2. Gershon RK, Cohen P, Hencin R, Liebhaber SA (1972) Suppressor T cells. J Immunol 108:586–590
3. Gershon RK, Kondo K (1971) Infectious immunological tolerance. Immunology 21:903–914
4. Hori S, Nomura T, Sakaguchi S (2003) Control of regulatory T cell development by the transcription factor Foxp3. Science 299:1057–1061
5. Fontenot JD, Gavin MA, Rudensky AY (2003) Foxp3 programs the development and function of CD4+CD25+ regulatory T cells. Nat Immunol 4:330–336
6. Sakaguchi S (2000) Regulatory T cells: key controllers of immunologic self-tolerance. Cell 101:455–458
7. Curotto de Lafaille MA, Lafaille JJ (2009) Natural and adaptive foxp3+ regulatory T cells: more of the same or a division of labor? Immunity 30:626–635
8. Moller G (1988) Do suppressor T cells exist? Scand J Immunol 27:247–250
9. Sakaguchi S, Wing K, Miyara M (2007) Regulatory T cells – a brief history and perspective. Eur J Immunol 37(Suppl 1):S116–S123
10. Kappler JW, Roehm N, Marrack P (1987) T cell tolerance by clonal elimination in the thymus. Cell 49:273–280
11. Goodnow CC, Cyster JG, Hartley SB, Bell SE, Cooke MP, Healy JI, Akkaraju S, Rathmell JC, Pogue SL, Shokat KP (1995) Self-tolerance checkpoints in B lymphocyte development. Adv Immunol 59:279–368
12. O'Garra A, Murphy K (1994) Role of cytokines in determining T-lymphocyte function. Curr Opin Immunol 6:458–466
13. Chen Y, Kuchroo VK, Inobe J, Hafler DA, Weiner HL (1994) Regulatory T cell clones induced by oral tolerance: suppression of autoimmune encephalomyelitis. Science 265:1237–1240
14. Nishizuka Y, Sakakura T (1969) Thymus and reproduction: sex-linked dysgenesis of the gonad after neonatal thymectomy in mice. Science 166:753–755
15. Sakaguchi S, Fukuma K, Kuribayashi K, Masuda T (1985) Organ-specific autoimmune diseases induced in mice by elimination of T cell subset. I. Evidence for the active participation of T cells in natural self-tolerance; deficit of a T cell subset as a possible cause of autoimmune disease. J Exp Med 161:72–87

16. Bloom BR, Salgame P, Diamond B (1992) Revisiting and revising suppressor T cells. Immunol Today 13:131–136

17. Brunkow ME, Jeffery EW, Hjerrild KA, Paeper B, Clark LB, Yasayko SA, Wilkinson JE, Galas D, Ziegler SF, Ramsdell F (2001) Disruption of a new forkhead/winged-helix protein, scurfin, results in the fatal lymphoproliferative disorder of the scurfy mouse. Nat Genet 27:68–73

18. Bennett CL, Christie J, Ramsdell F, Brunkow ME, Ferguson PJ, Whitesell L, Kelly TE, Saulsbury FT, Chance PF, Ochs HD (2001) The immune dysregulation, polyendocrinopathy, enteropathy, X-linked syndrome (IPEX) is caused by mutations of FOXP3. Nat Genet 27:20–21

19. Khattri R, Cox T, Yasayko SA, Ramsdell F (2003) An essential role for scurfin in CD4+CD25+ T regulatory cells. Nat Immunol 4:337–342

20. Weiner HL (2001) Induction and mechanism of action of transforming growth factor-beta-secreting Th3 regulatory cells. Immunol Rev 182:207–214

21. Kohm AP, Carpentier PA, Anger HA, Miller SD (2002) Cutting edge: CD4+CD25+ regulatory T cells suppress antigen-specific autoreactive immune responses and central nervous system inflammation during active experimental autoimmune encephalomyelitis. J Immunol 169:4712–4716

22. Zheng J, Chan PL, Liu Y, Qin G, Xiang Z, Lam KT, Lewis DB, Lau YL, Tu W (2013) ICOS regulates the generation and function of human CD4+ Treg in a CTLA-4 dependent manner. PLoS One 8:e82203

23. Currie AJ, Prosser A, McDonnell A, Cleaver AL, Robinson BW, Freeman GJ, van der Most RG (2009) Dual control of antitumor CD8 T cells through the programmed death-1/programmed death-ligand 1 pathway and immunosuppressive CD4 T cells: regulation and counterregulation. J Immunol 183:7898–7908

24. Harada H, Salama AD, Sho M, Izawa A, Sandner SE, Ito T, Akiba H, Yagita H, Sharpe AH, Freeman GJ, Sayegh MH (2003) The role of the ICOS-B7h T cell costimulatory pathway in transplantation immunity. J Clin Invest 112:234–243

25. Bertram EM, Tafuri A, Shahinian A, Chan VS, Hunziker L, Recher M, Ohashi PS, Mak TW, Watts TH (2002) Role of ICOS versus CD28 in antiviral immunity. Eur J Immunol 32:3376–3385

26. Maeda S, Fujimoto M, Matsushita T, Hamaguchi Y, Takehara K, Hasegawa M (2011) Inducible costimulator (ICOS) and ICOS ligand signaling has pivotal roles in skin wound healing via cytokine production. Am J Pathol 179:2360–2369

27. Haribhai D, Williams JB, Jia S, Nickerson D, Schmitt EG, Edwards B, Ziegelbauer J, Yassai M, Li SH, Relland LM, Wise PM, Chen A, Zheng YQ, Simpson PM, Gorski J, Salzman NH, Hessner MJ, Chatila TA, Williams CB (2011) A requisite role for induced regulatory T cells in tolerance based on expanding antigen receptor diversity. Immunity 35:109–122

28. Ray A, Khare A, Krishnamoorthy N, Qi Z, Ray P (2010) Regulatory T cells in many flavors control asthma. Mucosal Immunol 3:216–229

29. Sogut A, Yilmaz O, Kirmaz C, Ozbilgin K, Onur E, Celik O, Pinar E, Vatansever S, Dinc G, Yuksel H (2012) Regulatory-T, T-helper 1, and T-helper 2 cell differentiation in nasal mucosa of allergic rhinitis with olive pollen sensitivity. Int Arch Allergy Immunol 157:349–353

30. Wu K, Bi Y, Sun K, Wang C (2007) IL-10-producing type 1 regulatory T cells and allergy. Cell Mol Immunol 4:269–275

31. Noble A, Giorgini A, Leggat JA (2006) Cytokine-induced IL-10-secreting CD8 T cells represent a phenotypically distinct suppressor T-cell lineage. Blood 107:4475–4483

32. Cowan JE, Parnell SM, Nakamura K, Caamano JH, Lane PJ, Jenkinson EJ, Jenkinson WE, Anderson G (2013) The thymic medulla is required for Foxp3+ regulatory but not conventional CD4+ thymocyte development. J Exp Med 210:675–681

33. Itoh M, Takahashi T, Sakaguchi N, Kuniyasu Y, Shimizu J, Otsuka F, Sakaguchi S (1999) Thymus and autoimmunity: production of CD25+CD4+ naturally anergic and suppressive T cells as a key function of the thymus in maintaining immunologic self-tolerance. J Immunol 162:5317–5326

34. Salomon B, Lenschow DJ, Rhee L, Ashourian N, Singh B, Sharpe A, Bluestone JA (2000) B7/CD28 costimulation is essential for the homeostasis of the CD4+CD25+ immunoregulatory T cells that control autoimmune diabetes. Immunity 12:431–440

35. Cheng G, Yu A, Dee MJ, Malek TR (2013) IL-2R signaling is essential for functional maturation of regulatory T cells during thymic development. J Immunol 190:1567–1575

36. D'Cruz LM, Klein L (2005) Development and function of agonist-induced CD25+Foxp3+ regulatory T cells in the absence of interleukin 2 signaling. Nat Immunol 6:1152–1159

37. Tai X, Cowan M, Feigenbaum L, Singer A (2005) CD28 costimulation of developing thymocytes induces Foxp3 expression and regulatory T cell differentiation independently of interleukin 2. Nat Immunol 6:152–162

38. Zhang R, Huynh A, Whitcher G, Chang J, Maltzman JS, Turka LA (2013) An obligate cell-intrinsic function for CD28 in Tregs. J Clin Invest 123:580–593

39. Stritesky GL, Jameson SC, Hogquist KA (2011) Selection of self-reactive T cells in the thymus. Annu Rev Immunol 30:95–114

40. Martin-Gayo E, Sierra-Filardi E, Corbi AL, Toribio ML (2010) Plasmacytoid dendritic cells resident in human thymus drive natural Treg cell development. Blood 115:5366–5375

41. Srivastava RK, Tomar GB, Barhanpurkar AP, Gupta N, Pote ST, Mishra GC, Wani MR (2011) IL-3 attenuates collagen-induced arthritis by modulating the development of Foxp3+ regulatory T cells. J Immunol 186:2262–2272

42. Coquet JM, Middendorp S, van der Horst G, Kind J, Veraar EA, Xiao Y, Jacobs H, Borst J (2012) The CD27 and CD70 costimulatory pathway inhibits effector function of T helper 17 cells and attenuates associated autoimmunity. Immunity 38:53–65

43. Sun CM, Hall JA, Blank RB, Bouladoux N, Oukka M, Mora JR, Belkaid Y (2007) Small intestine lamina propria dendritic cells promote de novo generation of Foxp3 T reg cells via retinoic acid. J Exp Med 204:1775–1785

44. Lio CW, Hsieh CS (2008) A two-step process for thymic regulatory T cell development. Immunity 28:100–111

45. Burchill MA, Yang J, Vang KB, Moon JJ, Chu HH, Lio CW, Vegoe AL, Hsieh CS, Jenkins MK, Farrar MA (2008) Linked T cell receptor and cytokine signaling govern the development of the regulatory T cell repertoire. Immunity 28:112–121

46. Mahmud SA, Manlove LS, Schmitz HM, Xing Y, Wang Y, Owen DL, Schenkel JM, Boomer JS, Green JM, Yagita H, Chi H, Hogquist KA, Farrar MA (2014) Costimulation via the tumor-necrosis factor receptor superfamily couples TCR signal strength to the thymic differentiation of regulatory T cells. Nat Immunol 15:473–481

47. Burchill MA, Yang J, Vogtenhuber C, Blazar BR, Farrar MA (2007) IL-2 receptor beta-dependent STAT5 activation is required for the development of Foxp3+ regulatory T cells. J Immunol 178:280–290

48. Yao Z, Kanno Y, Kerenyi M, Stephens G, Durant L, Watford WT, Laurence A, Robinson GW, Shevach EM, Moriggl R, Hennighausen L, Wu C, O'Shea JJ (2007) Nonredundant roles for Stat5a/b in directly regulating Foxp3. Blood 109:4368–4375

49. Bluestone JA, Abbas AK (2003) Natural versus adaptive regulatory T cells. Nat Rev Immunol 3:253–257

50. Baecher-Allan C, Brown JA, Freeman GJ, Hafler DA (2003) CD4+CD25+ regulatory cells from human peripheral blood express very high levels of CD25 ex vivo. Novartis Found Symp 252:67–88; discussion 88–91, 106–14

51. Graca L, Le Moine A, Cobbold SP, Waldmann H (2003) Dominant transplantation tolerance. Opin Curr Opin Immunol 15:499–506

52. Bettini M, Vignali DA (2009) Regulatory T cells and inhibitory cytokines in autoimmunity. Curr Opin Immunol 21:612–618

53. La Cava A (2009) Natural Tregs and autoimmunity. Front Biosci (Landmark Ed) 14:333–343

54. Miyara M, Yoshioka Y, Kitoh A, Shima T, Wing K, Niwa A, Parizot C, Taflin C, Heike T, Valeyre D, Mathian A, Nakahata T, Yamaguchi T, Nomura T, Ono M, Amoura Z, Gorochov G, Sakaguchi S (2009) Functional delineation and differentiation dynamics of human CD4+ T cells expressing the FoxP3 transcription factor. Immunity 30:899–911

55. Liu W, Putnam AL, Xu-Yu Z, Szot GL, Lee MR, Zhu S, Gottlieb PA, Kapranov P, Gingeras TR, Fazekas de St Groth B, Clayberger C, Soper DM, Ziegler SF, Bluestone JA (2006) CD127 expression inversely correlates with FoxP3 and suppressive function of human CD4+ T reg cells. J Exp Med 203:1701–1711

56. Thornton AM, Korty PE, Tran DQ, Wohlfert EA, Murray PE, Belkaid Y, Shevach EM (2010) Expression of Helios, an Ikaros transcription factor family member, differentiates thymic-derived from peripherally induced Foxp3+ T regulatory cells. J Immunol 184:3433–3441

57. MacDonald KG, Han JM, Himmel ME, Huang Q, Kan B, Campbell AI, Lavoie PM, Levings MK (2013) Response to comment on "helios+ and helios- cells coexist within the natural FOXP3+ T regulatory cell subset in humans". J Immunol 190:4440–4441

58. Kim YC, Bhairavabhotla R, Yoon J, Golding A, Thornton AM, Tran DQ, Shevach EM (2012) Oligodeoxynucleotides stabilize Helios-expressing Foxp3+ human T regulatory cells during in vitro expansion. Blood 119:2810–2818

59. McClymont SA, Putnam AL, Lee MR, Esensten JH, Liu W, Hulme MA, Hoffmuller U, Baron U, Olek S, Bluestone JA, Brusko TM (2011) Plasticity of human regulatory T cells in healthy subjects and patients with type 1 diabetes. J Immunol 186:3918–3926

60. Ito T, Hanabuchi S, Wang YH, Park WR, Arima K, Bover L, Qin FX, Gilliet M, Liu YJ (2008) Two functional subsets of FOXP3+ regulatory T cells in human thymus and periphery. Immunity 28:870–880

61. Seddiki N, Santner-Nanan B, Tangye SG, Alexander SI, Solomon M, Lee S, Nanan R, Fazekas de Saint Groth B (2006) Persistence of naive CD45RA+ regulatory T cells in adult life. Blood 107:2830–2838

62. Apostolou I, Sarukhan A, Klein L, von Boehmer H (2002) Origin of regulatory T cells with known specificity for antigen. Nat Immunol 3:756–763

63. Thorstenson KM, Khoruts A (2001) Generation of anergic and potentially immunoregulatory CD25+CD4 T cells in vivo after induction of peripheral tolerance with intravenous or oral antigen. J Immunol 167:188–195

64. Yamazaki S, Dudziak D, Heidkamp GF, Fiorese C, Bonito AJ, Inaba K, Nussenzweig MC, Steinman RM (2008) CD8+ CD205+ splenic dendritic cells are specialized to induce Foxp3+ regulatory T cells. J Immunol 181:6923–6933

65. Gottschalk RA, Corse E, Allison JP (2010) TCR ligand density and affinity determine peripheral induction of Foxp3 in vivo. J Exp Med 207:1701–1711

66. Turner MS, Kane LP, Morel PA (2009) Dominant role of antigen dose in CD4+Foxp3+ regulatory T cell induction and expansion. J Immunol 183:4895–4903

67. Molinero LL, Miller ML, Evaristo C, Alegre ML (2011) High TCR stimuli prevent induced regulatory T cell differentiation in a NF-kappaB-dependent manner. J Immunol 186:4609–4617

68. Lathrop SK, Bloom SM, Rao SM, Nutsch K, Lio CW, Santacruz N, Peterson DA, Stappenbeck TS, Hsieh CS (2011) Peripheral education of the immune system by colonic commensal microbiota. Nature 478:250–254

69. Atarashi K, Tanoue T, Shima T, Imaoka A, Kuwahara T, Momose Y, Cheng G, Yamasaki S, Saito T, Ohba Y, Taniguchi T, Takeda K, Hori S, Ivanov II, Umesaki Y, Itoh K, Honda K (2011) Induction of colonic regulatory T cells by indigenous Clostridium species. Science 331:337–341

70. Dar HY, Pal S, Shukla P, Mishra PK, Tomar GB, Chattopadhyay N, Srivastava RK (2018) Bacillus clausii inhibits bone loss by skewing Treg-Th17 cell equilibrium in postmenopausal osteoporotic mice model. Nutrition 54:118–128

71. Dar HY, Shukla P, Mishra PK, Anupam R, Mondal RK, Tomar GB, Sharma V, Srivastava RK (2018) Lactobacillus acidophilus inhibits bone loss and increases bone heterogeneity in osteoporotic mice via modulating Treg-Th17 cell balance. Bone Rep 8:46–56

72. Round JL, Mazmanian SK (2010) Inducible Foxp3+ regulatory T-cell development by a commensal bacterium of the intestinal microbiota. Proc Natl Acad Sci U S A 107:12204–12209

73. Josefowicz SZ, Niec RE, Kim HY, Treuting P, Chinen T, Zheng Y, Umetsu DT, Rudensky AY (2012) Extrathymically generated regulatory T cells control mucosal TH2 inflammation. Nature 482:395–399

74. Chaudhry A, Rudensky AY (2013) Control of inflammation by integration of environmental cues by regulatory T cells. J Clin Invest 123:939–944

75. Zheng Y, Josefowicz S, Chaudhry A, Peng XP, Forbush K, Rudensky AY (2010) Role of conserved non-coding DNA elements in the Foxp3 gene in regulatory T-cell fate. Nature 463:808–812

76. Gottschalk RA, Corse E, Allison JP (2011) Expression of Helios in peripherally induced Foxp3+ regulatory T cells. J Immunol 188:976–980

77. Himmel ME, MacDonald KG, Garcia RV, Steiner TS, Levings MK (2013) Helios+ and Helios- cells coexist within the natural FOXP3+ T regulatory cell subset in humans. J Immunol 190:2001–2008

78. Abbas AK, Benoist C, Bluestone JA, Campbell DJ, Ghosh S, Hori S, Jiang S, Kuchroo VK, Mathis D, Roncarolo MG, Rudensky A, Sakaguchi S, Shevach EM, Vignali DA, Ziegler SF (2013) Regulatory T cells: recommendations to simplify the nomenclature. Nat Immunol 14:307–308

79. Murugaiyan G, Mittal A, Weiner HL (2010) Identification of an IL-27/osteopontin axis in dendritic cells and its modulation by IFN-gamma limits IL-17-mediated autoimmune inflammation. Proc Natl Acad Sci U S A 107:11495–11500

80. Maynard CL, Harrington LE, Janowski KM, Oliver JR, Zindl CL, Rudensky AY, Weaver CT (2007) Regulatory T cells expressing interleukin 10 develop from Foxp3+ and Foxp3- precursor cells in the absence of interleukin 10. Nat Immunol 8:931–941

81. Meiron M, Zohar Y, Anunu R, Wildbaum G, Karin N (2008) CXCL12 (SDF-1alpha) suppresses ongoing experimental autoimmune encephalomyelitis by selecting antigen-specific regulatory T cells. J Exp Med 205:2643–2655

82. Groux H, O'Garra A, Bigler M, Rouleau M, Antonenko S, de Vries JE, Roncarolo MG (1997) A CD4+ T-cell subset inhibits antigen-specific T-cell responses and prevents colitis. Nature 389:737–742

83. Bacchetta R, Sartirana C, Levings MK, Bordignon C, Narula S, Roncarolo MG (2002) Growth and expansion of human T regulatory type 1 cells are independent from TCR activation but require exogenous cytokines. Eur J Immunol 32:2237–2245

84. Kohyama M, Sugahara D, Sugiyama S, Yagita H, Okumura K, Hozumi N (2004) Inducible costimulator-dependent IL-10 production by regulatory T cells specific for self-antigen. Proc Natl Acad Sci U S A 101:4192–4197

85. Rahmoun M, Foussat A, Groux H, Pene J, Yssel H, Chanez P (2006) Enhanced frequency of CD18- and CD49b-expressing T cells in peripheral blood of asthmatic patients correlates with disease severity. Int Arch Allergy Immunol 140:139–149

86. Afonina IS, Zhong Z, Karin M, Beyaert R (2017) Limiting inflammation-the negative regulation of NF-kappaB and the NLRP3 inflammasome. Nat Immunol 18:861–869

87. Dinesh RK, Skaggs BJ, La Cava A, Hahn BH, Singh RP (2010) CD8+ Tregs in lupus, autoimmunity, and beyond. Autoimmun Rev 9:560–568

88. Emregul E, David A, Balthasar JP, Yang VC (2005) A GPIIb/IIIa bioreactor for specific treatment of immune thrombocytopenic purpura, an autoimmune disease. Preparation, in vitro characterization, and preliminary proof-of-concept animal studies. J Biomed Mater Res A 75:648–655

89. Chang CC, Ciubotariu R, Manavalan JS, Yuan J, Colovai AI, Piazza F, Lederman S, Colonna M, Cortesini R, Dalla-Favera R, Suciu-Foca N (2002) Tolerization of dendritic cells by T(S) cells: the crucial role of inhibitory receptors ILT3 and ILT4. Nat Immunol 3:237–243

90. Bin Dhuban K, Kornete M, Mason ES, Piccirillo CA (2014) Functional dynamics of Foxp3(+) regulatory T cells in mice and humans. Immunol Rev 259:140–158

91. Assadiasl S, Ahmadpoor P, Nafar M, Lessan Pezeshki M, Pourrezagholi F, Parvin M, Shahlaee A, Sepanjnia A, Nicknam MH, Amirzargar A (2014) Regulatory T cell sub-

types and TGF-beta1 gene expression in chronic allograft dysfunction. Iran J Immunol 11:139–152

92. Negrini S, Fenoglio D, Parodi A, Kalli F, Battaglia F, Nasi G, Curto M, Tardito S, Ferrera F, Filaci G (2017) Phenotypic alterations involved in CD8+ Treg impairment in systemic sclerosis. Front Immunol 8:18

93. Velasquez-Lopera MM, Correa LA, Garcia LF (2008) Human spleen contains different subsets of dendritic cells and regulatory T lymphocytes. Clin Exp Immunol 154:107–114

94. Wang B, Jiao Z, Shao X, Lu L, Yang N, Zhou X, Xin L, Zhou Y, Chou KY (2010) Phenotypic alterations of dendritic cells are involved in suppressive activity of trichosanthin-induced CD8+CD28- regulatory T cells. J Immunol 185:79–88

95. Nikoueinejad H, Amirzargar A, Sarrafnejad A, Einollahi B, Nafar M, Ahmadpour P, Pour-Reze-Gholi F, Sehat O, Lesanpezeshki M (2014) Dynamic changes of regulatory T cell and dendritic cell subsets in stable kidney transplant patients: a prospective analysis. Iran J Kidney Dis 8:130–138

96. Pierini A, Schneidawind D, Nishikii H, Negrin RS, Regulatory T (2016) Cell immunotherapy in immune-mediated diseases. Curr Stem Cell Rep 1:177–186

97. Long SA, Thorpe J, DeBerg HA, Gersuk V, Eddy J, Harris KM, Ehlers M, Herold KC, Nepom GT, Linsley PS (2017) Partial exhaustion of CD8 T cells and clinical response to teplizumab in new-onset type 1 diabetes. Sci Immunol 1

98. Chiswick EL, Mella JR, Bernardo J, Remick DG (2015) Acute-phase deaths from murine polymicrobial Sepsis are characterized by innate immune suppression rather than exhaustion. J Immunol 195:3793–3802

99. O'Leary S, Lloyd ML, Shellam GR, Robertson SA (2008) Immunization with recombinant murine cytomegalovirus expressing murine zona pellucida 3 causes permanent infertility in BALB/c mice due to follicle depletion and ovulation failure. Biol Reprod 79:849–860

100. Cheng MH, Nelson LM (2011) Mechanisms and models of immune tolerance breakdown in the ovary. Semin Reprod Med 29:308–316

101. Zhou C, Wu J, Borillo J, Torres L, McMahon J, Lou YH (2009) Potential roles of a special CD8 alpha alpha+ cell population and CC chemokine thymus-expressed chemokine in ovulation related inflammation. J Immunol 182:596–603

102. Xu H, Wang X, Malam N, Aye PP, Alvarez X, Lackner AA, Veazey RS (2015) Persistent simian immunodeficiency virus infection drives differentiation, aberrant accumulation, and latent infection of germinal center follicular T helper cells. J Virol 90:1578–1587

103. Bruno F, Fornara C, Zelini P, Furione M, Carrara E, Scaramuzzi L, Cane I, Mele F, Sallusto F, Lilleri D, Gerna G (2016) Follicular helper T-cells and virus-specific antibody response in primary and reactivated human cytomegalovirus infections of the immunocompetent and immunocompromised transplant patients. J Gen Virol 97:1928–1941

104. Muema DM, Macharia GN, Olusola BA, Hassan AS, Fegan GW, Berkley JA, Urban BC, Nduati EW (2017) Proportions of circulating follicular helper T cells are reduced and correlate with memory B cells in HIV-infected children. PLoS One 12:e0175570

105. Kurita D, Miyoshi H, Yoshida N, Sasaki Y, Kato S, Niino D, Sugita Y, Hatta Y, Takei M, Makishima M, Ohshima K (2016) A clinicopathologic study of Lennert lymphoma and possible prognostic factors: the importance of follicular helper T-cell markers and the association with angioimmunoblastic T-cell lymphoma. Am J Surg Pathol 40:1249–1260

106. Miles B, Miller SM, Folkvord JM, Levy DN, Rakasz EG, Skinner PJ, Connick E (2016) Follicular regulatory CD8 T cells impair the germinal center response in SIV and ex vivo HIV infection. PLoS Pathog 12:e1005924

107. Dar HY, Singh A, Shukla P, Anupam R, Mondal RK, Mishra PK, Srivastava RK (2018) High dietary salt intake correlates with modulated Th17-Treg cell balance resulting in enhanced bone loss and impaired bone-microarchitecture in male mice. Sci Rep 8:2503

108. Maggi L, Santarlasci V, Capone M, Peired A, Frosali F, Crome SQ, Querci V, Fambrini M, Liotta F, Levings MK, Maggi E, Cosmi L, Romagnani S, Annunziato F (2010) CD161 is a marker of all human IL-17-producing T-cell subsets and is induced by RORC. Eur J Immunol 40:2174–2181

109. Barbi J, Pardoll D, Pan F (2014) Treg functional stability and its responsiveness to the micro-environment. Immunol Rev 259:115–139

110. Beriou G, Costantino CM, Ashley CW, Yang L, Kuchroo VK, Baecher-Allan C, Hafler DA (2009) IL-17-producing human peripheral regulatory T cells retain suppressive function. Blood 113:4240–4249

111. Du R, Zhao H, Yan F, Li H (2014) IL-17+Foxp3+ T cells: an intermediate differentiation stage between Th17 cells and regulatory T cells. J Leukoc Biol 96:39–48

112. Duarte JH, Zelenay S, Bergman ML, Martins AC, Demengeot J (2009) Natural Treg cells spontaneously differentiate into pathogenic helper cells in lymphopenic conditions. Eur J Immunol 39:948–955

113. Tran DQ, Ramsey H, Shevach EM (2007) Induction of FOXP3 expression in naive human CD4+FOXP3 T cells by T-cell receptor stimulation is transforming growth factor-beta dependent but does not confer a regulatory phenotype. Blood 110:2983–2990

114. Bhaskaran N, Cohen S, Zhang Y, Weinberg A, Pandiyan P (2015) TLR-2 signaling promotes IL-17A production in CD4+CD25+Foxp3+ regulatory cells during oropharyngeal candidiasis. Pathogens 4:90–110

115. Saini C, Siddiqui A, Ramesh V, Nath I (2016) Leprosy reactions show increased Th17 cell activity and reduced FOXP3+ Tregs with concomitant decrease in TGF-beta and increase in IL-6. PLoS Negl Trop Dis 10:e0004592

116. Joller N, Lozano E, Burkett PR, Patel B, Xiao S, Zhu C, Xia J, Tan TG, Sefik E, Yajnik V, Sharpe AH, Quintana FJ, Mathis D, Benoist C, Hafler DA, Kuchroo VK (2014) Treg cells expressing the coinhibitory molecule TIGIT selectively inhibit proinflammatory Th1 and Th17 cell responses. Immunity 40:569–581

117. Komatsu N, Okamoto K, Sawa S, Nakashima T, Oh-hora M, Kodama T, Tanaka S, Bluestone JA, Takayanagi H (2013) Pathogenic conversion of Foxp3+ T cells into TH17 cells in autoimmune arthritis. Nat Med 20:62–68

118. Zhou X, Bailey-Bucktrout S, Jeker LT, Bluestone JA (2009) Plasticity of CD4(+) FoxP3(+) T cells. Curr Opin Immunol 21:281–285

119. Voo KS, Wang YH, Santori FR, Boggiano C, Arima K, Bover L, Hanabuchi S, Khalili J, Marinova E, Zheng B, Littman DR, Liu YJ (2009) Identification of IL-17-producing FOXP3+ regulatory T cells in humans. Proc Natl Acad Sci U S A 106:4793–4798

120. Kleinewietfeld M, Hafler DA (2013) The plasticity of human Treg and Th17 cells and its role in autoimmunity. Semin Immunol 25:305–312

121. Koenen HJ, Smeets RL, Vink PM, van Rijssen E, Boots AM, Joosten I (2008) Human CD25highFoxp3pos regulatory T cells differentiate into IL-17-producing cells. Blood 112:2340–2352

122. de Waal Malefyt R, Abrams J, Bennett B, Figdor CG, de Vries JE (1991) Interleukin 10(IL-10) inhibits cytokine synthesis by human monocytes: an autoregulatory role of IL-10 produced by monocytes. J Exp Med 174:1209–1220

123. de Waal Malefyt R, Yssel H, de Vries JE (1993) Direct effects of IL-10 on subsets of human CD4+ T cell clones and resting T cells. Specific inhibition of IL-2 production and proliferation. J Immunol 150:4754–4765

124. Chaudhry A, Samstein RM, Treuting P, Liang Y, Pils MC, Heinrich JM, Jack RS, Wunderlich FT, Bruning JC, Muller W, Rudensky AY (2011) Interleukin-10 signaling in regulatory T cells is required for suppression of Th17 cell-mediated inflammation. Immunity 34:566–578

125. Murai M, Turovskaya O, Kim G, Madan R, Karp CL, Cheroutre H, Kronenberg M (2009) Interleukin 10 acts on regulatory T cells to maintain expression of the transcription factor Foxp3 and suppressive function in mice with colitis. Nat Immunol 10:1178–1184

126. Shouval DS, Biswas A, Goettel JA, McCann K, Conaway E, Redhu NS, Mascanfroni ID, Al Adham Z, Lavoie S, Ibourk M, Nguyen DD, Samsom JN, Escher JC, Somech R, Weiss B, Beier R, Conklin LS, Ebens CL, Santos FG, Ferreira AR, Sherlock M, Bhan AK, Muller W, Mora JR, Quintana FJ, Klein C, Muise AM, Horwitz BH, Snapper SB (2014) Interleukin-10

receptor signaling in innate immune cells regulates mucosal immune tolerance and anti-inflammatory macrophage function. Immunity 40:706–719

127. Cottrez F, Groux H (2001) Regulation of TGF-beta response during T cell activation is modulated by IL-10. J Immunol 167:773–778

128. Donnelly RP, Dickensheets H, Finbloom DS (1999) The interleukin-10 signal transduction pathway and regulation of gene expression in mononuclear phagocytes. J Interf Cytokine Res 19:563–573

129. Goodman WA, Young AB, McCormick TS, Cooper KD, Levine AD (2011) Stat3 phosphorylation mediates resistance of primary human T cells to regulatory T cell suppression. J Immunol 186:3336–3345

130. Taylor A, Akdis M, Joss A, Akkoc T, Wenig R, Colonna M, Daigle I, Flory E, Blaser K, Akdis CA (2007) IL-10 inhibits CD28 and ICOS costimulations of T cells via src homology 2 domain-containing protein tyrosine phosphatase 1. J Allergy Clin Immunol 120:76–83

131. Kops GJ, Medema RH, Glassford J, Essers MA, Dijkers PF, Coffer PJ, Lam EW, Burgering BM (2002) Control of cell cycle exit and entry by protein kinase B-regulated forkhead transcription factors. Mol Cell Biol 22:2025–2036

132. Ouyang W, Liao W, Luo CT, Yin N, Huse M, Kim MV, Peng M, Chan P, Ma Q, Mo Y, Meijer D, Zhao K, Rudensky AY, Atwal G, Zhang MQ, Li MO (2012) Novel Foxo1-dependent transcriptional programs control T(reg) cell function. Nature 491:554–559

133. Letterio JJ, Roberts AB (1998) Regulation of immune responses by TGF-beta. Annu Rev Immunol 16:137–161

134. Kehrl JH, Wakefield LM, Roberts AB, Jakowlew S, Alvarez-Mon M, Derynck R, Sporn MB, Fauci AS (1986) Production of transforming growth factor beta by human T lymphocytes and its potential role in the regulation of T cell growth. J Exp Med 163:1037–1050

135. Gorelik L, Fields PE, Flavell RA (2000) Cutting edge: TGF-beta inhibits Th type 2 development through inhibition of GATA-3 expression. J Immunol 165:4773–4777

136. Heath VL, Murphy EE, Crain C, Tomlinson MG, O'Garra A (2000) TGF-beta1 down-regulates Th2 development and results in decreased IL-4-induced STAT6 activation and GATA-3 expression. Eur J Immunol 30:2639–2649

137. Ludviksson BR, Seegers D, Resnick AS, Strober W (2000) The effect of TGF-beta1 on immune responses of naive versus memory CD4+ Th1/Th2 T cells. Eur J Immunol 30:2101–2111

138. Zhang X, Giangreco L, Broome HE, Dargan CM, Swain SL (1995) Control of CD4 effector fate: transforming growth factor beta 1 and interleukin 2 synergize to prevent apoptosis and promote effector expansion. J Exp Med 182:699–709

139. Swain SL, Huston G, Tonkonogy S, Weinberg A (1991) Transforming growth factor-beta and IL-4 cause helper T cell precursors to develop into distinct effector helper cells that differ in lymphokine secretion pattern and cell surface phenotype. J Immunol 147:2991–3000

140. Massague J (1998) TGF-beta signal transduction. Annu Rev Biochem 67:753–791

141. Miyazono K (2000) TGF-beta signaling by Smad proteins. Cytokine Growth Factor Rev 11:15–22

142. Roberts AB (1999) TGF-beta signaling from receptors to the nucleus. Microbes Infect 1:1265–1273

143. Groux H, Bigler M, de Vries JE, Roncarolo MG (1996) Interleukin-10 induces a long-term antigen-specific anergic state in human CD4+ T cells. J Exp Med 184:19–29

144. Zeller JC, Panoskaltsis-Mortari A, Murphy WJ, Ruscetti FW, Narula S, Roncarolo MG, Blazar BR (1999) Induction of CD4+ T cell alloantigen-specific hyporesponsiveness by IL-10 and TGF-beta. J Immunol 163:3684–3691

145. Collison LW, Workman CJ, Kuo TT, Boyd K, Wang Y, Vignali KM, Cross R, Sehy D, Blumberg RS, Vignali DA (2007) The inhibitory cytokine IL-35 contributes to regulatory T-cell function. Nature 450:566–569

146. Niedbala W, Wei XQ, Cai B, Hueber AJ, Leung BP, McInnes IB, Liew FY (2007) IL-35 is a novel cytokine with therapeutic effects against collagen-induced arthritis through the expansion of regulatory T cells and suppression of Th17 cells. Eur J Immunol 37:3021–3029

147. Devergne O, Hummel M, Koeppen H, Le Beau MM, Nathanson EC, Kieff E, Birkenbach M (1996) A novel interleukin-12 p40-related protein induced by latent Epstein-Barr virus infection in B lymphocytes. J Virol 70:1143–1153

148. Oppmann B, Lesley R, Blom B, Timans JC, Xu Y, Hunte B, Vega F, Yu N, Wang J, Singh K, Zonin F, Vaisberg E, Churakova T, Liu M, Gorman D, Wagner J, Zurawski S, Liu Y, Abrams JS, Moore KW, Rennick D, de Waal-Malefyt R, Hannum C, Bazan JF, Kastelein RA (2000) Novel p19 protein engages IL-12p40 to form a cytokine, IL-23, with biological activities similar as well as distinct from IL-12. Immunity 13:715–725

149. Yen D, Cheung J, Scheerens H, Poulet F, McClanahan T, McKenzie B, Kleinschek MA, Owyang A, Mattson J, Blumenschein W, Murphy E, Sathe M, Cua DJ, Kastelein RA, Rennick D (2006) IL-23 is essential for T cell-mediated colitis and promotes inflammation via IL-17 and IL-6. J Clin Invest 116:1310–1316

150. Pflanz S, Timans JC, Cheung J, Rosales R, Kanzler H, Gilbert J, Hibbert L, Churakova T, Travis M, Vaisberg E, Blumenschein WM, Mattson JD, Wagner JL, To W, Zurawski S, McClanahan TK, Gorman DM, Bazan JF, de Waal Malefyt R, Rennick D, Kastelein RA (2002) IL-27, a heterodimeric cytokine composed of EBI3 and p28 protein, induces proliferation of naive CD4+ T cells. Immunity 16:779–790

151. Carl JW, Bai XF (2008) IL27: its roles in the induction and inhibition of inflammation. Int J Clin Exp Pathol 1:117–123

152. Bardel E, Larousserie F, Charlot-Rabiega P, Coulomb-L'Hermine A, Devergne O (2008) Human CD4+ CD25+ Foxp3+ regulatory T cells do not constitutively express IL-35. J Immunol 181:6898–6905

153. Collison LW, Chaturvedi V, Henderson AL, Giacomin PR, Guy C, Bankoti J, Finkelstein D, Forbes K, Workman CJ, Brown SA, Rehg JE, Jones ML, Ni HT, Artis D, Turk MJ, Vignali DA (2010) IL-35-mediated induction of a potent regulatory T cell population. Nat Immunol 11:1093–1101

154. Chaturvedi V, Collison LW, Guy CS, Workman CJ, Vignali DA (2011) Cutting edge: human regulatory T cells require IL-35 to mediate suppression and infectious tolerance. J Immunol 186:6661–6666

155. Schmidt A, Oberle N, Krammer PH (2012) Molecular mechanisms of Treg-mediated T cell suppression. Front Immunol 3:51

156. Grossman WJ, Revell PA, Lu ZH, Johnson H, Bredemeyer AJ, Ley TJ (2003) The orphan granzymes of humans and mice. Curr Opin Immunol 15:544–552

157. Appay V, Zaunders JJ, Papagno L, Sutton J, Jaramillo A, Waters A, Easterbrook P, Grey P, Smith D, McMichael AJ, Cooper DA, Rowland-Jones SL, Kelleher AD (2002) Characterization of CD4(+) CTLs ex vivo. J Immunol 168:5954–5958

158. Chtanova T, Kemp RA, Sutherland AP, Ronchese F, Mackay CR (2001) Gene microarrays reveal extensive differential gene expression in both CD4(+) and CD8(+) type 1 and type 2 T cells. J Immunol 167:3057–3063

159. Smyth MJ, Trapani JA (1995) Granzymes: exogenous proteinases that induce target cell apoptosis. Immunol Today 16:202–206

160. Trapani JA, Jans DA, Jans PJ, Smyth MJ, Browne KA, Sutton VR (1998) Efficient nuclear targeting of granzyme B and the nuclear consequences of apoptosis induced by granzyme B and perforin are caspase-dependent, but cell death is caspase-independent. J Biol Chem 273:27934–27938

161. Sarin A, Williams MS, Alexander-Miller MA, Berzofsky JA, Zacharchuk CM, Henkart PA (1997) Target cell lysis by CTL granule exocytosis is independent of ICE/Ced-3 family proteases. Immunity 6:209–215

162. Nagata S, Golstein P (1995) The Fas death factor. Science 267:1449–1456

163. Van Parijs L, Abbas AK (1996) Role of Fas-mediated cell death in the regulation of immune responses. Curr Opin Immunol 8:355–361

164. Motyka B, Korbutt G, Pinkoski MJ, Heibein JA, Caputo A, Hobman M, Barry M, Shostak I, Sawchuk T, Holmes CF, Gauldie J, Bleackley RC (2000) Mannose 6-phosphate/insulin-like

growth factor II receptor is a death receptor for granzyme B during cytotoxic T cell-induced apoptosis. Cell 103:491–500

165. Beresford PJ, Xia Z, Greenberg AH, Lieberman J (1999) Granzyme A loading induces rapid cytolysis and a novel form of DNA damage independently of caspase activation. Immunity 10:585–594

166. Grossman WJ, Verbsky JW, Barchet W, Colonna M, Atkinson JP, Ley TJ (2004) Human T regulatory cells can use the perforin pathway to cause autologous target cell death. Immunity 21:589–601

167. Bird CH, Sutton VR, Sun J, Hirst CE, Novak A, Kumar S, Trapani JA, Bird PI (1998) Selective regulation of apoptosis: the cytotoxic lymphocyte serpin proteinase inhibitor 9 protects against granzyme B-mediated apoptosis without perturbing the Fas cell death pathway. Mol Cell Biol 18:6387–6398

168. Chambers CA, Kuhns MS, Egen JG, Allison JP (2001) CTLA-4-mediated inhibition in regulation of T cell responses: mechanisms and manipulation in tumor immunotherapy. Annu Rev Immunol 19:565–594

169. Collins AV, Brodie DW, Gilbert RJ, Iaboni A, Manso-Sancho R, Walse B, Stuart DI, van der Merwe PA, Davis SJ (2002) The interaction properties of costimulatory molecules revisited. Immunity 17:201–210

170. Egen JG, Kuhns MS, Allison JP (2002) CTLA-4: new insights into its biological function and use in tumor immunotherapy. Nat Immunol 3:611–618

171. Parry RV, Chemnitz JM, Frauwirth KA, Lanfranco AR, Braunstein I, Kobayashi SV, Linsley PS, Thompson CB, Riley JL (2005) CTLA-4 and PD-1 receptors inhibit T-cell activation by distinct mechanisms. Mol Cell Biol 25:9543–9553

172. Krummel MF, Allison JP (1996) CTLA-4 engagement inhibits IL-2 accumulation and cell cycle progression upon activation of resting T cells. J Exp Med 183:2533–2540

173. Linsley PS, Bradshaw J, Greene J, Peach R, Bennett KL, Mittler RS (1996) Intracellular trafficking of CTLA-4 and focal localization towards sites of TCR engagement. Immunity 4:535–543

174. Tang F, Du X, Liu M, Zheng P, Liu Y (2018) Anti-CTLA-4 antibodies in cancer immunotherapy: selective depletion of intratumoral regulatory T cells or checkpoint blockade? Cell Biosci 8:30

175. Read S, Greenwald R, Izcue A, Robinson N, Mandelbrot D, Francisco L, Sharpe AH, Powrie F (2006) Blockade of CTLA-4 on CD4+CD25+ regulatory T cells abrogates their function in vivo. J Immunol 177:4376–4383

176. Piccirillo CA, Shevach EM (2004) Naturally-occurring CD4+CD25+ immunoregulatory T cells: central players in the arena of peripheral tolerance. Semin Immunol 16:81–88

177. Takahashi T, Tagami T, Yamazaki S, Uede T, Shimizu J, Sakaguchi N, Mak TW, Sakaguchi S (2000) Immunologic self-tolerance maintained by CD25(+)CD4(+) regulatory T cells constitutively expressing cytotoxic T lymphocyte-associated antigen 4. J Exp Med 192:303–310

178. Wing K, Onishi Y, Prieto-Martin P, Yamaguchi T, Miyara M, Fehervari Z, Nomura T, Sakaguchi S (2008) CTLA-4 control over Foxp3+ regulatory T cell function. Science 322:271–275

179. Tivol EA, Borriello F, Schweitzer AN, Lynch WP, Bluestone JA, Sharpe AH (1995) Loss of CTLA-4 leads to massive lymphoproliferation and fatal multiorgan tissue destruction, revealing a critical negative regulatory role of CTLA-4. Immunity 3:541–547

180. Waterhouse P, Penninger JM, Timms E, Wakeham A, Shahinian A, Lee KP, Thompson CB, Griesser H, Mak TW (1995) Lymphoproliferative disorders with early lethality in mice deficient in Ctla-4. Science 270:985–988

181. Karandikar NJ, Vanderlugt CL, Walunas TL, Miller SD, Bluestone JA (1996) CTLA-4: a negative regulator of autoimmune disease. J Exp Med 184:783–788

182. Luhder F, Hoglund P, Allison JP, Benoist C, Mathis D (1998) Cytotoxic T lymphocyte-associated antigen 4 (CTLA-4) regulates the unfolding of autoimmune diabetes. J Exp Med 187:427–432

183. Luhder F, Chambers C, Allison JP, Benoist C, Mathis D (2000) Pinpointing when T cell costimulatory receptor CTLA-4 must be engaged to dampen diabetogenic T cells. Proc Natl Acad Sci U S A 97:12204–12209

184. Nistico L, Buzzetti R, Pritchard LE, Van der Auwera B, Giovannini C, Bosi E, Larrad MT, Rios MS, Chow CC, Cockram CS, Jacobs K, Mijovic C, Bain SC, Barnett AH, Vandewalle CL, Schuit F, Gorus FK, Tosi R, Pozzilli P, Todd JA (1996) The CTLA-4 gene region of chromosome 2q33 is linked to, and associated with, type 1 diabetes. Belgian Diabetes Registry. Hum Mol Genet 5:1075–1080

185. Ueda H, Howson JM, Esposito L, Heward J, Snook H, Chamberlain G, Rainbow DB, Hunter KM, Smith AN, Di Genova G, Herr MH, Dahlman I, Payne F, Smyth D, Lowe C, Twells RC, Howlett S, Healy B, Nutland S, Rance HE, Everett V, Smink LJ, Lam AC, Cordell HJ, Walker NM, Bordin C, Hulme J, Motzo C, Cucca F, Hess JF, Metzker ML, Rogers J, Gregory S, Allahabadia A, Nithiyananthan R, Tuomilehto-Wolf E, Tuomilehto J, Bingley P, Gillespie KM, Undlien DE, Ronningen KS, Guja C, Ionescu-Tirgoviste C, Savage DA, Maxwell AP, Carson DJ, Patterson CC, Franklyn JA, Clayton DG, Peterson LB, Wicker LS, Todd JA, Gough SC (2003) Association of the T-cell regulatory gene CTLA4 with susceptibility to autoimmune disease. Nature 423:506–511

186. Gough SC, Walker LS, Sansom DM (2005) CTLA4 gene polymorphism and autoimmunity. Immunol Rev 204:102–115

187. Bachmann MF, Kohler G, Ecabert B, Mak TW, Kopf M (1999) Cutting edge: lymphoproliferative disease in the absence of CTLA-4 is not T cell autonomous. J Immunol 163:1128–1131

188. Cardona P, Cardona PJ (2019) Regulatory T cells in Mycobacterium tuberculosis infection. Front Immunol 10:2139

189. Saini C, Tarique M, Kumar S, Rao DN (2015) Role of FoxP3+ Tregs cells mediating immune suppression in leprosy. Curr Immunol Rev 11:66–72

190. Hansen DS, Schofield L (2010) Natural regulatory T cells in malaria: host or parasite allies? PLoS Pathog 6:e1000771

191. Kleinman AJ, Sivanandham R, Pandrea I, Chougnet CA, Apetrei C (2018) Regulatory T cells as potential targets for HIV cure research. Front Immunol 9:734

192. Okeke EB, Uzonna JE (2019) The pivotal role of regulatory T cells in the regulation of innate immune cells. Front Immunol 10:680

193. Kumar S, Naqvi RA, Khanna N, Pathak P, Rao DN (2011) Th3 immune responses in the progression of leprosy via molecular cross-talks of TGF-beta, CTLA-4 and Cbl-b. Clin Immunol 141:133–142

194. Saini C, Ramesh V, Nath I (2014) Increase in TGF-beta secreting CD4(+)CD25(+) FOXP3(+) T regulatory cells in anergic lepromatous leprosy patients. PLoS Negl Trop Dis 8:e2639

195. Kumar S, Naqvi RA, Bhat AAR, Ali R, Agnihotri A, Khanna N, Rao DN (2013) IL-10 production from dendritic cells is associated with DC SIGN in human leprosy. Immunobiology 218:1488–1496

196. Palermo ML, Pagliari C, Trindade MA, Yamashitafuji TM, Duarte AJ, Cacere CR, Benard G (2012) Increased expression of regulatory T cells and down-regulatory molecules in lepromatous leprosy. Am J Trop Med Hyg 86:878–883

197. Kumar S, Naqvi RA, Ali R, Rani R, Khanna N, Rao DN (2013) CD4+CD25+ T regs with acetylated FoxP3 are associated with immune suppression in human leprosy. Mol Immunol 56:513–520

198. Tarique M, Naqvi RA, Santosh KV, Kamal VK, Khanna N, Rao DN (2015) Association of TNF-alpha-(308(GG)), IL-10(-819(TT)), IL-10(-1082(GG)) and IL-1R1(+1970(CC)) genotypes with the susceptibility and progression of leprosy in North Indian population. Cytokine 73:61–65

199. Bobosha K, Wilson L, van Meijgaarden KE, Bekele Y, Zewdie M, van der Ploeg-van Schip JJ, Abebe M, Hussein J, Khadge S, Neupane KD, Hagge DA, Jordanova ES, Aseffa A, Ottenhoff TH, Geluk A (2014) T-cell regulation in lepromatous leprosy. PLoS Negl Trop Dis 8:e2773

200. Fernandes C, Goncalves HS, Cabral PB, Pinto HC, Pinto MI, Camara LM (2013) Increased frequency of CD4 and CD8 regulatory T cells in individuals under 15 years with multibacillary leprosy. PLoS One 8:e79072
201. Kumar S, Naqvi RA, Ali R, Rani R, Khanna N, Rao DN (2013) FoxP3 provides competitive fitness to CD4(+) CD25(+) T cells in leprosy patients via transcriptional regulation. Eur J Immunol 44:431–439
202. Tarique M, Saini C, Naqvi RA, Khanna N, Rao DN (2017) Increased IL-35 producing Tregs and CD19+IL-35+ cells are associated with disease progression in leprosy patients. Cytokine 91:82–88
203. Tarique M, Naz H, Kurra SV, Saini C, Naqvi RA, Rai R, Suhail M, Khanna N, Rao DN, Sharma A (2018) Interleukin-10 producing regulatory B cells transformed CD4(+) CD25(−) into Tregs and enhanced regulatory T cells function in human leprosy. Front Immunol 9:1636
204. Tarique M, Naqvi RA, Ali R, Khanna N, Rao DN (2017) CD4+ TCRgammadelta+ FoxP3+ cells: an unidentified population of immunosuppressive cells towards disease progression leprosy patients. Exp Dermatol
205. Saini C, Tarique M, Ramesh V, Khanna N, Sharma A (2018) Gammadelta T cells are associated with inflammation and immunopathogenesis of leprosy reactions. Immunol Lett 200:55–65
206. Singh I, Yadav AR, Mohanty KK, Katoch K, Bisht D, Sharma P, Sharma B, Gupta UD, Sengupta U (2012) Molecular mimicry between HSP 65 of Mycobacterium leprae and cytokeratin 10 of the host keratin; role in pathogenesis of leprosy. Cell Immunol 278:63–75
207. Singh I, Yadav AR, Mohanty KK, Katoch K, Sharma P, Mishra B, Bisht D, Gupta UD, Sengupta U (2015) Molecular mimicry between Mycobacterium leprae proteins (50S ribosomal protein L2 and Lysyl-tRNA synthetase) and myelin basic protein: a possible mechanism of nerve damage in leprosy. Microbes Infect 17:247–257
208. Singh I, Yadav AR, Mohanty KK, Katoch K, Sharma P, Pathak VK, Bisht D, Gupta UD, Sengupta U (2018) Autoimmunity to tropomyosin-specific peptides induced by Mycobacterium leprae in leprosy patients: identification of mimicking proteins. Front Immunol 9:642
209. Kursar M, Koch M, Mittrucker HW, Nouailles G, Bonhagen K, Kamradt T, Kaufmann SH (2007) Cutting edge: regulatory T cells prevent efficient clearance of Mycobacterium tuberculosis. J Immunol 178:2661–2665
210. Scott-Browne JP, Shafiani S, Tucker-Heard G, Ishida-Tsubota K, Fontenot JD, Rudensky AY, Bevan MJ, Urdahl KB (2007) Expansion and function of Foxp3-expressing T regulatory cells during tuberculosis. J Exp Med 204:2159–2169
211. Guyot-Revol V, Innes JA, Hackforth S, Hinks T, Lalvani A (2006) Regulatory T cells are expanded in blood and disease sites in patients with tuberculosis. Am J Respir Crit Care Med 173:803–810
212. Ribeiro-Rodrigues R, Resende Co T, Rojas R, Toossi Z, Dietze R, Boom WH, Maciel E, Hirsch CS (2006) A role for CD4+CD25+ T cells in regulation of the immune response during human tuberculosis. Clin Exp Immunol 144:25–34
213. Shafiani S, Tucker-Heard G, Kariyone A, Takatsu K, Urdahl KB (2010) Pathogen-specific regulatory T cells delay the arrival of effector T cells in the lung during early tuberculosis. J Exp Med 207:1409–1420
214. Saini C, Kumar P, Tarique M, Sharma A, Ramesh V (2018) Regulatory T cells antagonize proinflammatory response of IL-17 during cutaneous tuberculosis. J Inflamm Res 11:377–388
215. Belkaid Y, Piccirillo CA, Mendez S, Shevach EM, Sacks DL (2002) CD4+CD25+ regulatory T cells control Leishmania major persistence and immunity. Nature 420:502–507
216. Suffia IJ, Reckling SK, Piccirillo CA, Goldszmid RS, Belkaid Y (2006) Infected site-restricted Foxp3+ natural regulatory T cells are specific for microbial antigens. J Exp Med 203:777–788
217. Katara GK, Ansari NA, Verma S, Ramesh V, Salotra P (2011) Foxp3 and IL-10 expression correlates with parasite burden in lesional tissues of post kala azar dermal leishmaniasis (PKDL) patients. PLoS Negl Trop Dis 5:e1171

218. Katara GK, Raj A, Kumar R, Avishek K, Kaushal H, Ansari NA, Bumb RA, Salotra P (2013) Analysis of localized immune responses reveals presence of Th17 and Treg cells in cutaneous leishmaniasis due to Leishmania tropica. BMC Immunol 14:52

219. Mendez S, Reckling SK, Piccirillo CA, Sacks D, Belkaid Y (2004) Role for CD4(+) CD25(+) regulatory T cells in reactivation of persistent leishmaniasis and control of concomitant immunity. J Exp Med 200:201–210

220. Nylen S, Maurya R, Eidsmo L, Manandhar KD, Sundar S, Sacks D (2007) Splenic accumulation of IL-10 mRNA in T cells distinct from CD4+CD25+ (Foxp3) regulatory T cells in human visceral leishmaniasis. J Exp Med 204:805–817

221. Kemp K, Kemp M, Kharazmi A, Ismail A, Kurtzhals JA, Hviid L, Theander TG (1999) Leishmania-specific T cells expressing interferon-gamma (IFN-gamma) and IL-10 upon activation are expanded in individuals cured of visceral leishmaniasis. Clin Exp Immunol 116:500–504

222. Karp CL, el-Safi SH, Wynn TA, Satti MM, Kordofani AM, Hashim FA, Hag-Ali M, Neva FA, Nutman TB, Sacks DL (1993) In vivo cytokine profiles in patients with kala-azar. Marked elevation of both interleukin-10 and interferon-gamma. J Clin Invest 91:1644–1648

223. Ghalib HW, Piuvezam MR, Skeiky YA, Siddig M, Hashim FA, el-Hassan AM, Russo DM, Reed SG (1993) Interleukin 10 production correlates with pathology in human Leishmania donovani infections. J Clin Invest 92:324–329

224. Maizels RM, Smith KA (2011) Regulatory T cells in infection. Adv Immunol 112:73–136

225. Veiga-Parga T, Sehrawat S, Rouse BT (2013) Role of regulatory T cells during virus infection. Immunol Rev 255:182–196

226. Manangeeswaran M, Jacques J, Tami C, Konduru K, Amharref N, Perrella O, Casasnovas JM, Umetsu DT, Dekruyff RH, Freeman GJ, Perrella A, Kaplan GG (2012) Binding of hepatitis A virus to its cellular receptor 1 inhibits T-regulatory cell functions in humans. Gastroenterology 142:1516–1525. e3

227. Luhn K, Simmons CP, Moran E, Dung NT, Chau TN, Quyen NT, Thao le TT, Van Ngoc T, Dung NM, Wills B, Farrar J, McMichael AJ, Dong T, Rowland-Jones S (2007) Increased frequencies of CD4+ CD25(high) regulatory T cells in acute dengue infection. J Exp Med 204:979–985

228. Lund JM, Hsing L, Pham TT, Rudensky AY (2008) Coordination of early protective immunity to viral infection by regulatory T cells. Science 320:1220–1224

229. Kassiotis G, O'Garra A (2008) Immunology. Immunity benefits from a little suppression. Science 320:1168–1169

230. Moreno-Fernandez ME, Rueda CM, Rusie LK, Chougnet CA (2011) Regulatory T cells control HIV replication in activated T cells through a cAMP-dependent mechanism. Blood 117:5372–5380

231. Haase AT (2005) Perils at mucosal front lines for HIV and SIV and their hosts. Nat Rev Immunol 5:783–792

232. Graham JB, Da Costa A, Lund JM (2014) Regulatory T cells shape the resident memory T cell response to virus infection in the tissues. J Immunol 192:683–690

233. Aalaei-Andabili SH, Alavian SM (2012) Regulatory T cells are the most important determinant factor of hepatitis B infection prognosis: a systematic review and meta-analysis. Vaccine 30:5595–5602

234. Losikoff PT, Self AA, Gregory SH (2012) Dendritic cells, regulatory T cells and the pathogenesis of chronic hepatitis C. Virulence 3:610–620

235. Self AA, Losikoff PT, Gregory SH (2013) Divergent contributions of regulatory T cells to the pathogenesis of chronic hepatitis C. Hum Vaccin Immunother 9:1569–1576

236. Riezu-Boj JI, Larrea E, Aldabe R, Guembe L, Casares N, Galeano E, Echeverria I, Sarobe P, Herrero I, Sangro B, Prieto J, Lasarte JJ (2011) Hepatitis C virus induces the expression of CCL17 and CCL22 chemokines that attract regulatory T cells to the site of infection. J Hepatol 54:422–431

237. Chevalier MF, Weiss L (2013) The split personality of regulatory T cells in HIV infection. Blood 121:29–37

238. Moreno-Fernandez ME, Joedicke JJ, Chougnet CA (2014) Regulatory T cells diminish HIV infection in dendritic cells – conventional CD4(+) T cell clusters. Front Immunol 5:199

239. Nikolova M, Carriere M, Jenabian MA, Limou S, Younas M, Kok A, Hue S, Seddiki N, Hulin A, Delaneau O, Schuitemaker H, Herbeck JT, Mullins JI, Muhtarova M, Bensussan A, Zagury JF, Lelievre JD, Levy Y (2011) CD39/adenosine pathway is involved in AIDS progression. PLoS Pathog 7:e1002110

240. Schulze Zur Wiesch J, Thomssen A, Hartjen P, Toth I, Lehmann C, Meyer-Olson D, Colberg K, Frerk S, Babikir D, Schmiedel S, Degen O, Mauss S, Rockstroh J, Staszewski S, Khaykin P, Strasak A, Lohse AW, Fatkenheuer G, Hauber J, van Lunzen J (2011) Comprehensive analysis of frequency and phenotype of T regulatory cells in HIV infection: CD39 expression of FoxP3+ T regulatory cells correlates with progressive disease. J Virol 85:1287–1297

241. Cooper GS, Bynum ML, Somers EC (2009) Recent insights in the epidemiology of autoimmune diseases: improved prevalence estimates and understanding of clustering of diseases. J Autoimmun 33:197–207

242. Goodnow CC, Sprent J, Fazekas de St Groth B, Vinuesa CG (2005) Cellular and genetic mechanisms of self tolerance and autoimmunity. Nature 435:590–597

243. Gough SC, Simmonds MJ (2007) The HLA region and autoimmune disease: associations and mechanisms of action. Curr Genomics 8:453–465

244. Ohashi PS (2002) T-cell signalling and autoimmunity: molecular mechanisms of disease. Nat Rev Immunol 2:427–438

245. Torres-Aguilar H, Blank M, Jara LJ, Shoenfeld Y (2010) Tolerogenic dendritic cells in autoimmune diseases: crucial players in induction and prevention of autoimmunity. Autoimmun Rev 10:8–17

246. Torres-Aguilar H, Sanchez-Torres C, Jara LJ, Blank M, Shoenfeld Y (2010) IL-10/TGF-beta-treated dendritic cells, pulsed with insulin, specifically reduce the response to insulin of CD4+ effector/memory T cells from type 1 diabetic individuals. J Clin Immunol 30:659–668

247. Riella LV, Paterson AM, Sharpe AH, Chandraker A (2012) Role of the PD-1 pathway in the immune response. Am J Transplant 12:2575–2587

248. Wolfle SJ, Strebovsky J, Bartz H, Sahr A, Arnold C, Kaiser C, Dalpke AH, Heeg K (2011) PD-L1 expression on tolerogenic APCs is controlled by STAT-3. Eur J Immunol 41:413–424

249. Chikuma S, Terawaki S, Hayashi T, Nabeshima R, Yoshida T, Shibayama S, Okazaki T, Honjo T (2009) PD-1-mediated suppression of IL-2 production induces CD8+ T cell anergy in vivo. J Immunol 182:6682–6689

250. Pandiyan P, Zheng L, Ishihara S, Reed J, Lenardo MJ (2007) CD4+CD25+Foxp3+ regulatory T cells induce cytokine deprivation-mediated apoptosis of effector CD4+ T cells. Nat Immunol 8:1353–1362

251. Yanaba K, Bouaziz JD, Matsushita T, Magro CM, St Clair EW, Tedder TF (2008) B-lymphocyte contributions to human autoimmune disease. Immunol Rev 223:284–299

252. Yan W, Nguyen T, Yuki N, Ji Q, Yiannikas C, Pollard JD, Mathey EK (2014) Antibodies to neurofascin exacerbate adoptive transfer experimental autoimmune neuritis. J Neuroimmunol 277:13–17

253. Saadoun S, Waters P, Bell BA, Vincent A, Verkman AS, Papadopoulos MC (2010) Intracerebral injection of neuromyelitis optica immunoglobulin G and human complement produces neuromyelitis optica lesions in mice. Brain 133:349–361

254. Lim HW, Hillsamer P, Banham AH, Kim CH (2005) Cutting edge: direct suppression of B cells by CD4+ CD25+ regulatory T cells. J Immunol 175:4180–4183

255. Liu Y, Liu A, Iikuni N, Xu H, Shi FD, La Cava A (2014) Regulatory CD4+ T cells promote B cell anergy in murine lupus. J Immunol 192:4069–4073

256. van der Vliet HJ, Nieuwenhuis EE (2007) IPEX as a result of mutations in FOXP3. Clin Dev Immunol 2007:89017

257. Steinman L, Merrill JT, McInnes IB, Peakman M (2012) Optimization of current and future therapy for autoimmune diseases. Nat Med 18:59–65

258. von Boehmer H, Daniel C (2013) Therapeutic opportunities for manipulating T(Reg) cells in autoimmunity and cancer. Nat Rev Drug Discov 12:51–63

259. Akdis M, Verhagen J, Taylor A, Karamloo F, Karagiannidis C, Crameri R, Thunberg S, Deniz G, Valenta R, Fiebig H, Kegel C, Disch R, Schmidt-Weber CB, Blaser K, Akdis CA (2004) Immune responses in healthy and allergic individuals are characterized by a fine balance between allergen-specific T regulatory 1 and T helper 2 cells. J Exp Med 199:1567–1575
260. Akdis CA, Blesken T, Akdis M, Wuthrich B, Blaser K (1998) Role of interleukin 10 in specific immunotherapy. J Clin Invest 102:98–106
261. Jutel M, Akdis M, Budak F, Aebischer-Casaulta C, Wrzyszcz M, Blaser K, Akdis CA (2003) IL-10 and TGF-beta cooperate in the regulatory T cell response to mucosal allergens in normal immunity and specific immunotherapy. Eur J Immunol 33:1205–1214
262. Gri G, Piconese S, Frossi B, Manfroi V, Merluzzi S, Tripodo C, Viola A, Odom S, Rivera J, Colombo MP, Pucillo CE (2008) CD4+CD25+ regulatory T cells suppress mast cell degranulation and allergic responses through OX40-OX40L interaction. Immunity 29:771–781
263. Ring S, Schafer SC, Mahnke K, Lehr HA, Enk AH (2006) CD4+ CD25+ regulatory T cells suppress contact hypersensitivity reactions by blocking influx of effector T cells into inflamed tissue. Eur J Immunol 36:2981–2992
264. Trautmann A, Schmid-Grendelmeier P, Kruger K, Crameri R, Akdis M, Akkaya A, Brocker EB, Blaser K, Akdis CA (2002) T cells and eosinophils cooperate in the induction of bronchial epithelial cell apoptosis in asthma. J Allergy Clin Immunol 109:329–337
265. Meiler F, Klunker S, Zimmermann M, Akdis CA, Akdis M (2008) Distinct regulation of IgE, IgG4 and IgA by T regulatory cells and toll-like receptors. Allergy 63:1455–1463
266. Meiler F, Zumkehr J, Klunker S, Ruckert B, Akdis CA, Akdis M (2008) In vivo switch to IL-10-secreting T regulatory cells in high dose allergen exposure. J Exp Med 205:2887–2898
267. Platts-Mills T, Vaughan J, Squillace S, Woodfolk J, Sporik R (2001) Sensitisation, asthma, and a modified Th2 response in children exposed to cat allergen: a population-based cross-sectional study. Lancet 357:752–756
268. Verhasselt V, Milcent V, Cazareth J, Kanda A, Fleury S, Dombrowicz D, Glaichenhaus N, Julia V (2008) Breast milk-mediated transfer of an antigen induces tolerance and protection from allergic asthma. Nat Med 14:170–175
269. Karlsson MR, Rugtveit J, Brandtzaeg P (2004) Allergen-responsive CD4+CD25+ regulatory T cells in children who have outgrown cow's milk allergy. J Exp Med 199:1679–1688
270. Zou W (2006) Regulatory T cells, tumour immunity and immunotherapy. Nat Rev Immunol 6:295–307
271. Beyer M, Schultze JL (2006) Regulatory T cells in cancer. Blood 108:804–811
272. Curiel TJ (2007) Tregs and rethinking cancer immunotherapy. J Clin Invest 117:1167–1174
273. Ormandy LA, Hillemann T, Wedemeyer H, Manns MP, Greten TF, Korangy F (2005) Increased populations of regulatory T cells in peripheral blood of patients with hepatocellular carcinoma. Cancer Res 65:2457–2464
274. Woo EY, Chu CS, Goletz TJ, Schlienger K, Yeh H, Coukos G, Rubin SC, Kaiser LR, June CH (2001) Regulatory CD4(+)CD25(+) T cells in tumors from patients with early-stage non-small cell lung cancer and late-stage ovarian cancer. Cancer Res 61:4766–4772
275. Woo EY, Yeh H, Chu CS, Schlienger K, Carroll RG, Riley JL, Kaiser LR, June CH (2002) Cutting edge: regulatory T cells from lung cancer patients directly inhibit autologous T cell proliferation. J Immunol 168:4272–4276
276. Orentas RJ, Kohler ME, Johnson BD (2006) Suppression of anti-cancer immunity by regulatory T cells: back to the future. Semin Cancer Biol 16:137–149
277. Danese S, Rutella S (2007) The Janus face of CD4+CD25+ regulatory T cells in cancer and autoimmunity. Curr Med Chem 14:649–666
278. Barnett BG, Ruter J, Kryczek I, Brumlik MJ, Cheng PJ, Daniel BJ, Coukos G, Zou W, Curiel TJ (2008) Regulatory T cells: a new frontier in cancer immunotherapy. Adv Exp Med Biol 622:255–260
279. Lutsiak ME, Tagaya Y, Adams AJ, Schlom J, Sabzevari H (2008) Tumor-induced impairment of TCR signaling results in compromised functionality of tumor-infiltrating regulatory T cells. J Immunol 180:5871–5881

280. Sasada T, Kimura M, Yoshida Y, Kanai M, Takabayashi A (2003) CD4+CD25+ regulatory T cells in patients with gastrointestinal malignancies: possible involvement of regulatory T cells in disease progression. Cancer 98:1089–1099

281. Kawaida H, Kono K, Takahashi A, Sugai H, Mimura K, Miyagawa N, Omata H, Ooi A, Fujii H (2005) Distribution of CD4+CD25high regulatory T-cells in tumor-draining lymph nodes in patients with gastric cancer. J Surg Res 124:151–157

282. Kono K, Kawaida H, Takahashi A, Sugai H, Mimura K, Miyagawa N, Omata H, Fujii H (2006) CD4(+)CD25high regulatory T cells increase with tumor stage in patients with gastric and esophageal cancers. Cancer Immunol Immunother 55:1064–1071

283. Ichihara F, Kono K, Takahashi A, Kawaida H, Sugai H, Fujii H (2003) Increased populations of regulatory T cells in peripheral blood and tumor-infiltrating lymphocytes in patients with gastric and esophageal cancers. Clin Cancer Res 9:4404–4408

284. Tokuno K, Hazama S, Yoshino S, Yoshida S, Oka M (2009) Increased prevalence of regulatory T-cells in the peripheral blood of patients with gastrointestinal cancer. Anticancer Res 29:1527–1532

285. Shen LS, Wang J, Shen DF, Yuan XL, Dong P, Li MX, Xue J, Zhang FM, Ge HL, Xu D (2009) CD4(+)CD25(+)CD127(low/−) regulatory T cells express Foxp3 and suppress effector T cell proliferation and contribute to gastric cancers progression. Clin Immunol 131:109–118

286. Xu H, Mao Y, Dai Y, Wang Q, Zhang X (2009) CD4CD25+ regulatory T cells in patients with advanced gastrointestinal cancer treated with chemotherapy. Onkologie 32:246–252

287. Mizukami Y, Kono K, Kawaguchi Y, Akaike H, Kamimura K, Sugai H, Fujii H (2008) CCL17 and CCL22 chemokines within tumor microenvironment are related to accumulation of Foxp3+ regulatory T cells in gastric cancer. Int J Cancer 122:2286–2293

288. Gobert M, Treilleux I, Bendriss-Vermare N, Bachelot T, Goddard-Leon S, Arfi V, Biota C, Doffin AC, Durand I, Olive D, Perez S, Pasqual N, Faure C, Ray-Coquard I, Puisieux A, Caux C, Blay JY, Menetrier-Caux C (2009) Regulatory T cells recruited through CCL22/CCR4 are selectively activated in lymphoid infiltrates surrounding primary breast tumors and lead to an adverse clinical outcome. Cancer Res 69:2000–2009

289. Qin XJ, Shi HZ, Deng JM, Liang QL, Jiang J, Ye ZJ (2009) CCL22 recruits CD4-positive CD25-positive regulatory T cells into malignant pleural effusion. Clin Cancer Res 15:2231–2237

290. Wieczorek G, Asemissen A, Model F, Turbachova I, Floess S, Liebenberg V, Baron U, Stauch D, Kotsch K, Pratschke J, Hamann A, Loddenkemper C, Stein H, Volk HD, Hoffmuller U, Grutzkau A, Mustea A, Huehn J, Scheibenbogen C, Olek S (2009) Quantitative DNA methylation analysis of FOXP3 as a new method for counting regulatory T cells in peripheral blood and solid tissue. Cancer Res 69:599–608

291. Schaefer C, Kim GG, Albers A, Hoermann K, Myers EN, Whiteside TL (2005) Characteristics of CD4+CD25+ regulatory T cells in the peripheral circulation of patients with head and neck cancer. Br J Cancer 92:913–920

292. Strauss L, Bergmann C, Gooding W, Johnson JT, Whiteside TL (2007) The frequency and suppressor function of CD4+CD25highFoxp3+ T cells in the circulation of patients with squamous cell carcinoma of the head and neck. Clin Cancer Res 13:6301–6311

293. Gao Q, Qiu SJ, Fan J, Zhou J, Wang XY, Xiao YS, Xu Y, Li YW, Tang ZY (2007) Intratumoral balance of regulatory and cytotoxic T cells is associated with prognosis of hepatocellular carcinoma after resection. J Clin Oncol 25:2586–2593

294. Kobayashi N, Hiraoka N, Yamagami W, Ojima H, Kanai Y, Kosuge T, Nakajima A, Hirohashi S (2007) FOXP3+ regulatory T cells affect the development and progression of hepatocarcinogenesis. Clin Cancer Res 13:902–911

295. Miller AM, Lundberg K, Ozenci V, Banham AH, Hellstrom M, Egevad L, Pisa P (2006) CD4+CD25high T cells are enriched in the tumor and peripheral blood of prostate cancer patients. J Immunol 177:7398–7405

296. Carreras J, Lopez-Guillermo A, Fox BC, Colomo L, Martinez A, Roncador G, Montserrat E, Campo E, Banham AH (2006) High numbers of tumor-infiltrating FOXP3-positive regula-

tory T cells are associated with improved overall survival in follicular lymphoma. Blood 108:2957–2964

297. Rozkova D, Tiserova H, Fucikova J, Last'ovicka J, Podrazil M, Ulcova H, Budinsky V, Prausova J, Linke Z, Minarik I, Sediva A, Spisek R, Bartunkova J (2009) FOCUS on FOCIS: combined chemo-immunotherapy for the treatment of hormone-refractory metastatic prostate cancer. Clin Immunol 131:1–10

298. Ahmadzadeh M, Rosenberg SA (2006) IL-2 administration increases CD4+ CD25(hi) Foxp3+ regulatory T cells in cancer patients. Blood 107:2409–2414

299. Jensen HK, Donskov F, Nordsmark M, Marcussen N, von der Maase H (2009) Increased intratumoral FOXP3-positive regulatory immune cells during interleukin-2 treatment in metastatic renal cell carcinoma. Clin Cancer Res 15:1052–1058

300. Li X, Ye DF, Xie X, Chen HZ, Lu WG (2005) Proportion of CD4+CD25+ regulatory T cell is increased in the patients with ovarian carcinoma. Cancer Investig 23:399–403

301. Curiel TJ (2008) Regulatory T cells and treatment of cancer. Curr Opin Immunol 20:241–246

302. Wolf AM, Wolf D, Steurer M, Gastl G, Gunsilius E, Grubeck-Loebenstein B (2003) Increase of regulatory T cells in the peripheral blood of cancer patients. Clin Cancer Res 9:606–612

303. Liyanage UK, Moore TT, Joo HG, Tanaka Y, Herrmann V, Doherty G, Drebin JA, Strasberg SM, Eberlein TJ, Goedegebuure PS, Linehan DC (2002) Prevalence of regulatory T cells is increased in peripheral blood and tumor microenvironment of patients with pancreas or breast adenocarcinoma. J Immunol 169:2756–2761

304. Wang X, Zheng J, Liu J, Yao J, He Y, Li X, Yu J, Yang J, Liu Z, Huang S (2005) Increased population of CD4(+)CD25(high), regulatory T cells with their higher apoptotic and proliferating status in peripheral blood of acute myeloid leukemia patients. Eur J Haematol 75:468–476

305. Szczepanski MJ, Szajnik M, Czystowska M, Mandapathil M, Strauss L, Welsh A, Foon KA, Whiteside TL, Boyiadzis M (2009) Increased frequency and suppression by regulatory T cells in patients with acute myelogenous leukemia. Clin Cancer Res 15:3325–3332

306. Yang ZZ, Novak AJ, Ziesmer SC, Witzig TE, Ansell SM (2006) Attenuation of CD8(+) T-cell function by CD4(+)CD25(+) regulatory T cells in B-cell non-Hodgkin's lymphoma. Cancer Res 66:10145–10152

307. Jones D, Glimcher LH, Aliprantis AO (2011) Osteoimmunology at the nexus of arthritis, osteoporosis, cancer, and infection. J Clin Invest 121:2534–2542

308. Tai P, Wang J, Jin H, Song X, Yan J, Kang Y, Zhao L, An X, Du X, Chen X, Wang S, Xia G, Wang B (2008) Induction of regulatory T cells by physiological level estrogen. J Cell Physiol 214:456–464

309. Zaiss MM, Axmann R, Zwerina J, Polzer K, Guckel E, Skapenko A, Schulze-Koops H, Horwood N, Cope A, Schett G (2007) Treg cells suppress osteoclast formation: a new link between the immune system and bone. Arthritis Rheum 56:4104–4112

310. Luo CY, Wang L, Sun C, Li DJ (2011) Estrogen enhances the functions of CD4(+)CD25(+) Foxp3(+) regulatory T cells that suppress osteoclast differentiation and bone resorption in vitro. Cell Mol Immunol 8:50–58

311. Srivastava RK, Dar HY, Mishra PK (2018) Immunoporosis: Immunology of Osteoporosis-Role of T Cells. Front Immunol 9:657

312. Okamoto K, Nakashima T, Shinohara M, Negishi-Koga T, Komatsu N, Terashima A, Sawa S, Nitta T, Takayanagi H (2017) Osteoimmunology: the Conceptual Framework Unifying the Immune and Skeletal Systems. Physiol Rev 97:1295–1349

313. Morgan ME, Flierman R, van Duivenvoorde LM, Witteveen HJ, van Ewijk W, van Laar JM, de Vries RR, Toes RE (2005) Effective treatment of collagen-induced arthritis by adoptive transfer of CD25+ regulatory T cells. Arthritis Rheum 52:2212–2221

314. Morgan ME, Sutmuller RP, Witteveen HJ, van Duivenvoorde LM, Zanelli E, Melief CJ, Snijders A, Offringa R, de Vries RR, Toes RE (2003) CD25+ cell depletion hastens the onset of severe disease in collagen-induced arthritis. Arthritis Rheum 48:1452–1460

315. Kelchtermans H, Geboes L, Mitera T, Huskens D, Leclercq G, Matthys P (2009) Activated CD4+CD25+ regulatory T cells inhibit osteoclastogenesis and collagen-induced arthritis. Ann Rheum Dis 68:744–750

316. Zaiss MM, Frey B, Hess A, Zwerina J, Luther J, Nimmerjahn F, Engelke K, Kollias G, Hunig T, Schett G, David JP (2010) Regulatory T cells protect from local and systemic bone destruction in arthritis. J Immunol 184:7238–7246

317. Fischer L, Herkner C, Kitte R, Dohnke S, Riewaldt J, Kretschmer K, Garbe AI (2019) Foxp3(+) Regulatory T cells in bone and hematopoietic homeostasis. Front Endocrinol (Lausanne) 10:578

318. Tyagi AM, Yu M, Darby TM, Vaccaro C, Li JY, Owens JA, Hsu E, Adams J, Weitzmann MN, Jones RM, Pacifici R (2018) The Microbial metabolite butyrate stimulates bone formation via T regulatory cell-mediated regulation of WNT10B expression. Immunity 49:1116–1131. e7

Taj Mohammad, Rashmi Dahiya, and Md. Imtaiyaz Hassan

Abstract

Synthetic biology is an emerging field where biology, computer, chemistry and engineering are reciprocally revisited to manipulate, design, construct and develop new biological entities with novel functionalities. Synthetic biology has been widely used in pharmaceutical, chemical, agricultural and energy industries. It is used to design and build complex circuits inspired by electrical engineering for fast and effective solutions of biomedical challenges including antibiotic resistance, viral infections and cancer. Synthetic biology enables us to modify and reconstruct various cells and their components and even whole organisms precisely. Synthetic biology plays significant roles in biotherapeutic engineering for the development of diagnostics, drug designing, enzymes, tailoring tissues and synthetic organs. In addition, it has been widely used in the design and development of medical therapeutics including drugs, diagnostic devices and biocompatible materials to improve the living standards of individuals. Synthetic biology provides an enthusiastic platform to develop biotherapeutic engineering via providing precise molecular biology tools that allow us to manipulate living cells for beneficial use. In this chapter, we cover a detailed overview of recent advancements in molecular tools and approaches to engineer microbial biotherapeutics using synthetic biology especially in human therapeutics and biomedical engineering.

Keywords

Bio-therapeutics · Genome engineering · Synthetic biology · Engineered microbes

T. Mohammad · R. Dahiya · Md. I. Hassan (✉)
Centre for Interdisciplinary Research in Basic Sciences, Jamia Millia Islamia,
New Delhi, India
e-mail: mihassan@jmi.ac.in

© Springer Nature Singapore Pte Ltd. 2020
S. Singh (ed.), *Systems and Synthetic Immunology*,
https://doi.org/10.1007/978-981-15-3350-1_9

9.1   Introduction

Synthetic biology is an emerging discipline that covers vast scientific areas, especially biomedical engineering with the ultimate goal to improve human health. It deals with designing, programming cellular behaviour, modifications and creation of new biological parts, devices and biological systems [14]. Synthetic biology enables us to redesign the existing natural systems, even whole organisms, for beneficial resolution. Synthetic biology plays a vital role in the development of simple, fast and effective diagnostic [2].

Engineered bacteriophages are another example of synthetic biology which are used to detect specific bacterial strain producing bioluminescence [22] and can be designed to attack antibiotic-resistant bacteria by disrupting their defence mechanism [19]. Engineered enzymatic bacteriophages can degrade biofilms to inhibit bacterial pathogenesis. Engineered *E. coli* can screen gut microbiome for real-time biosensing to monitor changes in the cellular environment of organisms and can be used as a biosensor for whole-cell biosensing to detect pathogenic infections and even cancer [21]. Bacteria and phage are also being engineered selectively for a substance, for example, detecting arsenic in water.

RNA-based biosensor is another beautiful example of synthetic biology which can detect disease-specific RNA and metabolites. Taken together, synthetic biology has been widely used in biomedical engineering to develop human therapeutics and synthetic constructs for the treatment of bacterial infections and to improve pre-existing antibiotics [4].

Biotherapeutic engineering is another growing field where biological principles in complex with engineering tools are used to design and develop economically viable therapeutic products [23]. It has been widely used in designing and developing medical therapeutics including therapeutic drugs, diagnostic devices and biocompatible materials to improve the quality of life. For the development of biotherapeutics, synthetic biology provides an attractive platform via providing specific molecular biology tools to manipulate living cells, even whole organisms [25].

In this chapter, we cover a detailed overview of the use of synthetic biology especially in human therapeutics covering development of biotherapeutics, biological parts, and approaches in robust biotherapeutic engineering. We also discuss some state-of-the-art available tools of synthetic biology and their advancement in translational biology for human therapeutics using different approaches, mainly from the perception of biotherapeutic engineering.

9.2   Rise in Synthetic Biology Publications

Synthetic biology is one of the most growing interdisciplinary sciences in the developing world. We analysed the publication records of the Web of Science with keyword "Synthetic Biology" from 2000 to 2019 indicating a significant rise in the publications from synthetic biology (Fig. 9.1). Annual worldwide synthetic biology publication output grew from an average of about 150 publications per year from

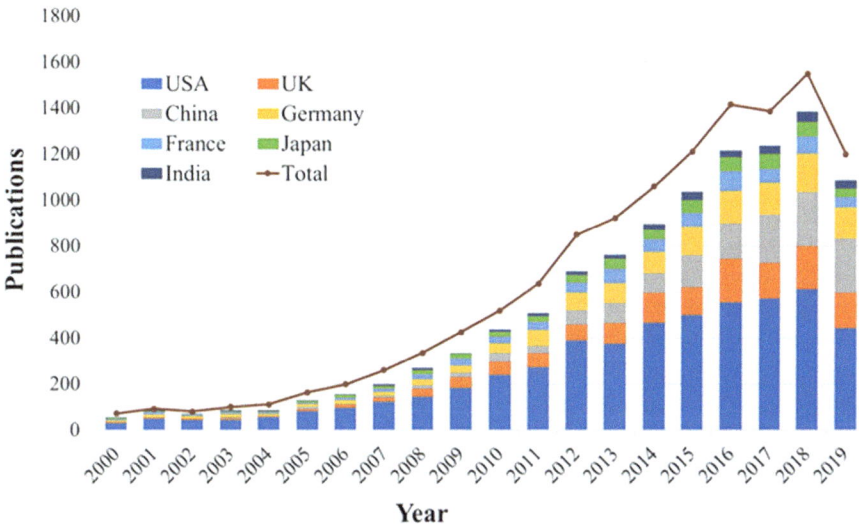

Fig. 9.1 A rise in synthetic biology publications globally and the contributions of leading nations. Source: Web of Sciences (http://apps.webofknowledge.com) data from January 2000 to September 2019. Marked lines showing worldwide annual publications while the stacked columns are showing annual publications for the six leading countries along with India

January 2000 to December 2010 to over 1200 publications per year in the period from January 2011 to September 2019. This record includes 12,500 publications by more than 35,000 authors at 5500 organizations located in 105 countries. The United States is the leading country in synthetic biology research contributing about 42.0% of total publications, while authors from the United Kingdom, China, Germany, France, Japan and India contribute 11.1%, 10.5%, 9.8%, 5.3%, 4.4% and 2.4%, respectively. Figure 9.1 clearly reflects the advancement in synthetic biology research globally.

9.3 Synthetic Biology in Development of Biotherapeutics

Synthetic biology is dramatically improving the existing production process of antibiotics, vitamins, enzymes, organic acids and synthetic organs. It has undergone considerable growth in scope, prospect and productivity and has become an extensively recognized branch of biomedical science [25]. It has already contributed significantly to modern medical science to encounter several global challenges, e.g. synthesis of artemisinin, an antimalarial drug through engineered *E. coli* and yeast. Biotherapeutics, for example, antibodies and therapeutic replacement enzymes, are the most successful and rapidly growing drugs for the treatment of complex diseases including cancer, neurodegeneration, inflammation, autoimmune diseases, infections and rare genetic disorders [13]. The approval and success rate of biotherapeutics is comparatively higher than small-molecule therapeutics.

Biotherapeutics such as monoclonal antibodies, large peptides and fusion proteins cannot be completely synthesized by chemical procedures [13]. Biotherapeutics can only be produced in living cells or organisms in predefined conditions to maintain product safety and efficacy using biotherapeutic engineering strategies. The integration of novel strategies and approaches of modern science such as synthetic biology tools are very useful in the modification of model organisms to produce the desired output such as therapeutic proteins with novel and improved efficacy [8]. Several biotherapeutic engineering platforms such as protein conjugation and derivatization approaches including generation of antibody-drug conjugates using synthetic biology are currently in use to improve half-life, efficacy, purity and production yield and to further limit toxicity of a drug [30]. A few examples of technological innovation and biotherapeutic engineering platforms are transgenic animal and plant, glyco-engineering, Fc fusion, antibody-drug conjugates and monoclonal antibody humanization/chimerism [29].

9.4 Engineering Microbial Therapeutics Using Synthetic Biology

Using microorganisms as small living factories to synthesize biologically active compounds is a systematic approach [26]. Different biotherapeutics such as monoclonal antibodies, small peptides, hormones, antigens, enzymes, vitamins and antibiotics are being produced by engineered microorganisms (bacteria and fungi) at industrial scale [9]. Microbes have probiotic and productive features which can be combined using a synthetic biology approach to protect from pathogens and to develop immunity and biotherapeutics against life-threatening diseases [26].

Engineered living cells including microbes are the future of biotherapeutics to treat complex diseases including cancer, neurodegeneration and metabolic disorders. They can be engineered to act like living therapeutics for defined actions within the human body. Recently, synthetic biology allowed us to develop microbial genetic tools for living therapeutics, biosensors, bio-switches and electrical-inspired circuits [26]. Engineered microbes are self-replicative, can detect abnormal conditions, and produce and transport therapeutics to the site of action inside the body. This approach will have numerous advantages over traditional therapeutics such as a significant reduction of cost for production and development. Using synthetic biology, microbes can be engineered to produce more than one biotherapeutic at a time making them more effective than currently used therapeutics to treat life-threatening diseases [20]. These engineered microbes can produce therapeutics directly in the human body thereby reducing many downstream processes; as a result, lowering dose, reduced side effects and much cheaper than traditional small-molecules therapeutics [15].

9.5   Overview of Biotherapeutic Discovery and Development

Biotherapeutics poses several challenges during their development as they require more complicated manufacturing and characterization process to produce them from living cells. With the advancement in synthetic biology applications and biotherapeutic engineering, it has been easy and accessible to produce a new generation of biotherapeutic designer drugs with increased efficacy and safety as compared to traditional and small-molecule therapeutics [17].

Since the first recombinant-DNA-derived human insulin was approved, more than 170 biotherapeutics have been marketed for medical applications [31]. Biotherapeutic drugs can be grouped into (1) peptides which include growth factors, hormones and cytokines represented by insulins, epoetin alpha and granulocyte colony-stimulating factor; (2) non-immune proteins which include therapeutic replacement enzymes, blood factors and anticoagulants represented by naglazyme, myozyme, elaprase, tissue plasminogen activator, recombinant hirudin and activated protein; and (3) antibodies and Fc fusion proteins including therapeutic antibodies and Fc-like fusion protein (rituximab, adalimumab, CD2-Fc, abatacept, Nplate, etc.) [12].

The success of biotherapeutic drugs cannot be described without the use of synthetic biology, providing a protein engineering platform that increased the stability and aggregation resistance of therapeutic candidates. Hybridoma technology, chimerization and humanization, human antibodies from transgenic mice and phage-display libraries, glycoengineering, multi-specific antibodies, intrabodies and protein engineering represent considerable examples of recent developments in biotherapeutic engineering with synthetic biology [16, 31]. A detailed discussion on using the synthetic biology approach in biotherapeutic engineering, available tools and development has been described previously [21, 22]. A pictorial representation of synthetic manipulation in living cells and its components using a biotherapeutic engineering approach to restore normal function is illustrated in Fig. 9.2.

Selection of potential drug target that drives a specific disease is a crucial step in biotherapeutic development. However, most tyrosine kinases and cytokine receptors fall in target categories for oncological and immunological disorders. Complex biotherapeutics such as rituximab – a genetically engineered chimeric monoclonal antibody targeting protein CD20 has been approved locally in India, China, and South Korea to treat autoimmune diseases and varying types of cancer – is a successful example [31].

9.6   Biotherapeutic Safety

Human diseases are complex and heterogeneous in nature and determined by different robust and diverse mechanisms that contribute to multifaceted pathologies with different symptoms. Biotherapeutic administration can cause adverse effects and

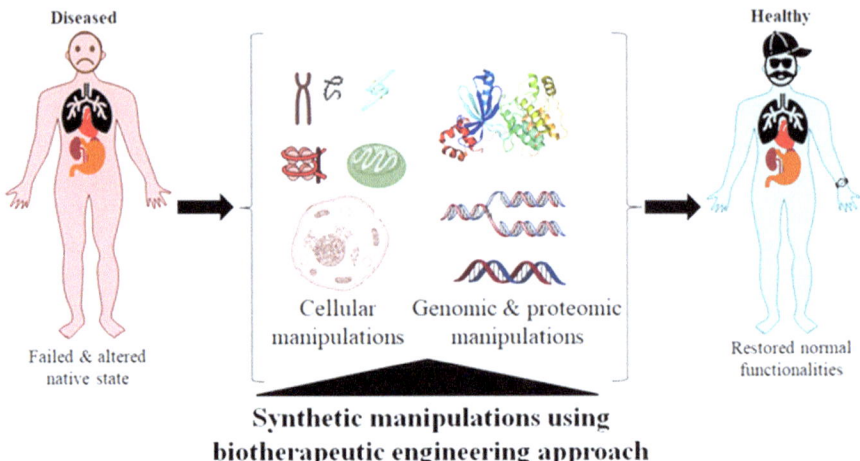

Fig. 9.2 A therapeutic approach of synthetic biology using biotherapeutic engineering approach to restore normal function in humans

immune response in the human body, for instance, induction of acute infusion reactions, immunogenicity, autoimmune diseases, tumorigenicity, platelet and thrombotic disorders, dermatitis, cardiotoxicity, hypercytokinemia, etc. [5]. Synthetic biology plays a crucial role in minimizing these side effects to develop safer and more effective biotherapeutics. Appropriate preclinical animal models are very important to develop safe and effective translational medicines targeting a specific pathway in complex disease. Synthetic biology helps to choose and even provide appropriate and customizable models for preclinical trials [25].

9.7 Synthetic Biology in Stabilizing Biotherapeutics

Synthetic applications in bioengineering have been extensively employed to stabilize recombinant biotherapeutics produced in engineered living cells [11]. To stabilize therapeutic proteins, modification of cysteine residues is used to form disulfide bridges which ultimately results in protein stability. Synthetic biology allow us to introduce precise alterations in proteins for their stabilization and resist them from degradation and formation of aggregates. Engineered *E. coli*-expressing non-glycosylated cytokine interleukin-2 is an aggregation-prone protein and showed decreased stability. Aldesleukin (recombinant interleukin-2) with Cys125Ser mutation is one of the wonderful examples of using classical protein engineering to stabilize proteins without altering their biological activity [6]. Likewise, interferon-β_{1b} is another biotherapeutic cytokine which contains three cysteines at 17th, 31st and 141st positions, which form one disulfide bond, leaving cysteine at 17th position in a reduced state. The Betaferon® and Betaseron® (recombinant interferon beta-1b) containing Cys17Ser mutation stabilize the protein without influencing its biological activity [24]. This approach with a fusion of advanced technologies has also been used to produce a stable variant of keratinocyte growth factor (KGF) by deleting N-terminal residues without influencing its native activity and is marketed as Biovitrum®, used in preventing chemotherapy- and radiotherapy-induced mucositis [3, 10]. Hence, altering basic constituents of biotherapeutics using advanced molecular editing tools of synthetic biology are beneficial for their stability and preservation without affecting their biological activity.

9.8 Synthetic Biology in the Pharmacoeconomics of Biotherapeutics

Biotherapeutics are relatively expensive due to high production costs and financial risk owing to their complex developmental process. Advancements using synthetic biology strategies can make these promising drugs more affordable with the aid of innovations. Improvement of trial designs, biomarker identification approaches and proper patient selection can increase their affordable biotherapeutic productivity by decreasing production cost. Recent advances in synthetic biology in biotherapeutic engineering are high-throughput production of mammalian cells to produce antibodies with shorter purification, formulation and production time [7]. Synthetic biology has become an effective tool in developing non-mammalian systems, e.g. engineered yeast and plant cells, to produce lower-cost biotherapeutics via eliminating costly viral inactivation validations step, used in production [18, 27].

9.9 Synthetic Biology in Vaccine Development

The developing world is in urgent demand for cost-effective vaccines to prevent growing infections. Still, the development of safe and new vaccines is a laborious task that requires precise identification (antigens, e.g. virus or microbial toxins) and development of immunogens to prevent or treat diseases [1]. Synthetic biology opens new avenues to develop precise molecular engineering tools required to read genetic information of different organisms to formulate vaccines in an effective and appropriate way, allowing scientists to save time and money. It enables researchers to engineer, produce and develop immunogens with high expression and improved efficacy. People can develop custom gene constructs for several vaccine candidates such as HIV and Ebola. The vaccines that come from such strategies are safer and well-tolerated which can trigger a stronger and long-lasting immune response in humans than ever existing vaccines [28].

9.10 Conclusions and Perspectives

Synthetic biology plays a very important role in the research and development of biotherapeutic engineering. Innovative technologies of synthetic biology are always emerging to address imminent challenges such as oral delivery of biotherapeutics, cost-saving production and development, and phase III success rate. With advancements in molecular biology tools and technologies, biotherapeutic engineering is extensively evolving over time. The existing biotherapies need to be optimized to achieve enhanced efficiency and functionalities with least adverse effects. The advanced engineering strategies along with synthetic biology tools allow us to modify existing entities for customization to introduce them for novel functionalities with precise clinical use. Using synthetic biology strategy represents a powerful approach to develop safe and effective biopharmaceuticals. It has been effectively used to improve stability and modulate native functionalities of biotherapeutics without unwanted side effects. Bioengineering is no longer limited to modifying and changing the genetic materials of living things. It is practical to expect that the existing biotherapeutics including proteins will be further studied and engineered in the near future. The rise in the scientific literature of synthetic biology and biotherapeutic engineering reflects an evocative meeting of advanced computational and high-throughput experimental methods for bioengineering platforms and has opened a new avenue to design and develop safe, effective and suitable biotherapeutics.

Conflict of Interest The authors declare no conflict of interest.

References

1. Andries O, Kitada T, Bodner K, Sanders NN, Weiss R (2015) Synthetic biology devices and circuits for RNA-based 'smart vaccines': a propositional review. Expert Rev Vaccines 14(2):313–331
2. Benner SA, Sismour AM (2005) Synthetic biology. Nat Rev Genet 6(7):533
3. Blijlevens N, Sonis S (2006) Palifermin (recombinant keratinocyte growth factor-1): a pleiotropic growth factor with multiple biological activities in preventing chemotherapy-and radiotherapy-induced mucositis. Ann Oncol 18(5):817–826
4. Chappell J, Watters KE, Takahashi MK, Lucks JB (2015) A renaissance in RNA synthetic biology: new mechanisms, applications and tools for the future. Curr Opin Chem Biol 28:47–56
5. Clarke JB (2010) Mechanisms of adverse drug reactions to biologics. In: Adverse drug reactions. Springer, pp 453–474
6. Doyle MV, Lee MT, Fong S (1985) Comparison of the biological activities of human recombinant interleukin-2 (125) and native interleukin-2. J Biol Response Mod 4(1):96–109
7. Fischbach MA, Bluestone JA, Lim WA (2013) Cell-based therapeutics: the next pillar of medicine. Sci Transl Med 5(179):179ps177–179ps177
8. Folcher M, Fussenegger M (2012) Synthetic biology advancing clinical applications. Curr Opin Chem Biol 16(3–4):345–354
9. Hays SG, Patrick WG, Ziesack M, Oxman N, Silver PA (2015) Better together: engineering and application of microbial symbioses. Curr Opin Biotechnol 36:40–49
10. Hsu E, Osslund T, Nybo R, Chen B-L, Kenney WC, Morris CF, Arakawa T, Narhi LO (2006) Enhanced stability of recombinant keratinocyte growth factor by mutagenesis. Protein Eng Des Sel 19(4):147–153
11. Huang L, Huang Y-Y, Mroz P, Tegos GP, Zhiyentayev T, Sharma SK, Lu Z, Balasubramanian T, Krayer M, Ruzié C (2010) Stable synthetic cationic bacteriochlorins as selective antimicrobial photosensitizers. Antimicrob Agents Chemother 54(9):3834–3841
12. Ishino T, Wang M, Mosyak L, Tam A, Duan W, Svenson K, Joyce A, O'Hara DM, Lin L, Somers WS (2013) Engineering a monomeric Fc domain modality by N-glycosylation for the half-life extension of biotherapeutics. J Biol Chem 288(23):16529–16537

13. Jones LH, McKnight AJ (2013) Biotherapeutics: recent developments using chemical and molecular biology, vol 36. Royal Society of Chemistry, Cambridge
14. Khalil AS, Collins JJ (2010) Synthetic biology: applications come of age. Nat Rev Genet 11(5):367
15. Kim JM (2010) Xylan-regulated delivery of human keratinocyte growth factor-2 to the inflamed colon by the human anaerobic commensal bacterium bacteroides ovatus. Gut 59:461–469. Korean J Gastroenterol 56(6):394–396
16. Li J, Zhu Z (2010) Research and development of next generation of antibody-based therapeutics. Acta Pharmacol Sin 31(9):1198
17. Lipsitz YY, Bedford P, Davies AH, Timmins NE, Zandstra PW (2017) Achieving efficient manufacturing and quality assurance through synthetic cell therapy design. Cell Stem Cell 20(1):13–17
18. Love KR, Dalvie NC, Love JC (2018) The yeast stands alone: the future of protein biologic production. Curr Opin Biotechnol 53:50–58
19. Lu TK, Collins JJ (2009) Engineered bacteriophage targeting gene networks as adjuvants for antibiotic therapy. Proc Natl Acad Sci 106(12):4629–4634
20. Mimee M, Citorik RJ, Lu TK (2016) Microbiome therapeutics—advances and challenges. Adv Drug Deliv Rev 105:44–54
21. Mohammad T, Hassan MI (2018a) Genome microbiology for synthetic applications. In: Synthetic biology. Springer, pp 75–86
22. Mohammad T, Hassan MI (2018b) Modern approaches in synthetic biology: genome editing, quorum sensing, and microbiome engineering. In: Synthetic biology. Springer, pp 189–205
23. Nielsen J, Keasling JD (2011) Synergies between synthetic biology and metabolic engineering. Nat Biotechnol 29(8):693
24. O'Rourke E, Drummond R, Creasey A (1984) Binding of 125I-labeled recombinant beta interferon (IFN-beta Ser17) to human cells. Mol Cell Biol 4(12):2745–2749
25. Ozdemir T, Fedorec AJ, Danino T, Barnes CP (2018) Synthetic biology and engineered live biotherapeutics: toward increasing system complexity. Cell Syst 7(1):5–16
26. Pedrolli DB, Ribeiro NV, Squizato PN, de Jesus VN, Cozetto DA, Unesp at iGEM TA (2018) Engineering microbial living therapeutics: the synthetic biology toolbox. Trend Biotechnol
27. Shahid N, Daniell H (2016) Plant-based oral vaccines against zoonotic and non-zoonotic diseases. Plant Biotechnol J 14(11):2079–2099
28. Skwarczynski M, Toth I (2016) Peptide-based synthetic vaccines. Chem Sci 7(2):842–854
29. Yu X, Marshall MJ, Cragg MS, Crispin M (2017) Improving antibody-based cancer therapeutics through glycan engineering. BioDrugs 31(3):151–166
30. Zhang Y-HP, Sun J, Ma Y (2017) Biomanufacturing: history and perspective. J Ind Microbiol Biotechnol 44(4–5):773–784
31. Zhong X, Neumann P, Corbo M, Loh E (2011) Recent advances in biotherapeutics drug discovery and development. In: Drug discovery and development-present and future. IntechOpen

Cytokines in Cancer Immunotherapy

10

Raki Sudan

Abstract

Cytokines are effector molecules of the immune system that act as messengers for cell to cell communications. Cytokines play an indispensable role in immune regulation and are involved in cell proliferation, cell death, inflammation, tissue repair, and cellular homeostasis. In recent years, with the advent of modern innovative technologies, our understanding of the immune system has expanded significantly. This increased understanding about our immune system has enabled us to target several immune mediators, including cytokines, in different diseases, ranging from autoimmunity to cancers. Recent success in the development of checkpoint blockade immunotherapies, targeting PD1 and CTLA-4, in treatment of cancers has revolutionized cancer treatment and sparked renewed interest among cancer immunologists for the discovery of new potential targets. Despite significant success, the response rate with checkpoint blockade therapies still remains limited to a fraction of patients and is often associated with several life-threatening side effects. Therefore, heightened efforts are being made to develop new and better therapies or improve current therapies for cancer treatment. Because of their pleiotropic effects on immune cells and their role in immune activation, cytokines have emerged as potential candidates for cancer immunotherapy and hold a central stage in this whole process of cancer immunotherapeutics. This chapter discusses about the major cytokines involved in cancer immunotherapy and their targeting strategies.

R. Sudan (✉)
Department of Pathology and Immunology, Washington University School of Medicine, St. Louis, MO, USA
e-mail: rakisudan@wustl.edu

© Springer Nature Singapore Pte Ltd. 2020
S. Singh (ed.), *Systems and Synthetic Immunology*,
https://doi.org/10.1007/978-981-15-3350-1_10

10.1   Introduction

Cytokines comprise a large family of regulatory proteins produced by various cell types of both immune and nonimmune origin that play a pivotal role in regulating and shaping both innate and adaptive immune responses. Cytokines could be secreted or membrane bound and act locally in an autocrine or over a distance in a paracrine manner. Membrane-bound cytokines act through cell to cell contact and communicate information between cells, often bidirectionally [1–5]. Cytokines exert their effects by binding to their specific receptors expressed on target cells. Cytokine receptors are often composed of two or more different receptor subunits that may be specific or shared between different cytokines. Among others, members of common gamma chain cytokine family play a crucial role in the development and functions of immune cells and are actively being explored for their antitumor potential either alone or in combination with other immunomodulatory agents. Common gamma chain cytokine family includes IL2, IL4, IL7, IL9, IL15, and IL21. As their name suggests, these cytokines share a common gamma chain receptor (γ_c or CD132) that is essential for signaling through JAK3. Other than common gamma chain, receptor complexes of IL4, IL7, IL9, and IL21 consist of a cytokine-specific alpha chain. IL2 and IL15, besides having a gamma chain and cytokine-specific alpha chains, also share IL2Rβ/IL15Rβ chain (CD122). IL2 and IL15 thus signal via JAK1/3 and STAT3/5 pathways leading to transcription of their target genes [3, 6–10].

Cytokines play an important role in the development of both innate and adaptive immune responses. Cytokines have emerged as attractive targets for cancer immunotherapy research because of their pleiotropic effects on immune cells and their role in shaping tumor microenvironment [2, 3, 5]. IL2 was the first cytokine that was FDA approved for use in patients with cancer. It proved the concept that cytokines can be used for cancer therapy and opened doors for a vast area of cytokine-based cancer immunotherapeutics [6–8]. Cytokines' ability to activate immune effector cells like CD8 T cells and NK cells is crucial for their immunotherapeutical potential. In modern-day cancer immunotherapy research, various cytokine-based immunotherapy strategies are used. Engineered versions of cytokines, like fusion products or agonists, where cytokines are fused to their receptor subunits and in some cases with Fc region of antibody or to tumor-antigen-specific antibodies, are being developed. These engineered cytokine products possess better activity and stability compared to recombinant parent cytokines. Cytokines or their engineered products are either used as monotherapy or in combination with other antitumor agents for cancer immunotherapy. Various cytokine-based combination strategies involve use of cytokines or their engineered products in combination with other antitumor agents like with chemotherapeutic drugs, with other cytokines, with cancer vaccines, with agonistic or tumor-antigen-specific antibodies, with checkpoint blockade antibodies like anti-PD1 and anti-CTLA4, and with adoptive cell therapy for cancer immunotherapy (Fig. 10.1) [1–3, 5, 9]. All these approaches will be covered briefly below [1–3, 5, 9]. This chapter discusses about IL15, IL21, and IL7 and their role in cancer immunotherapy.

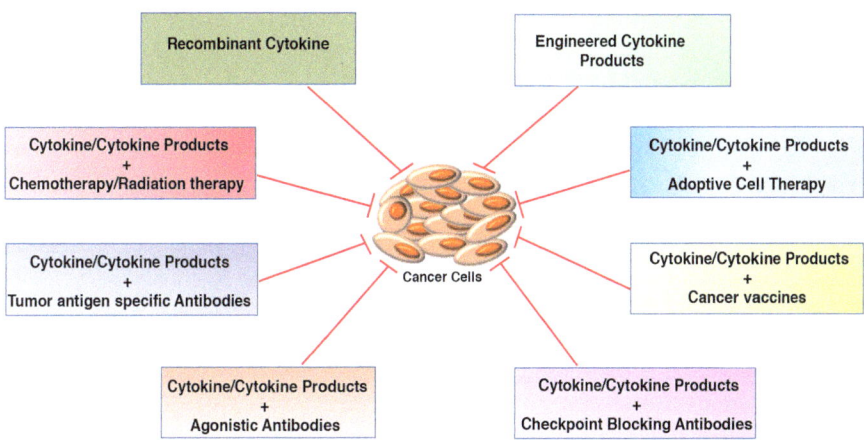

Fig. 10.1 Depiction of various cytokine-based approaches for cancer immunotherapy. Cytokines are either used as monotherapy or in combination with other anticancer agents for cancer immunotherapy. Different engineered cytokine-based products like cytokine-cytokine receptor subunit fusion proteins and cytokine-antibody fusion products provide better stability and activity for cancer immunotherapy. Cytokines are also being used in combination with other anticancer treatments like chemotherapy, antibody therapies, cancer vaccines, and with adoptive cell therapies in order to obtain better responses in patients with malignancies

10.2 IL15

10.2.1 Introduction

IL15 is a 14- to 15-kDa protein that was codiscovered in 1994 by two groups as a T-cell growth factor [11–13]. It is a member of the four α-helix cytokine bundle family. IL15 is primarily produced by cells of the innate immune system including dendritic cells, macrophages, and monocytes. IL15 expression is regulated tightly at the levels of transcription, mRNA splicing (post-transcription), translation, and intracellular trafficking. There are two isoforms of IL15 mRNA formed as a result of alternative splicing. These isoforms encode for IL15 protein either bearing a short signal peptide or long signal peptide. The signal peptides play a role in the intracellular trafficking of protein. The long signal peptide bearing IL15 is directed to endoplasmic reticulum (ER) secretory pathway and is exported outside the cell, whereas short signal peptide IL15 is not secreted and localizes to the cytoplasm and nucleus [13–17]. Also, multiple AUG present at 5′ UTR play a role in regulation of IL15 mRNA translation. This regulation at multiple points therefore ensures stringently controlled production of IL15, primarily by monocytes, macrophages, and dendritic cells, despite the fact that IL15 transcript is detected in multiple tissues and cell types. Probably this tight regulation of IL15 production is required because indiscriminant production of IL15 can induce inflammation and autoimmunity through the production of TNFα, IL1β, IFNγ, and other proinflammatory cytokines.

TLR stimulation, CD40 stimulation, and type I and type II interferons are known to induce IL15 transcription [13–18].

IL15 receptor complex is composed of three subunits and includes the common γc subunit shared by all γc cytokines, IL2Rβ/15Rβ (shared with IL2), and a specific IL15Rα subunit. Once IL15 is produced by cells of myeloid lineage (monocytes, macrophages, and DC), it binds to either secreted or membrane-bound form of its high-affinity unique receptor subunit, IL15Rα, that is almost ubiquitously expressed, and is presented in trans (trans-presentation) to the target cells expressing the dimeric IL2Rβ/γc receptor complex and signals through JAK1/3 and STAT3/5 pathways [14, 17–20]. Trans-presentation involves direct cell to cell contact and can stimulate neighboring or opposing cells, thus offering a tighter regulation by providing cell-directed delivery. The expression of IL2/15Rβ and γc is thought to be the major deciding attribute of a cell type to IL15 responsiveness. Also, soluble complexes of IL15 bound to IL15Rα (sIL15) are cleaved from IL15-expressing cell types in response to certain inflammatory signals such as TLR activation, CD40 ligation, and type 1 interferons, providing a sudden burst of IL15 activity [13, 14, 17, 18, 21]. IL15 alone, in the absence of high-affinity IL15R alpha, can also bind to the IL2Rβ/γc receptor complex, though with intermediate affinity, resulting in activation of Lyn, Lck, Fyn, and Syk tyrosine kinases and PI3K-MAPK pathway. However, soluble IL15 complexes are found to be superior to support an immune response compared to recombinant IL15 (rIL15). IL15 bound to its unique receptor, IL15Rα, is reported to undergo endosomal recycling, thus resulting in persistence of membrane-bound IL15 for longer durations of time. IL15 plays an important role in natural killer (NK) cell, innate lymphoid cell (ILC) 1, NKT-cell, B-cell, and T-cell development and function [13, 14, 17, 18, 21, 22].

10.2.2 IL15 in Immunomodulation and Cancer Therapy

10.2.2.1 Biology and Immunomodulatory Effects of IL15

IL15 was originally identified for its ability to induce T-cell proliferation, in a way similar to IL2 [11, 12]. Like IL-2, IL15 is reported to induce proliferation and differentiation of activated T cells. Further, IL15 also enhances the cytotoxic potential of CD8+ T cells. The similar functions of IL2 and 15 are attributed to their sharing of common receptor components, IL2/IL15Rβ and γc, and their signaling through JAK1/JAK3 and STAT3/STAT5 pathways. Despite these similarities with IL2, IL15 has certain unique functions too. IL15 is shown to have no major effects on T-regulatory cells and this is because of the inability of IL15 to bind to the IL2Rα chain. Also, unlike IL2 that induces activation-induced cell death of T cells, IL15 actually suppresses activation-induced cell death and is involved in maintenance of long-lasting CD8+CD44 [hi] memory T-cell phenotype [13, 14, 17, 18, 20].

As far as NK cells and ILC1 are concerned, IL15 is critical for their development and function. NK cells are innate lymphocytes that were first identified in the 1970s for their ability to kill leukemia cells without any prior sensitization. As their name indicates, NK cells exert cellular cytotoxicity through the release of granzyme and perforin and are known to kill tumor and virally infected cells. Also, NK cells secrete a number of cytokines and chemokines, thus playing an important role in shaping an immune response [14, 23]. Recent success of checkpoint blockade therapies has given a boost to cancer immunotherapy research and heightened efforts are being made to define and develop new targets for cancer therapy. The goal of immunotherapy in cancer patients is to stimulate the host's immune system to the point where it can attack cancer cells, and because of their cytolytic potential, NK cells have been subject of great interest in cancer immunotherapy research. IL15 regulates almost every aspect of NK cell's biology and has been shown to be essential for their development, proliferation, survival, and cytotoxic functions as evidenced from the phonotype of IL15Ra and IL15 KO mice. Because of these immune cell modulatory effects, IL15 holds a pivotal position in cancer immunotherapy research [13, 14, 18, 23].

10.2.2.2 Recombinant IL15 and IL15 Superagonists in Cancer Therapy

Prompted by studies showing immune-stimulatory effects of IL15, IL15 was tested for its antitumor potential in several preclinical experimental murine tumor models. IL15 protected MC38 colon carcinoma cell–injected mice, as IL15 transgenic mice survived longer than 6 months compared to wild-type mice. Wild-type mice died because of lung metastases in this study [24]. IL15 was shown to prolong survival or treatment with recombinant IL15 (rIL15) resulted in reduced tumor growth and decreased metastasis in several other transplantable murine tumor models like CT-26 colon cancer, B-16 melanoma, LA795 lung adenocarcinoma, P1A+, TRAMP-C2 prostate cancer, etc. [13–16, 18, 25, 26]. In all these murine tumor studies, the antitumor effects of IL15 were primarily due to its ability to enhance NK cell and CD8 T-cell responses and IFNγ production [13–16, 18, 25, 26]. Some of the mice treated with IL15 completely eradicated tumors, and upon rechallenge, these mice remained tumor-free suggesting that IL15 treatment can result in long-lasting antitumor immunity [15, 27]. Based upon the data showing immuno-stimulatory effects of IL15 and promising results from preclinical animal tumor model studies, the National Cancer Institute (NCI) ranked IL15 as the most promising immunotherapeutic agent to be brought to clinical trials [28]. Potential of toxicity remains a concern for use of IL15 as a therapeutic agent in humans. In this regard, the safety of recombinant human IL15 (rhIL15) and macaque was evaluated in primates (rhesus macaques) by several groups. It was observed that treatment with IL15 resulted in significant expansion of CD8 T cells and NK cells and in some cases CD4 T cells also. Higher dose and continuous delivery of cytokine resulted in greater response but there were clinical toxicities associated. These toxicities included diarrhea, reduced appetite, weight loss, anemia, rash, and reversible grade 3–4 neutropenia. Despite these transient toxicities, treatment with rhIL15 was well tolerated and no severe autoimmune reactions were observed [15, 29]. These observations led to use of rhIL15 as an alternative treatment in patients with metastatic melanoma and refractory metastatic renal cell cancer as a first human phase I clinical trial. The results from this recently completed phase I clinical trial reveal that similar to animal model studies, IL15 treatment resulted in efflux of NK cell and CD8 memory T cell as early as in 30 mins of treatment. Also, expansion of NK cells, CD8 cells, and gamma delta cells was observed as evident from their increased count. Serum levels of multiple proinflammatory cytokines like IL6, IL8, IFNγ, and TNFα were markedly elevated in treated patients. A half-life of 2.5 h was reported by pharmacokinetic studies. Treatment with rhIL15 resulted in decrease and/or clearance of lung lesions in melanoma patients; however, there were clinical toxicities like grade 3 hypotension, thrombocytopenia, fever, and increased transaminase levels. Elevated cytokine levels were thought to be responsible for these toxicities. Disease stabilization was reported as the best response in this first phase I trial. Toxicities associated with this trial suggested need to reconsider dosing and route of administration in future studies [15, 30]. Several other trials using different routes and schedules of rhIL15 treatment are still going on.

Despite being a promising immunotherapeutic candidate for cancer treatment, short in vivo half- life of rIL15 is a limiting factor for its applications. Because of this short half-life, often high doses of rIL15 are required to achieve the maximum response and this in turn results in toxicities. Therefore, several strategies have been used to develop new and better formulations of IL15 to address these limitations. As mentioned earlier, IL15 is predominantly present bound to IL15Rα, under physiological conditions, for its long-term persistence and is presented in trans to NK and CD8 T cells for their activation and function [13–16]. Therefore, these new strategies focused on designing protein complexes of IL15/IL15Rα, with better half-life and activity, often termed as IL15 superagonists. Also, these strategies address the issue of low expression of IL15Rα when rIL15 is used as monomer as soluble complexes of IL15/IL15Rα need not to be presented in trans. These new approaches are briefly mentioned in coming lines. One approach involved combining rIL15 with recombinant soluble murine IL15Rα linked to the human IgG1 Fc portion (sIL15/IL15Rα-Fc). This soluble IL15/IL15Rα-Fc complex had an impressive half-life of 20 h and was more potent compared to rIL15 and induced expansion of NK cells, NKT cells, and memory CD8 T cells (more than 50-fold). Importantly, treatment with IL15/IL15Rα-Fc in mice prevented tumor growth in a B16 melanoma model or resulted in regression of pancreatic (RIP1-Tag2) tumors and B16F10 melanoma tumors without significant toxicities [14–16, 31]. Another strategy involved development of a fusion protein consisting of a cytokine-binding amino-terminal domain of IL15Rα (Sushi domain) coupled to IL15 via an amino acid linker. This fusion protein termed as RLI (receptor-linker-IL15) has potent NK- and T-cell stimulatory activity and increased half-life compared to rIL15. RLI also showed strong antitumor activity in preclinical tumor models (B16F10 melanoma and in an orthotopic colon cancer model) [14–16, 18, 32]. Another fusion protein known as ALT-803 consisting of a mutated version of IL15 (N72D) linked to the Sushi domain of IL15Rα fused to the Fc region of human IgG1 also showed enhanced NK and CD8 stimulatory activity and antitumor potential in a mouse myeloma model [15, 33]. All these superagonists are currently being tested in several clinical trials and it will be interesting to know how these different superagonists will behave.

10.2.2.3 IL15 in Combination Cancer Therapy

Tumor heterogeneity and the presence of multiple immunological barriers at tumor sites, among many others, are the major detrimental factors for the success of a single immunotherapeutic agent to be effective in achieving complete remission. Therefore, modern-day approaches for cancer treatment involve combining different antitumor agents to achieve better results. In this regard, IL15 and its different agonists have been investigated in combination cancer therapy. The different combinatorial approaches involve combining IL15 therapy with another cytokine, with chemotherapeutic agents, with checkpoint blockade antibody therapy using anti-CTLA4 and anti-PD1/PD-L1, and with agonistic antibodies like CD40 and other tumor antigen–directed antibodies for successful treatment of different types of cancers [14, 15, 25, 34–38].

Combining IL15 treatment with multiple chemotherapeutic agents resulted in increased survival as well as tumor regression in multiple experimental murine tumor models compared to a single agent alone. This enhanced antitumor effect could be due to the ability of chemotherapy to induce tumor cell death resulting in less tumor burden and tumor-mediated immunosuppression and subsequently induction of increased CD8 T-cell and NK cell activity by IL15 because of its stimulatory effects on these cell types. IL15 when used in conjunction with other cytokines showed more potent antitumor effects. In mouse models, combined IL15 and IL12 therapy acted synergistically to achieve maximum tumor clearance mostly through stimulation of CTLs and IFNγ production. Despite having various immune-stimulatory effects, IL15 is associated with induction of PD-1, TIGIT, and IL10, thus suggesting that combining checkpoint blockade therapy with IL15 treatment will provide enhanced antitumor activity. In view of this, IL15 or its agonists mentioned above were tested in combination with different checkpoint blocking antibodies (anti-CTLA4, PD1, PD-L1) in experimental murine tumor models. Results from these studies were promising as combined treatment showed enhanced efficacy as observed by prolonged survival and reduced tumor burden in mice receiving combination therapy [14, 15, 17, 25, 34–39]. Currently, this approach of using IL15 or its agonists with checkpoint blockade therapy is tested in clinical trials for treatment of several advanced malignancies. In same lines based upon promising results from murine studies, IL15 or its agonists are being tested in combination with agonistic antibodies or antibodies directed against specific tumor antigens. In another approach, IL15 or IL15 agonists are being fused with proteins (e.g., Apo-A1) or antibodies (anti-CD20 antibody, anti-GD2 ganglioside antibody) that specifically direct IL15 to tumors and are tested for their clinical efficacy. Also, IL15 is being tested as an adjuvant either alone or in combination with other agents in NK-cell- and T-cell-based adoptive cell therapies [14, 15, 17, 25, 34–40]. Overall, it is expected that these ongoing studies will help us in expanding our understanding of IL15 and tell us how and which combination therapy works best.

10.3 IL21

10.3.1 Introduction

IL21 is another cytokine from the common γ-chain cytokine family that is being actively investigated for its antitumor potential. IL21 is primarily produced by CD4 T cells, both Th1 and Th17, NKT cells, and follicular helper T cells (Tfh). The IL21 receptor is a heterodimeric receptor composed of IL21Rα and common γ-chain subunits and signals through JAK1 and JAK3 resulting in activation of STATs (STAT3, STAT1, and STAT5). Involvement of PI3K as well as MAPK signaling pathway is also reported in IL21 signaling [41–44]. IL21 has pleiotropic effects on a variety of immune cells including T cells, B cells, NK cells, NKT cells, macrophages, and dendritic cells. It plays an important role in B cell differentiation into plasma cells and in Th17 development. IL21 enhances cytotoxicity of NK cells, CD8 T cells, and NKT cells. It has suppressive effects on FOXP3 expression and expansion of T-reg cells. Because of its ability to enhance NK and CD8 cell function and suppress T-regs, IL21 has been evaluated for its antitumor potential [41–46].

10.3.2 IL21 in Cancer Immunotherapy

In several experimental murine tumor models, treatment with IL21 or IL21 gene transfer successfully inhibited tumor growth. The various murine tumor models tested included melanoma, mammary adenocarcinoma, colon cancer, renal cell carcinoma, bladder cancer, pancreatic carcinoma, fibrosarcoma, etc. IL21-producing cells were also used as whole-cell vaccines in certain mouse tumor studies (TS/A mammary adenocarcinoma, glioblastoma, myeloma) to treat mice with established tumors from wild type tumor cells. Significant tumor regression was observed in these studies because of enhanced CTL and NK cell responses including cytotoxicity and cytokine (IFNγ) production [41, 42, 45, 47–51]. In other studies, IL21 treatment when combined with another antitumor agent (cytokine, antibodies, adoptive cell transfer) mediated enhanced antitumor activity compared to the treatment with single agent alone. IL21 and IL15 when coadministered acted synergistically resulting in enhanced expansion and function of CD8 cells and clearance/regression of established B16 melanomas [52]. IL21 is known to induce cell death in certain B cell lymphomas like diffuse large B-cell lymphoma and mantle cell lymphoma [41–44, 53]. Also, when used in combination with anti-CD20 monoclonal antibody rituximab, enhanced killing of cancer cells was reported. Further effective tumor regression or clearance was reported when IL21 was combined with antibody-mediated depletion of CD4 T-regs. When combined with checkpoint blockade therapy or agonistic antibody therapy (anti-CD40, anti-DR5, anti-CD137), IL21 showed cooperative antitumor activity [41–43, 54–56]. Further, IL21 enhances NK-cell-mediated ADCC and is reported to enhance the therapeutic activity of tumor antigen–directed monoclonal antibodies [41–43, 50, 57]. It is important to note that treatment with rIL21 was not associated with significant toxicity in mice as unlike

IL2 no vascular leak syndrome was observed thus suggesting it being a safe candidate to be potentially used for clinical cancer immunotherapy.

Prompted by its success in preclinical murine tumor models, several clinical trials were initiated using IL21 as monotherapy or in combination with other agents for cancer immunotherapy. First clinical trials using recombinant IL21 in patients with metastatic melanoma and renal cell carcinoma reported favorable antitumor response. At higher doses of treatment, some reversible adverse effects including pruritus, neutropenia, thrombocytopenia, fatigue, and liver toxicity were observed. However, no vascular leak syndrome and significant autoimmune reactions were observed suggesting treatment to be safe and well tolerated within a maximum tolerated dose of 30 micrograms per kg [41, 42, 58]. IL21 therapy in combination with several other agents has been evaluated in several other phase I/II clinical trials. The combination treatment of rIL21 and sorafenib (a kinase inhibitor) in patients with metastatic renal cell carcinoma was evaluated and partial response and disease stabilization were the main response reported [59]. Complete or partial response was also observed in a phase I trial when IL21 treatment combined with anti-CD20 antibody (rituximab) was tested in patients with relapsed and refractory B cell lymphoproliferative disorders [60]. In other phase I trials, IL21 treatment is actively being tested in combination with checkpoint blockade therapy and other tumor antigen–directed antibody therapies for its safety and efficacy [41, 42, 48, 50, 61]. Also, there is a great interest in using IL21 in adoptive cell therapies and this area is also being actively explored. Overall data from clinical studies suggest that IL21 is a promising candidate for cancer immunotherapy as IL21 therapy is well tolerated and should be evaluated further.

10.4   IL7

10.4.1 Introduction

IL7 is another member of the common γc cytokine family discovered in 1980 that is being actively investigated in anticancer therapies. IL7 is primarily produced by non-hematopoietic cells including fibroblastic stromal cells, endothelial cells, keratinocytes, and epithelial cells. Dendritic cells are also reported to produce small amounts of IL7. IL7 signals through IL7 receptor, which is a heterodimeric receptor consisting of IL7Rα (CD127) and common γc (CD132) subunits [62–67]. Expression of IL7Rα is reported on a variety of immune cells including T and B cell precursors, most mature T-cell types, and innate lymphoid cells (ILCs). A soluble form of IL7R is also reported which competes with the cell-bound form of the receptor for IL7, thus regulating its availability and activity. Binding of IL7 to IL7R results in the activation of JAK-STAT as well as PI3K-AKT pathway. JAK1 and JAK3 are the two JAKs associated with IL7R complex and their activation results in activation of STAT5 and transcription of STAT5-regulated genes. IL7 is critical for the development and maintenance of T cells, B cells, and innate lymphoid cells (ILC1, ILC2, ILC3). Both αβ and γδ T-cell lineages require IL7 for their development. It is required for the survival of naive T cells as well as for the generation and maintenance of both CD4 and CD8 memory T cells. IL7R is down-regulated on the activation of naive T cells. T-regs express very low levels of IL7R, and unlike IL2, IL7 does not induce their proliferation. Although being a critical factor for B cell development, mature B cells lack IL7R and are not dependent on it. Il7 is also implicated in the development of thymic NK cells which produce IFNγ. IL7 is required for the proper development of lymph nodes and Peyer's patches in the gut by regulating LTi cells (a subset of ILCs), thus playing an important role in immune regulation at barrier sites. There is evidence that both ILCs and T cells compete for the IL7 pool and this competition appears to have an effect on the size of ILC compartment. Due to all these immune modulatory effects, IL7 is being explored in cancer immunotherapy [62–70].

10.4.2 IL7 in Cancer Immunotherapy

In preclinical mouse studies, IL7 administration significantly expanded the T-cell compartment and improved their function. IL7 augmented antigen-specific T-cell responses to tumor vaccination resulting in recognition of weak subdominant tumor antigens. IL7 administration either alone or in combination with another antitumor agent resulted in reduced tumor burden and prolonged survival in several murine tumor studies [62–66]. For adoptive T-cell therapy, when tumor-specific T cells were expanded in the presence of IL7 plus IL15, greater tumor regression was observed in a melanoma and 4T1 mammary carcinoma models. IL7 when combined with IFNγ enhanced its antitumor effects in rat glioma tumor models. In another study, intratumoral injection of adenoviral transduced IL7-expressing DC resulted in complete tumor regression in two murine lung cancer models. In another approach, IL7-producing whole cell vaccines were found effective in a prostate cancer model. IL7 when used as an adjuvant after a vaccine-induced response significantly improved survival and induced enhanced antitumor immune responses in another murine tumor model. The improved immune response was associated with increased IL6 production and augmented Th17 differentiation. Also IL-7 was shown to inhibit PD1 expression as well as antagonize the effects of TGFβ on CD8 T cells. Thus, all these preclinical studies employing different strategies for use of IL7 in different preclinical tumor models strongly supported the application of this cytokine in clinical cancer therapy [62–77].

In early clinical trials using recombinant human IL7 (rhIL7) as monotherapy for the treatment of patients with advanced malignancies, IL7 was found to be safe and well tolerated with limited toxicity. Treatment with hIL7 resulted in sustained increase in both CD4 and CD8 cells along with decrease in percentage of T-regs in these patients. Significant increase in TCR receptor diversity was also observed indicating that IL7 can broaden an immune response by selective expansion of naïve T cells. However, no significant anticancer response was observed in these two trials suggesting that rIL7 monotherapy may not be sufficient to achieve significant response and hence must be combined with other anticancer therapies [62, 65, 78,

79]. Patients with relapsed or refractory pediatric sarcoma are subject to antineo-plastic regimens resulting in lymphocyte depletion. Recombinant IL7 when used as adjuvant therapy in these patients promoted immune recovery (as measured by CD4 counts) and enhanced immune response. In another approach, IL7 was also used in CAR T-cell therapy in combination with other cytokines like IL15 and IL2 for expansion of these cells. CAR T cells expanded in the presence of IL7 along with IL4 and IL21 expressed less inhibitory receptors. CAR T cells expanded well and persisted for longer duration and showed enhanced antitumor responses when IL7 was combined with another cytokine (IL15/IL21), thus signifying an important and beneficial role of IL7 in adoptive cell therapy. Currently, IL7 is being actively tested in clinical trials using CAR T cells [62, 63, 65, 67, 80]. IL7 when administered with a prostate cancer vaccine Provenge resulted in increased PSA (prostate-specific antigen)-specific T cells. Recently, a hybrid version of IL7 consisting of recombi-nant human IL-7 fused with hybrid Fc (rhIL-7-hyFc) was developed that showed enhanced antitumor effects in preclinical models. This rhIL-7-hyFc addresses the limitation of short half-life and stability of rhIL7. Currently, rhIL-7-hyFc is being tested in several clinical trials combined with other anticancer agents including pembrolizumab (anti-PD-1) in triple-negative breast cancer; atezolizumab (anti-PD-L1) for treatment of melanoma, Merkel cell carcinoma, cutaneous squamous cell carcinoma; in combination with temozolomide in glioblastoma [62–65, 67]. In another clinical trial, glycosylated recombinant human interleukin-7 (CYT107) is tested with atezolizumab for treatment of advanced urothelial carcinoma [62, 63, 65, 67, 80]. Results from all these ongoing studies will expand our understanding of IL7 biology and will tell us how effective it is as an agent for cancer immunotherapy.

10.5 Conclusion

Cytokines play a critical role in shaping an immune response against tumors because of their pleiotropic effects on immune cells. Their ability to stimulate NK cells and CD8+ cytotoxic T lymphocytes is vital for their antitumor potential. IL15, IL21, and IL7, all three cytokines discussed above, have some degree of immuno-stimulatory activity that is crucial for antitumor immunity. These three cytokines are currently being tested for clinical cancer therapy in several malignancies. As early data from clinical trials suggest that these cytokines are not very effective when used as monotherapy, the current approaches using these cytokines in clinical trials involve various combinatorial approaches, where these cytokines are used in com-bination with another anticancer agent for cancer immunotherapy. There are cur-rently several challenges in developing successful cytokine-based immunotherapies. Short in vivo half-life, low availability at tumor site, potential toxicities associated with systemic administration, autoimmunity, deciding maximum tolerable dose, and best route of administration are few among the many challenges. Also because of their effect on inducing proliferation of lymphoid cells, chronic cytokine stimula-tion sometime results in development of lymphoid tumors. Therefore, it is crucial to use the right dose for the right duration of time for the right type of tumors to achieve a finely tuned and calibrated antitumor immune response for any cytokine immunotherapy to be effective, which itself seems a challenging job. Modern-day approaches employing new and novel strategies to engineer better versions of cytokines with enhanced half-life or designing cytokine-based fusion products, which combine cytokine with some antibody or protein, hold promise and are currently being tested for anticancer potential. It is believed that with all these new approaches, cytokines will ultimately hold an important position in cancer immunotherapy, but only the future can tell that.

References

1. Berraondo P et al (2019) Cytokines in clinical cancer immunotherapy. Br J Cancer 120(1):6–15
2. Conlon KC et al (2019) Cytokines in the treatment of cancer. J Interf Cytokine Res 39(1):6–21
3. Dwyer CJ et al (2019) Fueling cancer immunotherapy with common gamma chain cytokines. Front Immunol 10:263
4. Smyth MJ et al (2004) Cytokines in cancer immunity and immunotherapy. Immunol Rev 202:275–293
5. Waldmann TA (2018) Cytokines in cancer immunotherapy. Cold Spring Harb Perspect Biol 10(12):a028472
6. Malek TR (2008) The biology of interleukin-2. Annu Rev Immunol 26:453–479
7. Arenas-Ramirez N et al (2015) Interleukin-2: biology, design and application. Trends Immunol 36(12):763–777
8. Wrangle JM et al (2018) IL-2 and beyond in cancer immunotherapy. J Interf Cytokine Res 38(2):45–68
9. Shourian M et al (2019) Common gamma chain cytokines and CD8 T cells in cancer. Semin Immunol 42(101307):101307
10. Pol JG et al (2020) Effects of interleukin-2 in immunostimulation and immunosuppression. J Exp Med 217(1)
11. Burton JD et al (1994) A lymphokine, provisionally designated interleukin T and produced by a human adult T-cell leukemia line, stimulates T-cell proliferation and the induction of lymphokine-activated killer cells. Proc Natl Acad Sci U S A 91(11):4935–4939
12. Grabstein KH et al (1994) Cloning of a T cell growth factor that interacts with the beta chain of the interleukin-2 receptor. Science 264(5161):965–968
13. Steel JC et al (2012) Interleukin-15 biology and its therapeutic implications in cancer. Trends Pharmacol Sci 33(1):35–41
14. Rautela J, Huntington ND (2017) IL-15 signaling in NK cell cancer immunotherapy. Curr Opin Immunol 44:1–6
15. Robinson TO, Schluns KS (2017) The potential and promise of IL-15 in immuno-oncogenic therapies. Immunol Lett 190:159–168
16. Waldmann TA (2014) Interleukin-15 in the treatment of cancer. Expert Rev Clin Immunol 10(12):1689–1701
17. Waldmann TA et al (2020) Interleukin-15 (dys)regulation of lymphoid homeostasis: implications for therapy of autoimmunity and cancer. J Exp Med 217(1)
18. Mishra A et al (2014) Molecular pathways: interleukin-15 signaling in health and in cancer. Clin Cancer Res 20(8):2044–2050
19. Sim GC, Radvanyi L (2014) The IL-2 cytokine family in cancer immunotherapy. Cytokine Growth Factor Rev 25(4):377–390
20. Waldmann TA (2015) The shared and contrasting roles of IL2 and IL15 in the life and death of normal and neoplastic lymphocytes: implications for cancer therapy. Cancer Immunol Res 3(3):219–227

21. Guo Y et al (2017) Immunobiology of the IL-15/IL-15Ralpha complex as an antitumor and antiviral agent. Cytokine Growth Factor Rev 38:10–21
22. Zhang M et al (2012) Augmented IL-15Ralpha expression by CD40 activation is critical in synergistic CD8 T cell-mediated antitumor activity of anti-CD40 antibody with IL-15 in TRAMP-C2 tumors in mice. J Immunol 188(12):6156–6164
23. Chiossone L et al (2018) Natural killer cells and other innate lymphoid cells in cancer. Nat Rev Immunol 18(11):671–688
24. Kobayashi H et al (2005) Role of trans-cellular IL-15 presentation in the activation of NK cell-mediated killing, which leads to enhanced tumor immunosurveillance. Blood 105(2):721–727
25. Thi VAD et al (2019) Cell-based IL-15:IL-15Ralpha secreting vaccine as an effective therapy for CT26 colon cancer in mice. Mol Cells 42(12):869
26. Tamzalit F et al (2014) IL-15.IL-15Ralpha complex shedding following trans-presentation is essential for the survival of IL-15 responding NK and T cells. Proc Natl Acad Sci U S A 111(23):8565–8570
27. Tang F et al (2008) Activity of recombinant human interleukin-15 against tumor recurrence and metastasis in mice. Cell Mol Immunol 5(3):189–196
28. Cheever MA (2008) Twelve immunotherapy drugs that could cure cancers. Immunol Rev 222:357–368
29. Waldmann TA et al (2011) Safety (toxicity), pharmacokinetics, immunogenicity, and impact on elements of the normal immune system of recombinant human IL-15 in rhesus macaques. Blood 117(18):4787–4795
30. Conlon KC et al (2015) Redistribution, hyperproliferation, activation of natural killer cells and CD8 T cells, and cytokine production during first-in-human clinical trial of recombinant human interleukin-15 in patients with cancer. J Clin Oncol 33(1):74–82
31. Epardaud M et al (2008) Interleukin-15/interleukin-15R alpha complexes promote destruction of established tumors by reviving tumor-resident CD8+ T cells. Cancer Res 68(8):2972–2983
32. Mortier E et al (2006) Soluble interleukin-15 receptor alpha (IL-15R alpha)-sushi as a selective and potent agonist of IL-15 action through IL-15R beta/gamma. Hyperagonist IL-15 x IL-15R alpha fusion proteins. J Biol Chem 281(3):1612–1619
33. Xu W et al (2013) Efficacy and mechanism-of-action of a novel superagonist interleukin-15: interleukin-15 receptor alphaSu/Fc fusion complex in syngeneic murine models of multiple myeloma. Cancer Res 73(10):3075–3086
34. Alizadeh D et al (2019) IL15 enhances CAR-T cell antitumor activity by reducing mTORC1 activity and preserving their stem cell memory phenotype. Cancer Immunol Res 7(5):759–772
35. Sanseviero E et al (2019) Anti-CTLA-4 activates Intratumoral NK cells and combined with IL15/IL15Ralpha complexes enhances tumor control. Cancer Immunol Res 7(8):1371–1380
36. Vallera DA et al (2016) IL15 Trispecific Killer Engagers (TriKE) make natural killer cells specific to CD33+ targets while also inducing persistence, in vivo expansion, and enhanced function. Clin Cancer Res 22(14):3440–3450
37. Zhang M et al (2018) IL-15 enhanced antibody-dependent cellular cytotoxicity mediated by NK cells and macrophages. Proc Natl Acad Sci U S A 115(46):E10915–E10924
38. Zhang M et al (2009) Interleukin-15 combined with an anti-CD40 antibody provides enhanced therapeutic efficacy for murine models of colon cancer. Proc Natl Acad Sci U S A 106(18):7513–7518
39. Cooley S et al (2019) First-in-human trial of rhIL-15 and haploidentical natural killer cell therapy for advanced acute myeloid leukemia. Blood Adv 3(13):1970–1980
40. Nguyen R et al (2019) Interleukin-15 enhances anti-GD2 antibody-mediated cytotoxicity in an orthotopic PDX model of neuroblastoma. Clin Cancer Res 25(24):7554–7564
41. Croce M et al (2015) IL-21: a pleiotropic cytokine with potential applications in oncology. J Immunol Res 2015:696578
42. Davis MR et al (2015) The role of IL-21 in immunity and cancer. Cancer Lett 358(2):107–114
43. Leonard WJ, Wan CK (2016) IL-21 signaling in immunity. F1000Res 5:224
44. Stolfi C et al (2012) Interleukin-21 in cancer immunotherapy: friend or foe? Oncoimmunology 1(3):351–354

45. Santegoets SJ et al (2013) IL-21 in Cancer immunotherapy: at the right place at the right time. Oncoimmunology 2(6):e24522
46. Skak K et al (2008) Interleukin 21: combination strategies for cancer therapy. Nat Rev Drug Discov 7(3):231–240
47. Aravindaram K et al (2014) Tumor-associated antigen/IL-21-transduced dendritic cell vaccines enhance immunity and inhibit immunosuppressive cells in metastatic melanoma. Gene Ther 21(5):457–467
48. Chapuis AG et al (2016) Combined IL-21-primed polyclonal CTL plus CTLA4 blockade controls refractory metastatic melanoma in a patient. J Exp Med 213(7):1133–1139
49. Croce M et al (2010) Transient depletion of CD4(+) T cells augments IL-21-based immunotherapy of disseminated neuroblastoma in syngeneic mice. Int J Cancer 127(5):1141–1150
50. Tangye SG, Ma CS (2020) Regulation of the germinal center and humoral immunity by interleukin-21. J Exp Med 217(1):e20191638
51. Wang G et al (2003) In vivo antitumor activity of interleukin 21 mediated by natural killer cells. Cancer Res 63(24):9016–9022
52. Zeng R et al (2005) Synergy of IL-21 and IL-15 in regulating CD8+ T cell expansion and function. J Exp Med 201(1):139–148
53. Bhatt S et al (2015) Direct and immune-mediated cytotoxicity of interleukin-21 contributes to antitumor effects in mantle cell lymphoma. Blood 126(13):1555–1564
54. Lewis KE et al (2017) Interleukin-21 combined with PD-1 or CTLA-4 blockade enhances antitumor immunity in mouse tumor models. Oncoimmunology 7(1):e1377873
55. Pan XC et al (2013) Synergistic effects of soluble PD-1 and IL-21 on antitumor immunity against H22 murine hepatocellular carcinoma. Oncol Lett 5(1):90–96
56. Rigo V et al (2014) Recombinant IL-21 and anti-CD4 antibodies cooperate in syngeneic neuroblastoma immunotherapy and mediate long-lasting immunity. Cancer Immunol Immunother 63(5):501–511
57. Chapuis AG et al (2016) T-cell therapy using Interleukin-21-primed cytotoxic T-cell lymphocytes combined with cytotoxic T-cell lymphocyte antigen-4 blockade results in long-term cell persistence and durable tumor regression. J Clin Oncol 34(31):3787–3795
58. Davis ID et al (2007) An open-label, two-arm, phase I trial of recombinant human interleukin-21 in patients with metastatic melanoma. Clin Cancer Res 13(12):3630–3636
59. Bhatia S et al (2014) Recombinant interleukin-21 plus sorafenib for metastatic renal cell carcinoma: a phase 1/2 study. J Immunother Cancer 2:2
60. Timmerman JM et al (2012) A phase I dose-finding trial of recombinant interleukin-21 and rituximab in relapsed and refractory low grade B-cell lymphoproliferative disorders. Clin Cancer Res 18(20):5752–5760
61. Hashmi MH, Van Veldhuizen PJ (2010) Interleukin-21: updated review of Phase I and II clinical trials in metastatic renal cell carcinoma, metastatic melanoma and relapsed/refractory indolent non-Hodgkin's lymphoma. Expert Opin Biol Ther 10(5):807–817
62. Barata JT et al (2019) Flip the coin: IL-7 and IL-7R in health and disease. Nat Immunol 20(12):1584–1593
63. Gao J et al (2015) Mechanism of action of IL-7 and its potential applications and limitations in cancer immunotherapy. Int J Mol Sci 16(5):10267–10280
64. Zarogoulidis P et al (2014) Interleukin-7 and interleukin-15 for cancer. J Cancer 5(9):765–773
65. Mackall CL et al (2011) Harnessing the biology of IL-7 for therapeutic application. Nat Rev Immunol 11(5):330–342
66. ElKassar N, Gress RE (2010) An overview of IL-7 biology and its use in immunotherapy. J Immunotoxicol 7(1):1–7
67. Sportes C, Gress RE (2007) Interleukin-7 immunotherapy. Adv Exp Med Biol 601:321–333
68. Shi LZ et al (2016) Interdependent IL-7 and IFN-gamma signalling in T-cell controls tumour eradication by combined alpha-CTLA-4+alpha-PD-1 therapy. Nat Commun 7:12335
69. Ding ZC et al (2016) IL-7 signaling imparts polyfunctionality and stemness potential to CD4(+) T cells. Oncoimmunology 5(6):e1171445

70. Gunnarsson S et al (2010) Intratumoral IL-7 delivery by mesenchymal stromal cells potentiates IFNgamma-transduced tumor cell immunotherapy of experimental glioma. J Neuroimmunol 218(1–2):140–144

71. Zoon CK et al (2017) Expansion of T cells with Interleukin-21 for adoptive immunotherapy of murine mammary carcinoma. Int J Mol Sci 18(2):270

72. Shum T et al (2017) Constitutive signaling from an engineered IL7 receptor promotes durable tumor elimination by tumor-redirected T cells. Cancer Discov 7(11):1238–1247

73. Suzuki T et al (2016) Requirement of interleukin 7 signaling for anti-tumor immune response under lymphopenic conditions in a murine lung carcinoma model. Cancer Immunol Immunother 65(3):341–354

74. Deiser K et al (2016) Interleukin-7 modulates anti-tumor CD8+ T cell responses via its action on host cells. PLoS One 11(7):e0159690

75. Choi YW et al (2016) Intravaginal administration of fc-Fused IL7 suppresses the cervico-vaginal tumor by recruiting HPV DNA vaccine-induced CD8 T cells. Clin Cancer Res 22(23):5898–5908

76. Toyota H et al (2015) Vaccination with OVA-bound nanoparticles encapsulating IL-7 inhibits the growth of OVA-expressing E.G7 tumor cells in vivo. Oncol Rep 33(1):292–296

77. Fritzell S et al (2013) IFNgamma in combination with IL-7 enhances immunotherapy in two rat glioma models. J Neuroimmunol 258(1–2):91–95

78. Rosenberg SA et al (2006) IL-7 administration to humans leads to expansion of CD8+ and CD4+ cells but a relative decrease of CD4+ T-regulatory cells. J Immunother 29(3):313–319

79. Sportes C et al (2008) Administration of rhIL-7 in humans increases in vivo TCR repertoire diversity by preferential expansion of naive T cell subsets. J Exp Med 205(7):1701–1714

80. Ding ZC et al (2017) Adjuvant IL-7 potentiates adoptive T cell therapy by amplifying and sustaining polyfunctional antitumor CD4+ T cells. Sci Rep 7(1):12168